Advances in Soil Science

Advances in Soil Science

B.A. Stewart, Editor

Advances in Soil Science

Volume 20

Edited by B.A. Stewart

With Contributions by
M. Becker, B.T. Christensen, D.J. Horne,
C.S. Khind, J.K. Ladha, R.P. Pareek, T.B. Parkin,
J.A. Robinson, Bijay-Singh, Yadvinder-Singh, M.G. Wallis

With 67 Illustrations

Springer-Verlag
New York Berlin Heidelberg London Paris
Tokyo Hong Kong Barcelona Budapest

B.A. Stewart
USDA Conservation & Production Research Laboratory
Bushland, TX 79012, USA

ISSN: 0176-9340

Printed on acid-free paper.

Softcover reprint of the hardcover 1st edition 1992

Production managed by Dimitry L. Loseff; Manufacturing supervised by Jacqui Ashri.
Typeset by Best-set Typesetter Ltd., Hong Kong.

9 8 7 6 5 4 3 2 1

ISBN-13: 978-1-4612-7724-8 e-ISBN-13: 978-1-4612-2930-8
DOI: 10.1007/978-1-4612-2930-8

Preface

The key to sustaining the soil resource base is to maintain, or enhance, soil quality. Soil quality cannot be seen or measured directly from the soil alone but is inferred from soil characteristics and soil behavior under defined conditions. In essence, the quality of soils is analogous to the health of humans, and just as there is no single characteristic that can be measured to quantify a person's health, there is no single measurement that can quantify soil quality. However, there are certain characteristics, particularly when considered together, that are good indicators. Soil quality, just as human health, can be maintained or enhanced by good management practices; and seriously degraded—sometimes irreversibly—with poor practices.

Soil quality is also important because it has direct and indirect effects on air quality and water quality. While the enhancement of soil quality does not always assure parallel improvements in the quality of air and, particularly, water resources, this is often the case. However, soil degradation is invariably accompanied by degraded qualities of both air and water resources.

The consensus among many scientists is that the greatest challenge is not increasing production, but preventing serious deterioration of the soil and water resource base so that the production level can be sustained. Soil degradation is a complex phenomenon produced by a strong interaction between socioeconomic and biophysical factors, which today include the effects of increasing population, a fragile economy, and dismal farm policies, as well as the fragility of soil itself and the harshness of climate. Soil degradation can be subtle and slow until a certain threshold is reached, and then it deteriorates rapidly.

Advances in Soil Science provides a forum for leading scientists to analyze and summarize the available information on a topic, assessing its importance and identifying additional research needs. Most importantly, contributors are asked to develop principles that have practical applications to both developing and developed agriculture. *Advances in Soil Science* was formulated to be international in scope and cover all subjects relating to soil science. This volume certainly fulfills those goals in that authors are from Denmark, the Philippines, New Zealand, India, United States, and Australia; and the topics cover organic matter, green manures, legumes, soil water repellency, and statistics.

The reception of *Advances in Soil Science* by both authors and readers has been very gratifying. I want to thank the the authors for their excellent contributions, the Editorial Board for their help in selecting such competent contributors, and the Springer-Verlag staff for their kind and efficient assistance. Last, and most important, I want to thank the readers for their acceptance and use of *Advances in Soil Science*.

B.A. Stewart

Contents

Contributors

M. Becker, International Rice Research Institute, P.O. Box 933, Manila, Philippines.

Bent T. Christensen, Askov Experimental Station, Vejenvej 55, DK-6600 Vejen, Denmark.

D.J. Horne, Department of Soil Science, Massey University, Palmerston North, New Zealand.

C.S. Khind, Department of Soils, Punjab Agricultural University, Ludhiana 141 004, India.

J.K. Ladha, International Rice Research Institute, P.O. Box 933, Manila, Philippines.

R.P. Pareek, Department of Soil Science, G.B. Pant University for Agriculture and Technology, Pantnagar, India.

T.B. Parkin, U.S. Department of Agriculture, Agricultural Research Service, National Soil Tilth Laboratory, Ames, Iowa 50011, U.S.A.

J.A. Robinson, The Upjohn Company, Kalamazoo, Michigan, U.S.A.

Bijay-Singh, Department of Soils, Punjab Agricultural University, Ludhiana 141 004, India.

Yadvinder-Singh, Department of Soils, Punjab Agricultural University, Ludhiana 141 004, India.

M.G. Wallis, Australian Turfgrass Research Institute, P.O. Box 190, Concord West, New South Wales 2138, Australia.

Physical Fractionation of Soil and Organic Matter in Primary Particle Size and Density Separates

Bent T. Christensen

Advances in Soil Science, Volume 20

I. Introduction

A. Soil Organic Matter (SOM)

The soil organic matter (SOM) pool encompasses plant, animal, and microbial residues in all stages of decay and a diversity of heterogeneous organic substances intimately associated with inorganic soil components. The soil microbiota and fine roots may also be considered part of the SOM pool. The turnover of the different SOM components varies continuously due to the complex interaction of biological, chemical, and physical processes in soil. The complexity of SOM and its importance to soil fertility have challenged generations of soil scientists, and numerous studies, of which some date back more than two centuries (see historical review in Kononova, 1961), have covered a vast array of aspects of SOM.

Different SOM concepts have prevailed in different historical periods, and, accordingly, methods used in SOM research have developed under the influence of very diverse scientific disciplines. However, chemical disciplines have historically had the most apparent impact on the methodology applied in SOM research, and much experimental and theoretical effort has been allocated to studies on the chemical structure of SOM. The functional aspects of SOM have traditionally been explored by scientists associated with agricultural research, because SOM plays a key role in soil productivity. Knowledge about the nature and turnover of SOM is a prerequisite for understanding the structure, chemical reactivity, and inherent fertility of soils, and for predicting the fate of mineral fertilizers, animal manure, and crop residues added to soil.

Various aspects of SOM have been reviewed regularly in recent years (e.g., Swift et al., 1979; Jenkinson and Ladd, 1981; Stotzky and Burns, 1982; Tisdall and Oades, 1982; Vaughan and Malcolm, 1985; Huang and Schnitzer, 1986; Oades, 1989; Paul and Clark, 1989), and SOM has been in focus at many seminars (e.g., IAEA, 1977; Tinsley and Darbyshire, 1984; Jensen et al., 1986; Cooley, 1987; Coleman et al., 1989). It appears that chemical aspects of SOM and microbially mediated processes are especially well covered. In contrast, reviews of SOM research based on physical soil fractionation and organomineral complexes occupy a much less prominent position.

Throughout most of the history of SOM research, it has been recognized that the interaction of SOM with clay minerals is crucial to SOM behavior (Greenland, 1965a, 1965b; Huang and Schnitzer, 1986). A number of studies have, for example, addressed the effect of various species of mined clay minerals on SOM stabilization in soils incubated with organic substrates (Allison et al., 1949; Lynch et al., 1956; Sørensen, 1967, 1972), but due to the complexity of organomineral associations and the lack of adequate methods, the organic and the mineral parts of soils have most often been treated in separate studies. The mineral part has been studied by mineralogists after careful removal of SOM, while organic chemists studying various SOM extracts have struggled to minimize the ash content.

B. Physical Fractionation of Soils

Before 1960, relatively few studies had addressed the interaction between SOM and soil minerals directly by attempting to isolate naturally occurring primary organomineral complexes. This is in contrast to the large number of studies on aggregates, which in the present context may be termed secondary organomineral complexes. Although dispersion of soils by shaking with water was used to isolate primary organomineral-size separates, it was realized that simple shaking in water would not ensure a complete disintegration of microaggregates. Size separates obtained after limited dispersion may not represent the soil fraction which the results are intended to refer to. The recovered fraction may contain only the most easily dispersed part of the entire size fraction, causing a selective isolation of SOM. Furthermore, the recovered size fraction may consist of an unknown mixture of primary particles and microaggregates of the same size but made up of particles belonging to smaller size classes. The major progress in the study of primary organomineral complexes was to devise soil dispersion techniques that would allow a quantitative isolation of size and density separates without major effects on the configuration and biological properties of SOM.

The most significant single factor affecting this progress was the application of ultrasonic vibrations in soil dispersion. The use of ultrasonic vibrations for dispersion of soils dates back to 1923 (see review by Watson, 1971), but the studies of Edwards and Bremner (1967) and Ford et al. (1969), and the review by Watson (1971) appear to be the most important milestones, triggering a dramatic increase in the number of studies on natural soil organomineral complexes.

The concept behind physical fractionation of soil emphasizes the role of soil minerals in SOM stabilization and turnover. The physical fractionation techniques are considered chemically less destructive, and the results obtained from physical soil fractions are anticipated to relate more directly to the structure and function of SOM in situ.

This contribution considers physical fractionation of soils and its use in SOM studies. Physical fractionation includes density and size separation of primary organomineral particles in whole soil. Thus, aggregates (secondary organomineral complexes) and aggregation will not be considered. Various aspects of physical soil fractionation in relation to SOM studies have been included in the papers by Tiessen et al. (1984b), Christensen (1985, 1987b), Oades (1989), Stevenson et al. (1989), and Elliott and Cambardella (1991), but the methods used in physical soil fractionation have not been reviewed in detail.

Density fractionation is applied to isolate SOM which is not firmly associated with the mineral part of the soil. Soil is dispersed in heavy organic liquids or inorganic salt solutions with specific densities typically ranging from 1.6 to 2.2 g/ml. The light fraction is taken to be less decomposed plant and animal residues, whereas the sediment (or heavy fraction) is expected to encompass "true" organomineral complexes in which SOM is more processed.

Particle size fractionation is based on the concept that SOM fractions associated with particles of different size (and therefore also of different mineralogical composition) differ in structure and function, and therefore play different roles in SOM turnover. Size fractionation may be applied to whole soil samples or to heavy fractions following density fractionation. Soil is usually dispersed in water, the fraction >63 μm being recovered and further separated by wet or dry sieving. Organomineral complexes <63 μm are isolated by gravity sedimentation in water, and the <2 μm fraction may be subdivided by centrifugation. Size separates are occasionally exposed to subsequent density fractionation.

II. Soil Fractionation Methodology

The effectiveness of soil dispersion procedures is crucial for density as well as particle size fractionation. In standard methods of analysis of soil particle size distribution, samples may be exposed to chemical pretreatments in order to remove SOM, sesquioxides, and carbonates, and other cementing agents before being dispersed by a combination of chemical and physical means (Gee and Bauder, 1986). Chemical dispersion treatments are not considered feasible for isolation of intact organomineral complexes, although some studies have applied SOM extractions or included chemicals at various stages of the dispersion procedure (Kyuma et al., 1969; McGill and Paul, 1976; Ladd et al., 1977a; Paul and McGill, 1977; Busacca et al., 1984; Balesdent et al., 1987; Vitorello et al., 1989).

Most studies on organomineral complexes have relied on ultrasonic vibrations to accomplish soil dispersion, but alternative physical procedures have also been employed (e.g., shaking and resins). A number of studies have combined shaking and ultrasonic treatments (Ladd et al., 1977a; Adams, 1982; Balesdent et al., 1987). While standard methods of particle size analysis have been subject to detailed evaluations, procedures used for isolation of intact organomineral complexes have received less attention. This is especially true for techniques used in density fractionation. Soil dispersion methodology will therefore be discussed in relation to isolation of size separates. The rather limited information on dispersion procedures used in density fractionation will be treated in Section II.C.

A. Ultrasonic Dispersion of Soil

Ultrasound, its chemical, physical, and biological effects, and the mechanisms behind acoustic cavitation have been reviewed in detail by Suslick (1988a). The review considers sonochemistry, and industrial, biological, and clinical applications, but reference to soil studies is not included.

Early information on ultrasonic dispersion of soils was reviewed by Watson (1971). The use of ultrasonic vibrations for soil dispersion dates back more than 50 years, but it was not until the beginning of the 1960s that soil dispersion by ultrasonics began to be studied more intensively. During this period much work focused on the use of tank-type ultrasonic disintegrators.

1. Instrument Features

A range of different tank-type vibrators operating at different frequencies has been tested by various workers (Barkoff, 1960; Edwards and Bremner, 1967; Saly, 1967; Bourget, 1968; Parasher and Lowe, 1970; Genrich and Bremner, 1972a). Tank-type ultrasonic vibrators were found to be inefficient for soil dispersion in water even when prolonged treatments were applied. Satisfactory dispersion could be achieved after relatively short treatment periods if a peptizing agent (e.g., Calgon) was added to the soil suspension. In contrast, Edwards and Bremner (1964, 1967) tested sonic (9 kHz) sample-cup and ultrasonic (18 to 20 kHz) probe-type disintegrators and reported that a complete dispersion could be obtained for a range of soils without the use of any oxidants or peptizers. This observation was subsequently confirmed (Watson, 1970; Elonen, 1971; Emerson, 1971), although Emerson (1971) found that some soils showed flocculation of clay particles shortly after dispersion.

Genrich and Bremner (1972a) compared tank-type and probe-type disintegrators. Tank-type ultrasonic equipment did not provide a complete dispersion of soil in water, neither when the soil solutions were vibrated directly in the tank nor when they were kept in flasks and immersed into the water-filled transducer tank. It has been noted, however, (Pritchard, 1974; Suslick, 1988b) that in order to obtain effective dispersion with tank-type vibrators, the position of the sample vessel and the depth of liquid in the tank in relation to equipment frequency are of critical importance. A substantial spatial variability in cavitation intensity produced by tank-type ultrasonic vibrators was clearly indicated by Weissler and Hine (1962). They found greatest cavitation intensity in the center of the tank while intensities decreased sharply towards the sides of the tank. Cavitation in tank corners was 100 times lower than in the center. They also observed a declining intensity with increasing time of continuous operation. Reports on studies using tank-type vibrators for soil dispersion do not, however, indicate whether experimental conditions were optimized, and the importance of reproducible sample positioning and heights of the liquid in the tank and sample flasks was recognized.

Several potential drawbacks (Suslick, 1988b), including the ones mentioned above, cause tank-type ultrasonic vibrators to be viewed as equipment of limited capability for isolation of intact primary organomineral complexes, and following the appearance of the paper by Genrich and Bremner (1972a), soil dispersion studies have almost exclusively relied on probe-type disintegrators.

More recently, Busacca et al. (1984) and Zhang et al. (1988) applied cup-horn ultrasonic disintegrators for dispersion of soils in water. The cup-horn device replaces the probe tip of standard probe-type disintegrators and radiates energy through the base of the container enclosing the soil suspension. Thereby dispersion is achieved without any direct contact between the ultrasonic probe and the sample. This feature may be essential in studies sensitive to contamination by elements shed from the alloy of probe tips. Significant probe wear has been reported for standard probes after prolonged use (Genrich and Bremner, 1972b; Hinds and Lowe, 1980a), and Busacca et al. (1984) found that 15 min of ultrasonic treatment increased the Ti content of silt by 33% due to Ti shed from a conventional probe. The cup-horn configuration is, however, claimed to be sensitive to liquid levels and to shape of reaction vessel (Suslick, 1988b). Also, the acoustic intensities may be lower than for probe-type disintegrators because the vibrating device is not in direct contact with the sample.

The relationship between working frequency of sonic/ultrasonic vibrators and their efficiency in dispersing soils has not been examined in detail. While tank-type disintegrators may operate at widely different frequencies, most probe-type disintegrators now in use operate at 18 to 22 kHz. Working with two tank-type vibrators operating at 20 kHz and 1 MHz, respectively, Saly (1967) concluded that the lower frequency was more efficient in dispersing soils. Edwards and Bremner (1967) found equipment operating at 9 and 18 to 20 kHz to be equally satisfactory if the dispersion procedures of the instruments were optimized individually. A screening of various types of sonic and ultrasonic vibrators showed that common characteristics of vibrators effective in soil dispersion were high power intensity (power input per unit mass of sample treated) and the absence of a transmitting medium (air or liquid) between vibrating mechanisms and soil suspension. Suslick (1988b) indicates that in the range of kHz to a few MHz, frequency is of minor importance for sonochemical reactions to take place as long as the cavitation threshold is reached and the cavitation intensity is adequate.

Excessive erosion of the probe tip has been found to significantly influence the performance of high-intensity probe-type ultrasonic disintegrators (Watson, 1970; Genrich and Bremner, 1972a,b; Hinds and

Lowe, 1980a). Normal probe wear (gray tip surface) is usually detected after only a few hours use, whereas prolonged use may cause significant probe pitting. Genrich and Bremner (1972b) observed a 19% reduction in clay yield when testing a probe, previously exposed to 40 h of use without maintenance. Hinds and Lowe (1980a) employed a probe severely eroded after 70 h of use, and found its energy output reduced by 60%. Although longer treatment times occasionally could compensate for excessive probe wear, regular polishing of the probe tip was recommended in order to prevent probe pitting and to eliminate reductions in energy outputs. Probe tip maintenance is clearly important, and the maintenance procedures recommended in operating manuals should be followed closely.

The performance of probe-type disintegrators in soil dispersion depends heavily on the characteristics of the instrument. Using a cup-horn probe attached to transducers of two different manufacturers, Busacca et al. (1984) found that both the rated power output and treatment time needed to successfully disperse soils differed substantially between instruments. To describe the effective output from ultrasonic vibrators, Saly (1967) suggested the amount of energy dissipated per unit of irradiated area (W/cm^2) as a basic index for exposure conditions. This was challenged by Pritchard (1974), who suggested that the effect produced on standard materials under specified conditions be assessed and used to standardize experimental setups. Genrich and Bremner (1972b) described a calibration method in which the power output of ultrasonic probes was assessed from the radiation pressure exerted by ultrasonic waves when travelling through liquids. The power output was calculated from the difference in the weight of a balance pan with the probe turned on and off. Probe output power has also been estimated from the heat buildup in water exposed to ultrasonics for a given period (North, 1976; Gregorich et al., 1988; Gregorich, 1989; Morra et al., 1991). Probe output power estimated this way was found to be much lower than the rated energy supply displayed by instrument output control. A determination of the "true" probe power output may provide a useful check on the performance of ultrasonic vibrators engaged for longer periods in a particular experimental setup. The significance of this calibration is less obvious for interstudy comparisons because the efficiency of ultrasonic dispersion procedures depends heavily on noninstrument factors, as will be discussed later.

North (1976) tried to estimate the dispersive component of the total amount of energy conveyed from the ultrasonic probe to the soil by using the difference in heat buildup between water and soil suspension. North (1976) claimed that his calibrations enabled a controlled application to soil of a range of known dispersive energies, a point later critically commented upon by Koenigs (1978).

The disruptive effect of ultrasonics is related to the process of cavitation. Cavitation involves three steps: nucleation (bubble formation), bubble growth, and implosion (bubble collapse). In pure liquids bubbles remain spherical while collapsing, whereas bubbles near a solid surface implode asymmetrically, generating a liquid jet directed at the surface of the solid (Suslick, 1988a). The very high pressure waves released upon cavitation are held responsible for soil dispersion. The visual assessment of cavitation intensity may be used as a guide for chosing the power output of optimum dispersion. Too vigorous cavitation should be avoided as it may cause a buildup of bubbles at the interface between probe tip surface and liquid (bubble shrouding). The ultrasound may become highly attenuated as it passes through this region, and sonification of the bulk of the liquid may become greatly reduced (Atchley and Crum, 1988).

Effective cavitation depends not only on instrumental specifications such as wattage, frequency, probe dimensions, and probe wear, but also on heat buildup in instrument components and on properties of the irradiated solution. Ultrasonic treatments cause heat buildup in the soil solution (Saly, 1967; Chichester, 1969; Watson, 1970; Elonen, 1971; Hinds and Lowe, 1980a), and it is recommended that suspension temperature be kept at ambient levels e.g., by use of water cooling jackets or ice baths. Heat is also formed in the ultrasonic equipment and may reduce probe performance. Heat formation occurs in the transducer which converts the electrical high-frequency signal into mechanical vibrations, and in the probe which focuses and transmits the ultrasonic vibrations to the soil suspension. Although most ultrasonic equipment has built-in cooling fans, it is recommended to leave the ultrasonic instrument to cool off between consecutive treatments (Watson, 1970; Hinds and Lowe, 1980a; Tan and Troth, 1981; Christensen, 1985). Watson (1970) found that clay yields were reduced by 50% when the ultrasonic vibrator was operated continuously. Similarly, Hinds and Lowe (1980a) recorded a 30 to 40% reduction in energy output in successive runs without intervening rest periods.

2. Dispersion Procedures

The outcome of ultrasonic treatments of soils depends on instrument specifications, but also on actual experimental procedures and soil characteristics. To assess the effectiveness of ultrasonic dispersion, the complete experimental setup must therefore be considered as a whole. Table 1 is a compilation of the experimental conditions of 27 studies using probe-type ultrasonic disintegrators for soil dispersion in water. Sample weights range from 5 to 100 g, and vibrated soil volumes vary between 29 and 538 ml. The soil/water ratios were between 1:20 and 1:2.5, and the rated energy supply and vibration periods cover 60 to 600 W and 3 to 30 min, respectively.

Although the effective probe output power remains unknown, it appears that studies applying more than 1500 J/ml often produce ade-

Table 1. Experimental conditions of 27 studies using ultrasonic probe-type disintegrators for dispersion of soil in water. A particle density of 2.6 g/cm^3 was used in calculating the volume of soil suspension and amount of energy dissipated per ml of soil suspension

Reference	Soil clay content (%)	Sample weight (g)	Volume of soil suspension (ml)	Soil/water ratio	Vibration period (min)	Rated energy supply (W)	Energy dissipated per ml of soil suspension (J/ml)	Degree of dispersion[a]
Healy and Claridge (1974)	9–24	25	260	1/10	5	75	90	–
Hamblin (1977)	20–45	20	408	1/20	3	250	110	–
Chichester (1969)	38–44	50	519	1/10	15	75	130	?
Ahmed and Oades (1984)	16–47	50	169	1/3	5	80	142	–
Tate and Churchman (1978)	6–26	20	208	1/10	3	180	155	–
Zhang et al. (1988)	15–31	100	438	1/4	15	140	288	–
Turchenek and Oades (1979)	17–46	20	58	1/2.5	5	80	420	+/–
Cheshire and Mundie (1981)	10–15	50	144	1/2.5	15	100	625	?
Baldock et al. (1990)	10	20	58	1/2.5	5	125	650	+
Catroux and Schnitzer (1987)	22	100	538	1/5	15	400	669	–
Anderson et al. (1981)	24–32	20	208	1/10	8	300	695	+
Tiessen and Stewart (1983)	16–65	20	208	1/10	8	300	695	?
Reuter and Leinweber (1988)	1–69	100	288	1/2.5	10	400	832	+/–
Broersma and Lavkulich (1980)	6–16	20	108	1/5	20	80	890	?
Genrich and Bremner (1974)	15–42	60	173	1/2.5	30	100	1040	+
Armour et al. (1990)	6–65	50	219	1/4	15	300	1232	–
Gregorich et al. (1988)	20	15	81	1/5	16	120	1500	+
Christensen (1985)	3–46	30	162	1/5	15	300	1670	+
Angers and Mehuys (1990)	47	40	215	1/5	25	240	1674	–
Elonen (1971)	0–76	10	54	1/5	30	60	2010	+
Hinds and Lowe (1980a)	21–50	25	135	1/5	20	240	2140	+
Edwards and Bremner (1967)	6–75	10	29	1/2.5	20	60	2500	+
Genrich and Bremner (1972a)	8–72	10	29	1/2.5	15	80	2500	+
Tan and Troth (1981)	11–80	25	85	1/3	20	180	2555	?
McKeague (1971)	8–44	40	115	1/2.5	15	400	3120	–
Dong et al. (1983)	16–61	10	34	1/3	15	125	3324	+
Watson and Parsons (1974a)	4–60	25	135	1/5	20	600	5350	+/–

[a] Degree of sample dispersion according to reference: adequate (+), incomplete (–), or unknown (?), while +/– indicate that adequate dispersion was achieved for some soils, but not for others. Conventional textural analyses used as standards.

quate soil dispersion but also that these studies occasionally fail to reach complete dispersion. It should be recognized, however, that the degree of dispersion stated in the references may reflect subsequent flocculation of clay particles, and that the dispersion procedure as such may have accomplished a complete dispersion. The stability of ultrasonically dispersed soil solutions is discussed in Section II.A.3.

Soil samples are usually dried and passed through a 2 mm mesh before being exposed to ultrasonic dispersion. Although dried samples are often presoaked before dispersion (Watson, 1970; Watson and Parsons, 1974a; Hamblin, 1977; Hinds and Lowe, 1980a; Tan and Troth, 1981; Dong et al., 1983; Christensen, 1985; Reuter and Leinweber, 1988), the necessity of this step is not clear. Genrich and Bremner (1972a) found increased clay yields when samples were allowed to presoak for 24 h before a 3 min ultrasonic dispersion. When longer ultrasonic treatments were used, no effect of presoaking was observed. Churchman and Tate (1986) found that limited ultrasonic treatments (260 J/ml) produced higher clay contents for field-moist than for air-dry (20°C) subsamples of four grassland soils. For two soils, air-drying produced similar or higher clay contents.

Compared with a soil/water ratio of 1:5, Edwards and Bremner (1967) found reduced clay yields when using ratios of 1:1.7 and 1:2.5. Hinds and Lowe (1980a) noted that clay recoveries were similar for soil/water ratios of 1:5 and 1:10, whereas a ratio of 1:20 produced significantly lower clay recoveries. Similarly, Watson (1970) and Christensen (1985) found insignificant differences in clay and silt yields for ratios between 1:3 and 1:8. A soil/water ratio of 1:5 to 1:10 therefore seems preferable.

Watson (1970) found higher clay yields for 25 g of soil in 125 ml of water than for 50 g in 250 ml. The relationship between soil suspension volume and degree of dispersion is probably related to the time of exposure of microaggregates to zones of intense cavitation. To increase the contact period, stirring of soil suspensions during ultrasonic treatments is often applied (Emerson, 1971; McKeague, 1971; Genrich and Bremner, 1974; Watson and Parsons, 1974a; Turchenek and Oades, 1979; Cheshire and Mundie, 1981; Catroux and Schnitzer, 1987; Armour et al., 1990). The pressure waves created by cavitation may, however, cause sufficient mixing of the soil suspension (North, 1976; Hinds and Lowe, 1980a; Christensen, 1985; Gregorich et al., 1988), and the need for additional mixing is probably related to the volume of the soil suspension, the shape and dimensions of the containers enclosing the soil suspension, and to the intensity of cavitation. It is envisaged that probe tips with small radiating surface areas used in combination with larger volumes of soil suspension will provide less efficient mixing than larger probes used on smaller volumes. As the same factors probably affect the overall intensity of cavitation, probes with relatively large diameters should be used in combination with moderate soil-suspension volumes.

Given the instrument features (probe dimensions, power output intensity) and sample properties (soil type, soil/water ratio, soil-suspension volume), the outcome of ultrasonic dispersion procedures depends on the length of treatment time. Limited ultrasonic treatments provide incompletely dispersed soils, causing microaggregates of smaller size particles to be included in the silt and sand size separates. Excessive treatments may lead to abrasion of primary organomineral complexes (see Section II.A.3).

The influence of ultrasonic treatment time on the degree of soil dispersion has been studied regularly (Edwards and Bremner, 1967; Chichester, 1970; Watson, 1970; Emerson, 1971; Genrich and Bremner, 1972a; North, 1976; Busacca et al., 1984; Christensen, 1985; Gregorich et al., 1988; Morra et al., 1991), and Figure 1 illustrates the basic relationship between clay and silt yields and treatment period. The yield of clay increases and that of silt decreases gradually with increased treatment time, demonstrating that a progressive disruption of silt size aggregates takes place. In accordance with other studies (Edwards and Bremner, 1967; Watson, 1970; Gregorich et al., 1988), the gain in clay exceeds the concomitant loss of silt particles showing that sand size aggregates are being dispersed too. The rate of change in clay and silt yields levels off with increased time of treatment, and after 15 min of ultrasonic treatment yields of clay and silt reach a plateau, indicating that complete dispersion has been achieved. Similar dispersion patterns have been demonstrated in other studies (Edwards and Bremner, 1967; Chichester, 1970; Watson, 1970; Emerson, 1971; North, 1976; Gregorich et al., 1988). It was also observed that treatment times exceeding what was needed for complete dispersion caused but a slight further increase in clay yields. This suggests that a "true" plateau of complete dispersion exists, but also that excessive inputs of ultrasonic energy may lead to further dispersion of strongly cemented microaggregates or to a slow gradual breakdown of primary particles. However, as noted by Edwards and Bremner (1967), the lack of an absolute standard of complete dispersion is an adverse feature which studies of primary organomineral complexes share with conventional textural analyses.

Given an optimized experimental setup, the time of ultrasonic treatment required to achieve complete dispersion will vary between soil types. Ideally, the optimum treatment time should be determined for each soil under study in order to ensure that ultrasonic dispersion is neither too limited nor excessive. Treatment times adopted in practice, however, will have to be tuned to cover a range of different soils.

Table 1 demonstrates that a relatively large number of studies have failed to disperse soils completely. Watson (1971) and North (1976) stressed the lack of standard procedures and indicated that standardization of ultrasonic dispersion methods was desirable. Differences between instrument specifications, and the interaction between soil type and the treatment time required to achieve adequate dispersion do not,

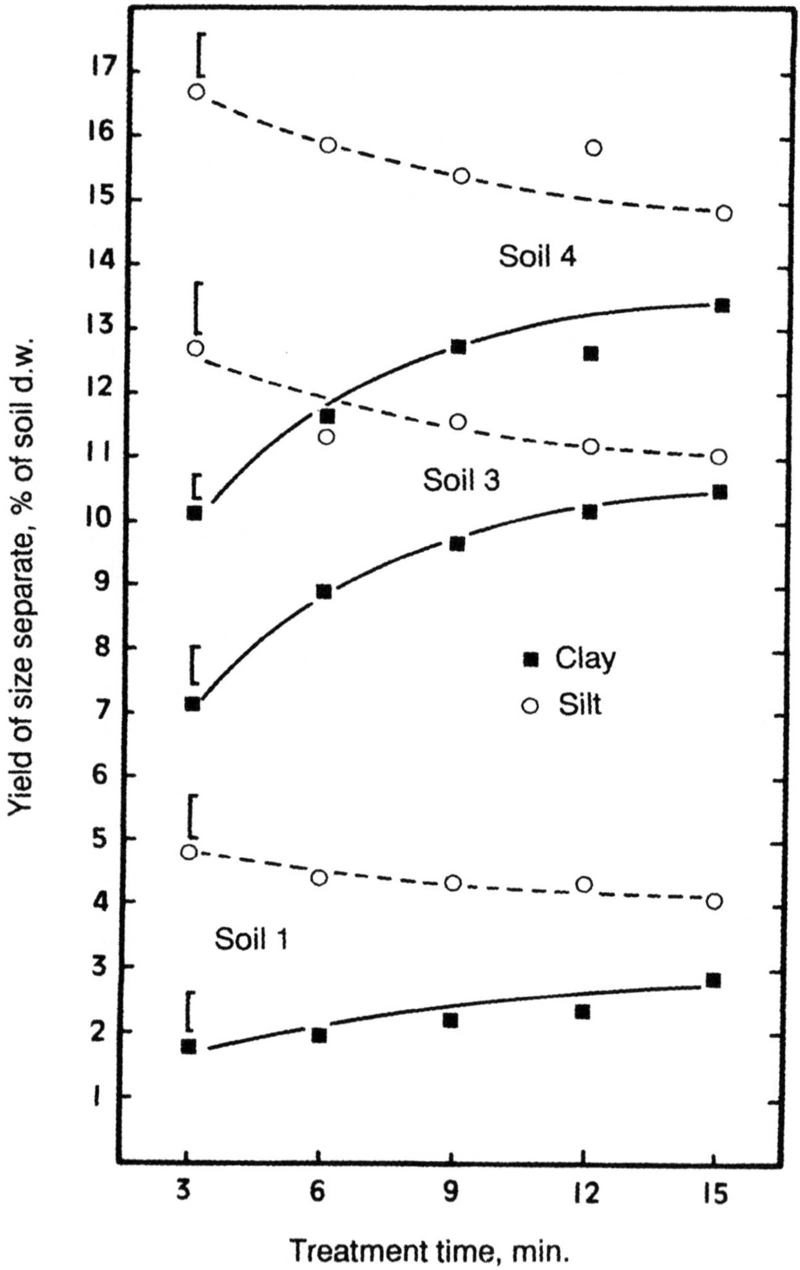

Figure 1. Influence of ultrasonic treatment period at 300 W (30 g soil in 150 ml water) on yields of clay (<2 μm) and silt (2 to 20 μm) from samples of three Danish sandy soils with <1.4% C (from Christensen, 1985).

however, suggest that a universally applicable standard procedure for isolation of intact primary organomineral complexes is feasible. For each particular study, the experimental procedure should be optimized for representative soil types and the particle size distribution calibrated against accepted standards such as conventional textural analyses.

3. Effect of Dispersion on Soil

It has been claimed that the ultrasonic dispersion of soil causes the breakdown of primary particles (Elonen, 1971; Watson, 1971), but the significance of particle abrasion in studies of primary organomineral complexes is not known. For very fragile minerals there may be an overlap between the cohesive strength of microaggregates and the stability of primary particles (Pritchard, 1974).

Although the possibility of some breakdown of primary particles cannot be excluded, the existence of a plateau of complete dispersion suggests that for most mineral soils a quantitative disruption of microaggregates takes place much more readily than does the breakdown of primary particles. Watson (1971) considered that SOM and clay would provide some buffering against the abrasion of coarse particles, and that abrasion would be more serious for well-defined minerals free of SOM and other amorphous material. Gregorich et al. (1988) found that abrasion of clay minerals was unlikely because the bond strength in the clay lattice is in the order of kJ per mol, an energy level much higher than provided by ultrasonic treatments. However, ultrasonic cavitation creates transient domains of much higher energy intensity than indicated by average probe output power (Suslick and Doktycz, 1990). Such locally very high temperature and pressure gradients may promote cavitational erosion, but the significance of these ultrasonic "hot spots" on the structure of organomineral complexes is not known.

Most minerals tested in the study of Edwards and Bremner (1967) suffered less breakdown when exposed to ultrasonics than when shaken in water for 10 h or dispersed by H_2O_2 and Calgon treatments. Walker and Hutka (1973) compared a number of different dispersion treatments (end-over-end shaking, high-speed mixing, ultrasonic vibrations) and showed that the probe-type ultrasonic vibrator caused the least fragmentation. Fragile materials such as biotite, mica, and sedimentary rocks (shales) have been shown to suffer breakdown when dispersed ultrasonically (Edwards and Bremner, 1967; Adams and Stewart, 1969; Pritchard, 1974), and some abrasion of quartz has also been indicated (Busacca et al., 1984; McTainsh and Duhaylungsod, 1989).

Busacca et al. (1984) found that 6 to 11% of the 20 to 2000 μm fraction obtained by conventional textural analyses was moved into smaller size fractions after 25 min of ultrasonic dispersion. However, X-ray diffraction analyses demonstrated that the material gained in the clay size separate was phyllosilicate materials, and it was concluded that the ultrasonic treatment cleaned sand particles for clay coatings left after conventional dispersion. Gregorich et al. (1988) found similar surface areas in clay isolated by ultrasonic dispersion of increasing intensity, suggesting that structural changes in the clay were insignificant. This observation agrees with previous reports (Saly, 1967; Watson, 1970). Particle abrasion induced by ultrasonics deserves more attention, however, especially when the biological properties and chemical reactions of organomineral complexes are assessed.

The possibility of SOM being detached from organomineral complexes during ultrasonic dispersion, and subsequently comminuted and redistributed between different size separates has not been directly addressed experimentally. Electron microscopy suggests that the SOM is intimately associated with the mineral part of the soil (Turchenek and Oades, 1978; Tiessen and Stewart, 1983; Spycher et al., 1983; Christensen, 1985; Tiessen and Stewart, 1988), but sample pretreatments associated with electron microscopy and uncertainty with regard to the representativeness of information obtained from extremely small subsamples hamper conclusive statements on SOM detachment and redistribution. Circumstantial evidence suggests that detachment of SOM from organomineral complexes may be, quantitatively at least, relatively insignificant. Due to a low specific density and, thus, a slow sedimentation rate, SOM released during ultrasonic dispersion would tend to accumulate in the clay and water-soluble fractions, and the silt would appear low in SOM. The relatively high SOM content of silt found for most soils indicates, however, that SOM associated with silt does occur in intact organomineral complexes. Assuming similar intracomplex stabilities for clay- and silt-size separates, the SOM in clay would also be expected to be present in organomineral combination. Moreover, if a significant redistribution of detached or solubilized SOM takes place during dispersion, the readsorption of SOM has to be selective with regard to particle size since difference in the quality of SOM associated with different size fractions is well established (see below).

Even limited ultrasonic treatments cause release of microorganisms from soil (Stevenson, 1958; Jenkinson et al., 1976; Amato and Ladd, 1980; Ahmed and Oades, 1984; Ramsay, 1984). The intensity of ultrasonic treatments required to disperse soils completely most probably causes a substantial release of microbial SOM components, which may subsequently be redistributed among size separates. Part of the microbial cell content most probably appears in the water-soluble fraction (SOM recovered in the supernatant of centrifuged clay suspensions), which has been found to contain <2% of the SOM pool (Cameron and Posner, 1979). Christensen (1987a) found that 1 to 2% of native SOM was water soluble, whereas 4 to 9% of labeled SOM was in the water-soluble fraction of soils incubated for 5 to 6 years with ^{14}C-labeled substrates (Christensen and Sørensen, 1985).

Soil pH is not affected significantly by ultrasonic dispersion treatments (Barkoff, 1960; Saly, 1967; Edwards and Bremner, 1967), although a slight but consistent increase (0.1 to 0.5 pH units) has been noted for Gleysolic soils (Hinds and Lowe, 1980c). Compared with simple shaking ultrasonic dispersion in water may increase the solubility of SOM and various elements (Hinds and Lowe, 1980c; Dong et al., 1985). Hinds and Lowe (1980c) observed that ultrasonic dispersion in electrolytes substantially suppressed the dissolution effects and concluded that ultrasonic dispersion of Gleysolic soils in electrolytes (0.01 *M* $CaCl_2$ or $BaCl_2$, 0.1 *M* NaCl) would produce higher yields of less-altered organoclay complexes.

The resistance against flocculation of clay particles in soil suspensions dispersed ultrasonically in water is crucial for the quantitative recovery of primary organomineral size separates. It is generally reported that soil suspensions are very stable and do not show signs of flocculation even when left to stand undiluted for several days (Edwards and Bremner, 1967; Watson, 1970; Genrich and Bremner, 1972a; Busacca et al., 1984; Christensen, 1985). Suspensions of soils low in pH, subsoils low in SOM, and soils with high contents of Ca^{2+} and Mg^{2+} may, however, be unstable unless dispersants are added (Saly, 1967; Emerson, 1971; Watson and Parsons, 1974a). Genrich and Bremner (1972a) examined the stability of 28 soils dispersed ultrasonically in water, including soils differing widely in pH and content of SOM, clay, and carbonates, and representing different soil types and pedogenic horizons. Except for one soil extremely high in carbonates, flocculation was not observed when suspensions were allowed to stand for 4 days. Genrich and Bremner (1972a) stated that some of the soils used in their study resembled soils previously reported to flocculate readily upon ultrasonic dispersion in water (Saly, 1967; Emerson, 1971). No explanation for the discrepancy between results was, however, offered and the particular mechanisms responsible remain unknown.

B. Alternative Dispersion Methods

Alternative dispersion methods range in intensity from short-term stirring or shaking by hand in water (Arshad and Lowe, 1966; Parasher and Lowe, 1970; Cameron and Posner, 1979; Schnitzer and Ivarson, 1982) to prolonged chemically assisted shaking in rotary, reciprocating, and end-over-end shakers (Shiel, 1986; Balesdent et al., 1987; Vitorello et al., 1989). Various shaking procedures in water or dispersants have been combined with limited or prolonged ultrasonic treatments (Ladd et al., 1977a; Adams, 1982; Balesdent et al., 1988). High-speed mixers (Kanazawa and Filip, 1986; Livens and Baker, 1988) and homogenizers (Spycher et al., 1983; Janzen, 1987) have been employed too. Essington and Mattigood (1990) dispersed soil by vigorous shaking in ethanol containing 10% polyvinyl pyrrolidone. This dispersant is absorbed by soil particles (Francis, 1973) and its removal before analysis for SOM content is therefore essential.

High-speed mixers may cause unacceptable abrasion of primary particles (Lemieux, 1964; Thornburn and Shaw, 1987), and simple shaking in water has been found to provide incomplete dispersion even after prolonged treatment periods (Edwards and Bremner, 1967; Genrich and Bremner, 1974; Young and Spycher, 1979; Zhu et al., 1983). Genrich and Bremner (1974) found that clay isolated by shake procedures with limited dispersion capability differed significantly in content of C, N, P, and S from clay isolated by an ultrasonic dispersion procedure that gave complete dispersion. Gregorich et al. (1988) found that the C content of clay, silt, and sand size separates varied with the degree of soil dispersion. Simple shake methods are therefore not recommended for isolation of primary organomineral complexes.

Chemically assisted dispersion procedures may introduce unintended in-process changes of SOM structure and distribution (e.g., Scharpenseel et al., 1986). Such changes are liable to conflict with the concept of intact organomineral complexes, and the introduction of chemicals in various soil dispersion procedures should be avoided unless the action of the chemical is specific and well documented.

The dispersive effect of prolonged shaking may be enhanced by introducing glass or agate balls (usually five balls with 10 mm diameter) to the plastic container enclosing the soil suspension during shaking. Andreaux et al. (1980) found that the number of balls (two to eight balls examined) appeared less critical. The highest proportion of whole soil C was found in the <50 μm fraction after 15 h of shaking with five balls. The recovery of <50 μm particles increased linearly from 1 to 15 h treatment and no plateau of complete dispersion was reached.

The soil suspension obtained by the ball-assisted shake procedure may subsequently be fractionated according to particle size or density (Feller, 1979; Andreux et al., 1980; Bruckert and Kilbertus, 1980; Ahl et al., 1983; Ahl, 1984) or may be dispersed further by ultrasonic treatments before being fractionated (Cerri et al., 1985; Balesdent et al., 1988; Martin et al., 1990). The ball-assisted shake procedure for dispersing soils has also been followed by further chemically assisted dispersion treatments (Balesdent et al., 1987; Vitorello et al., 1989).

Most studies employing the ball-assisted shake method do not include a standard textural analysis of the soils under study, and accordingly it is difficult to evaluate the degree of dispersion obtained. Using the procedure of Bruckert and Kilbertus (1980), Ahl et al. (1985) found however that 51 to 76% of the clay in a number of temperate arable soils was recovered in silt size separates. These results suggest that the ball-assisted shake method most probably provides limited soil dispersion, and unless additional dispersion measures are applied, it may not be recommended for studies on primary organomineral complexes. Balesdent et al. (1988) found that an additional 10 min of ultrasonic treatment caused soils to disperse completely.

Shaking with resins has been found to disperse soils completely (Edwards and Bremner, 1965, 1967) and this method has been used to isolate primary organomineral complexes (Dalal and Mayer, 1986c; Dalal and Henry, 1988). Resins may, however, extract substantial amounts of SOM (Edwards and Bremner, 1965), and this led Watson (1970) to conclude that the use of resins had serious limitations in studies of organomineral complexes. Clearly the resin shake method should be examined in much greater detail before it is adopted for isolation of intact organomineral complexes.

C. Density Fractionation

1. Concepts of Light/Heavy Fractions

Although the bonding strength between SOM and soil inorganics varies continuously, the original concept behind density fractionation assumes that SOM can be divided into two pools differing in structure and function. The so-called light fraction ("free" or noncomplexed SOM) is considered to be decomposing plant and animal residues with a relatively high C:N ratio, a rapid turnover, and a specific density considerably lower than that of soil minerals. The heavy fraction includes the organomineral complexed SOM which is taken to be comparatively more processed decomposition products ("true humus") with a narrow C:N ratio, a slower turnover rate, and a higher specific density due to its intimate association with soil minerals (Monnier et al., 1962; Greenland and Ford, 1964; Greenland, 1965a, 1971). Experimentally the light and heavy fractions are most often obtained by introducing the soil into a liquid with a specific density between 1.6 and 2.2 g/cm^3, but sieving and flotation in water has also been used. Early information on density fractionation was reviewed by Greenland (1965a, 1971).

The original two-fraction concept has been kept in most studies, but some workers have fractionated soil into light, medium, and heavy fractions by subsequently exposing the sample to liquids of increasing density. Shaymukhametov et al. (1985) summarized that a successful density fractionation divides SOM into three distinct pools: (1) a light fraction (density below 1.8 to 2.0 g/cm^3) with little mineral content and including slightly decomposed macroorganic matter of plant and animal origin, (2) a medium density fraction (density 2.0 to 2.4 g/cm^3) including the "true" organoclay complexes, and (3) a heavy fraction (density above 2.4 g/cm^3) containing mineral grains coated with SOM. This three-compartment concept has been expanded further to cover a range of density fractions isolated from whole soils (Richter et al., 1975; Ladd and Amato, 1980; Leuschner et al., 1981; Dalal and Mayer,1986d, 1987, Skjemstad and

Dalal, 1987; Essington and Mattigood, 1990) or size separates (Spycher and Young, 1977, 1979; Turchenek and Oades, 1974, 1978, 1979; Young and Spycher, 1979; Oades et al., 1987). The various density fractions may differ in one or more properties, but the interpretation of results is far less straightforward than when the orginal two-fraction concept is applied. Lichtfuss and Brümmer (1981) attempted to characterize and conceptually define five separate fractions of different density (density range, 1.6 to 2.8 g/cm^3), whereas Spycher and Young (1979) concluded that a continuum existed in properties and composition of fractions between density extremes. Clearly a theoretically more consistent concept justifying the use of a range of density fractions should precede further work in this direction.

2. Methodological Aspects

Table 2 presents examples of the very diverse experimental procedures used to isolate light fractions in various studies. The feasibility of a particular fractionation scheme is not easily evaluated, and comparison between studies is not possible because descriptions of dispersion treatments are incomplete.

In contrast to particle size fractionation, the outcome of which can be compared to standard textural analysis, density fractionation can only be subject to intrinsic evaluation, e.g., by making material and element balances or by examining the ash content of the light fractions. According to the concept of light fractions being made up of noncomplexed SOM, the ash content of light fractions should be low and confined to minerals inherent in plant and animal material. Most studies do, however, find ash contents of light fractions exceeding 50% (e.g., Greenland and Ford, 1964; Ford et al., 1969; Spycher et al., 1983) or carbon contents considerably below 25% (e.g., Turchenek and Oades, 1979; Dalal and Mayer, 1986d).

One reason for the wide variation in methodology (Table 2) is probably the very limited research on various aspects of the fractionation methods. Few studies have evaluated in any detail the methods applied, and many aspects of the experimental procedures remain obscure.

During size fractionation, light fraction material is often recovered along with the sand size separates. This SOM, referred to as macroorganic matter, may be isolated from the sand by flotation in water (Hamblin, 1977; Cameron and Posner, 1979; Feller, 1979; Tiessen and Stewart, 1983; Balesdent et al., 1988; Zhang et al., 1988), or by exposing the dried sample to winnowing (Theodorou, 1990). Adams (1980) isolated macroorganic matter from a number of grassland and woodland soils by washing soil through a 0.25 mm sieve, followed by flotation in water.

Density fractionation has historically relied on organic liquids, but aqueous solutions of inorganic salts have become increasingly popular,

Table 2. Experimental procedures used for isolation of light fractions (noncomplexed or "free" SOM)

Reference	Dispersion (treatment time)	Specific density (g/cm^3)	Heavy liquid mixture	Surfactant (additives)
Monnier et al. (1962)	B/SH–?	1.8–2.3	Bromoform/benzene/ethanol	None
Greenland and Ford (1964)	UP–3 min	2.0	Bromoform/petroleumspirit	None
Ford et al. (1969)	UP–?	2.0	Nemagon/petroleumspirit	0.1% w/w Aerosol-OT-100
McKeague (1971)	?	2.0	Tetrabromoethane/benzene	None
Richter et al. (1975)	SH–?	1.6–2.5	Bromoform/ethanol	None
Ladd et al. (1977a)	UP–3 min	1.6	Tetrachloromethane	0.1% w/v Aerosol-OT-100
Ladd et al. (1977a)	UP–2 min	2.1	Nemagon	0.1% w/v Aerosol-OT-100
Spycher and Young (1977)	SM–10 h	1.7–2.3	Tetrabromoethane/ethanol	None
Murayama et al. (1979)	UT–5 min	2.0	Bromoform/chloroform	0.5% w/w Aerosol-OT-100
Turchenek and Oades (1979)	UP–?	1.8–2.6	Decalin/Nemagon/bromoform	0.2% w/v Aerosol-OT-100
Ladd and Amato (1980)	UP–3 min	1.6–2.1	Tetrachloromethane/Nemagon	0.1% w/v Aerosol-OT-100
Leuschner et al. (1981)	UT–3 min	1.6–2.2	Bromoform/petroleumether	None
Lichtfuss and Brümmer (1981)	G	1.6–2.8	Tetrabromoethane/ethanol	None
Ahl et al. (1983)	SM–24 h	2.2	Bromoform/ethanol	None
Spycher et al. (1983)	ST–30 s	1.6	NaI/water	None
Sollins et al. (1984)	?	1.4–1.6	NaI/water	None
Dalal and Mayer (1986d)	SM–1 h	2.0–2.4	Bromoform/ethanol	0.1% v/v Triton-X
Janzen (1987)	ST–1 min	1.6	NaI/water	None
Oades et al. (1987)	UP–?	2.0–2.2	$ZnBr_2$/water	None
Baldock et al. (1990)	?	2.0	Sodium-polytungstate/water ($Na_3WO_4 \cdot 9WO_3 \cdot H_2O$)	None
Essington and Mattigod (1990)	?–48 h	1.9–3.3	Tetrabromoethane/ethanol/ diiodomethane	20% w/w PVP
Skjemstad et al. (1990)	G	1.6	$ZnBr_2$/water	None

UT = ultrasonic dispersion, tank type
UP = ultrasonic dispersion, probe-type
S = shaking (by hand (H) or mechanically (M))
G = ground sample
ST = vigorous stirring
B = boiling in water
? = details not stated in reference

Bromoform = $CHBr_3$
Tetrabromoethane = $C_2H_2Br_4$
Diiodomethane = CH_2I_2
Tetrachloromethane = CCl_4
Chloroform = $CHCl_3$
Ethanol = CT_2H_5OH
Nemagon = 1,2-dibromo-3-chloropropane

Aerosol OT-100 = di-iso-octyl-sodium-sulfosuccinate
Triton-X = octylphenoxy-polyethoxy-ethanol
PVP = polyvinylpyrrolidone

one reason for this being the high potential toxicity of halogenated hydrocarbons. The most frequently used organic liquids are tetrabromoethane ($C_2H_2Br_4$, 2.96 g/cm^3), bromoform ($CHBr_3$, 2.88 g/cm^3), and tetrachloromethane (CCl_4, 1.59 g/cm^3). The densities of the bromated hydrocarbons are most often adjusted by adding ethanol, chloroform, or CCl_4. Because residual water in dried samples may significantly affect the yield of light fractions when nonpolar mixtures of organic liquids are used, the admixture of polar liquids to adjust densities of heavy liquids may be advantageous (Ford et al., 1969; Richter et al., 1975). Dalal and Mayer (1986d) found that sample water contents <32% did not affect yields of light fractions in bromoform/ethanol mixtures, whereas separate ethanol-water and bromoform layers were formed when samples contained 65% water.

The inclusion of a surfactant may reduce effects of residual sample water content in nonpolar liquids (Ford et al., 1969). Ford and Greenland (1968) introduced the surfactant Aerosol OT-100 and found increased yields of light fractions especially for samples ground to less than 0.25 mm (Ford et al., 1969). It was stated that 0.1% of surfactant gave the maximum yield of light fractions. Dalal and Mayer (1986d) examined various concentrations of Triton-X in bromoform/ethanol mixtures and observed that 0.1% (vol/vol) of surfactant maximized light-fraction yields. A concentration of 0.2% caused reduced yields, probably due to precipitation of flocculated light-fraction material. Turchenek and Oades (1979) found that 0.2% (wt/vol) of the surfactant Aerosol OT-100 also promoted flocculation of clay particles.

Using mixtures of bromoform and Nemagon (1,2-dibromo-3-chloropropane) with 0.1% (wt/vol) of surfactant (Aerosol OT-100), Turchenek and Oades (1974) observed that the density fractions obtained were 20 to 30% too high in carbon. Turchenek and Oades (1979) later examined this point in greater detail and found that bromoform, Nemagon, and Aerosol OT-100 all were adsorbed to clay. Several washings with acetone and ethanol, either alone or in combination, would not remove the excess carbon, whereas washings with acetone/water mixtures returned the carbon level to that of unprocessed clay.

Inorganic salts used for density fractionation include Mg_2SO_4, $ZnBr_2$, NaI, and more recently also sodium polytungstate (Baldock et al., 1990). Solutions of sodium polytungstate are less viscous and less toxic than solutions of $ZnBr_2$ (Plewinsky and Kamps, 1984). Aqueous solutions of inorganic salts eliminate the adverse effects of residual water in samples and of surfactant and organic liquid adsorption to fractions. The effect of concentrated salt solutions on the nature of SOM has, however, received little attention. Sollins et al. (1984) found that nitrogen mineralization was decreased by 70% when soil was washed in NaI solution. Vigorous agitation caused a further 53% reduction in mineralization potential. The reduced mineralization was ascribed to extraction of mineralizable nitrogen, the amount of nitrogen extracted by NaI accounting for 2.2% of soil total-N. Nevertheless, mineralization of nitrogen from heavy fractions was found to be similar to or greater than mineralization from whole soils, and it was concluded that flotation in water was advantageous when isolating noncomplexed SOM for subsequent microbiological experiments.

For organic as well as inorganic heavy liquids, information is obviously needed with regard to their effects on chemical, physical, and biological properties of light and heavy fractions. When halogenated hydrocarbons are used, a kill-off of the soil biomass is to be expected, the effectiveness of chloroform in assays of soil microbial biomass being well known.

Light-fraction yields are very sensitive to changes in the density of fractionation liquids (Richter et al., 1975; Ladd and Amato, 1980; Dalal and Mayer, 1986d), indicating that choice of density, exact density adjustment, and temperature control during processing are all critical factors for the outcome of density separations. Minor deviations in liquid density may produce significant differences in the carbon concentration of light fractions, especially in the density range of 1.9 to 2.4 g/cm^3 (Richter et al., 1975). For example, the proportion of whole soil carbon found in light fractions more than doubled when liquid density was raised from 1.9 to 2.0 g/cm^3.

The specific density of heavy liquids differs widely among studies. Originally, the concept of light fractions was experimentally based on a liquid density near 2.0 g/cm^3 (Monnier et al., 1962; Greenland and Ford, 1964), and this density has been kept in much subsequent work (Ford et al., 1969; McKeague, 1971; Murayama et al., 1979; Dalal and Mayer, 1987; Oades et al., 1987; Dalal and Henry, 1988; Baldock et al., 1990). Other studies have favored a density of about 1.6 g/cm^3 (Ladd et al., 1977a; Spycher et al., 1983; Ahmed and Oades, 1984; Sollins et al., 1984; Janzen, 1987; Skjemstad et al., 1990). Ladd and Amato (1980) summarized, from work based on density limits of 1.59 and 2.06 g/cm^3 (Ladd et al., 1977a,b,c) that dispersion of soils at densities below 2.0 g/cm^3 probably yields light fractions with higher proportions of macroorganic matter. A similar conclusion was reached by Turchenek and Oades (1979), and Oades (1989) stated that no critical density would completely separate out the free SOM from the mineral-complexed SOM.

As for size separation, the degree of sample dispersion is critical to the outcome of density fractionation. Dispersion procedures used in density fractionation include boiling in water, high-speed stirring, manual and mechanical shaking, and ultrasonic dispersion (Table 2). Compared to size fractionation, density fractionation methods appear to rely on less vigorous dispersions. However, very little can be concluded on the degree of dispersion obtained in various studies because procedures are incompletely described and not evaluated for effectiveness. Limited dispersion causes density fractions to include a mixture of noncomplexed SOM,

SOM in organomineral complexes, and SOM in aggregates of finer particles. Microaggregates become included in light fractions because entrapped air and adsorbed water lower their effective density. Results based on limited soil dispersion are difficult to interprete because the nature and proportion of each SOM component in a particular density fraction remains unknown.

Turchenek and Oades (1979) summarized the problems encountered in their density fractionation study as: (1) removal of water and air from microaggregate surfaces and cavities, (2) disruption of aggregates and floccules, and maintenance of dispersion during the fractionation, (3) removal of adsorbed heavy liquids and surfactant from particles, and (4) evaluation of the changes in organomineral complexes arising from repeated ultrasonic treatments and from solution effects of heavy liquids and solvents. Undoubtedly, this list of problems is valid for most studies involving density fractionation.

Despite the frequent use of density fractionation in SOM studies, a number of fundamental conceptual and experimental problems appear to remain unsettled. Although different densities may apply to different soil types, the high ash content of most light fractions suggests that the dispersion procedures are inadequate or that too high critical densities are used to isolate noncomplexed SOM. The rationale of separating SOM into a range of density fractions needs to be clarified, and an adequate and consistent conceptual framework should be developed and tested critically as a prerequisite to further work in this direction. If a satisfactory concept can be outlined, it should be followed by detailed examinations of the fundamental aspects of the fractionation procedures.

Specific examples of laboratory procedures used in density fractionation can be found in Ford et al. (1969), Richter et al. (1975), Turchenek and Oades (1979), and Dalal and Mayer (1986d).

D. Isolation of Size Separates

Following dispersion, the whole soils or heavy fractions "freed" of noncomplexed SOM may be divided into different size separates by sieving and sedimentation. The particle size limits basically follow the USDA or the ISSS classification schemes (see Gee and Bauder, 1986), but minor modifications of size limits and subdivisions of clay and silt separates are often adopted. Clay is generally taken to be particles $<2\,\mu m$ in diameter, while silt is 2 to $20\,\mu m$ in the ISSS and 2 to $50\,\mu m$ in the USDA classification. The coarser material is termed "sand" and falls in the range 20 to $2000\,\mu m$ (ISSS) or 50 to $2000\,\mu m$ (USDA).

The use of different size limits restricts the comparability of results from different studies. Although most studies specify the adopted size limits, discussions of organomineral size separates regularly fail to appreciate the interpretative restrictions attached to different size limits, especially when comparing data on silt and clay subfractions. Often reference is made to clay and silt in general, disregarding that silt in one study may include material which in another study is isolated along with the sand size separates. While clay separates are dominated by clay-type minerals, the silt may be considered a transistory fraction containing clay-type minerals, mica, and various proportions of the less weathered minerals that normally dominate the sand separate (quartz and feldspars). The size limits chosen for silt are therefore critical to the interpretation of results obtained on SOM structure and dynamics, especially since a large proportion of whole soil SOM is associated with the silt (see below).

Dry and wet sieving has been applied to separate out fractions >50 to 60 µm before or after isolation of clay and silt. Size separates between 60 and 2 µm are isolated by gravity sedimentation, while centrifugation is applied for subdividing separates <2 µm. Extended sedimentation times and effects of Brownian movement make gravity sedimentation less feasible for isolation of small clay particles. Centrifugation greatly speeds up processing time compared with gravity sedimentation, and this may be advantageous in order to reduce the possible effects of microbial activities in soil suspensions.

1. Application of Stokes Law to Sedimenting Soil Particles

The sedimentation dynamics under gravity of chemically dispersed and SOM-free soil particles have been discussed in detail by Elonen (1971), but little information can be traced on the sedimentation behavior of intact primary organomineral particles. Sedimentation under gravity can be described by Stokes Law:

$$v = \frac{h}{t} = \frac{GD^2(d_p - d_1)}{18n}$$

where v is particle settling velocity, t is the time required for particles to fall through a fixed distance h, G is the acceleration induced by gravity, D is particle diameter, d_p is particle density, and d_1 is the density, and n the viscosity of the liquid in which sedimentation takes place. The density and viscosity of liquids depends on the temperature, which makes temperature control during sedimentation important.

Values for d_1 and n of water are usually applied, but the viscosity of soil suspensions may deviate significantly from that of water. Elonen (1971) found a substantial increase in soil-solution viscosity when clay concentrations were increased from a few to 90 g of clay per L. The relationship was almost linear, with every 10 g increase in clay causing viscosity to increase by 5% relative to that of water. According to Stokes Law this corresponds to a 5% decrease in the settling velocity of particles. Applying the viscosity of pure water to a soil solution containing 50 g of clay per L will therefore cause a given particle size class to include 13% of particles larger than the stated size. The error introduced by applying the viscosity of water may be significant during initial sedimentation cycles only because the concentration of small particles dereases rapidly during successive cycles. However, the results indicate that soil-to-water ratios should be kept as low as possible during sedimentation. Genrich and Bremner (1974) stated that clay and silt contents should be less than 5% by weight, as sedimentation in more concentrated suspensions was considered not to obey Stokes Law.

Stokes Law applies to particles exposed to laminar flow. Based on calculations using Reynolds Number, Elonen (1971) concluded that Stokes Law was not applicable to particles >60 μm. Because of Brownian movement, Stokes Law would not apply to <0.1 μm particles sedimenting under gravity. Centrifugal forces of several times gravity may be used for smaller particle size classes, the equations used to calculate sedimentation times being derived from Stokes Law (e.g., Tanner and Jackson, 1947; Ahl, 1984).

Basic assumptions underlying Stokes Law include that particles are spherical and smooth and do not interact with each other or with the wall of the sedimentation cylinder. Elonen (1971) discussed the effect of particle-particle and particle-wall interactions. The particle-wall interaction was considered unimportant if the ratio particle/cylinder diameter is kept below 1:100. The effect of particle-particle interaction is not straightforward as deceleration and acceleration forces are both involved. Another exception to Stokes Law is the result of the nonspherical shape of soil particles. Organomineral particles may be porous and water absorbed in the particles should be included when considering their density. Plate-shaped mica particles are common in soils. The particle size limits used in size fractionation, therefore, refer to equivalent spherical diameter (ESD), signifying the diameter of a spherical particle with a similar density and settling velocity as the actual particle under analysis.

The particle density used in sedimentation is taken to be in the range 2.5 to 2.7 g/cm^3. Densities for minerals free of SOM are 2.65 for quartz, 2.5 to 2.8 for feldspars, 2.7 to 3.3 for micas, and 3.1 to 3.3 for apatite. The density of SOM is taken to be <1.5 g/cm^3, and the densities for organomineral complexes will therefore be lower than those of pure minerals. The settling velocity calculated from Stokes Law is directly related to the difference in density of the sedimenting particle and the sedimentation liquid. For an organomineral particle containing 20% SOM with a density of 1.2 g/cm^3, and assuming a mineral density of 2.65 g/cm^3, the calculated density of the organomineral particle will be 2.36 g/cm^3. A 0.3 g/cm^3 decrease in particle density reduces the settling velocity by 18%, corresponding to a change in measured particle size of <10%. Most particles will contain much less than 20% SOM, and using a particle density of 2.5 to 2.7 g/cm^3 is considered feasible.

It may be summarized that practically all the assumptions underlying Stokes Law are violated to some extent when it is applied to sedimenting soil particles, and that the isolation of clay and silt separates is by no means theoretically straightforward. The concept of equivalent spherical diameters is a practical but crude approximation, and it appears difficult to justify its use in attempts to isolate several size fractions over a limited range of particle sizes.

2. Experimental Procedures

Wet and dry sieving have been used to isolate sand size separates. When applied to whole soil samples, most of the macroorganic matter (noncomplexed SOM) will be recovered in the sand. Christensen (1985) compared wet and dry sieving procedures applied before and after isolation of clay and silt, respectively, and found no significant differences in dry weight and carbon recovery. The 20 to 63 μm and 63 to 200 μm separates did, however, tend to be slightly lower and higher, respectively, in dry weight after wet sieving.

Following dispersion (and sometimes also removal of noncomplexed SOM by density fractionation and sand size separates by wet sieving), the soil suspension may be transferred to sedimentation cylinders to separate out clay and silt by gravity sedimentation. After addition of water and throughout mixing, the cylinders are left to stand undisturbed, preferably at constant temperature (e.g., 20°C). Clay and silt size separates are isolated in turn by successive sedimentation-decanting (or siphoning) cycles, fresh water being added to the sediment, and the suspension carefully remixed after each decanting. For complete separation of a given size fraction, the sedimentation-decanting process must be repeated several times. The number of sedimentation cycles needed to achieve a given accuracy depends on the content of that fraction in the soil (Christensen, 1985). The visual assessment of a clear supernatant has proved to be an adequate criterion for a sufficiently complete and reproducible separation. Typically, 5 to 15 sedimentation cycles are needed (Watson, 1970; Cheshire and Mundie, 1981; Christensen, 1985; Catroux and Schnitzer, 1987; Gregorich et al., 1988). For clay size particles, each sedimentation cycle may last more than 8 h. To completely separate out the clay, the soil may therefore be submerged in water for a period of 2 to 3 weeks. In contrast, the separation of silt can be achieved within less than 2 h.

Generally no precautions are taken against microbial activity during sedimentation. The significance of microbial activity during the extended processing time involved in gravity sedimentation of clay size separates has not been examined in detail, although there are studies indicating metabolic activity during fractionations (Cheshire and Mundie, 1981; Baldock et al., 1990). It may be envisaged that the microbial population present initially is relatively small due to the kill-off of microorganisms during ultrasonic dispersion. Moreover, the microbial population is rapidly diluted by the successive sedimentation-decanting cycles which also remove clay particles and soluble carbon and nutrients. The lack of more detailed information may, however, warrant the addition of microbial inhibitors to soil suspensions. Baldock et al. (1990) proposed the use of sodium azide or mercuric chloride. Alternatively, isolation of clay by centrifugation should be favored.

Clay particles in the decanted suspensions are generally flocculated by adding $CaCl_2$, $MgCl_2$, $MgSO_4$, or NaCl and then concentrated by centrifugation. The concentrated clay separate may be dialyzed or washed with acetone to remove excessive salt contents, and then oven- (e.g., 40°C) or freeze-dried. Silt suspensions are also concentrated by centrifugation but without the use of flocculants. Solublized SOM may be included in the finest size separate or may be isolated separately from the supernatant left after centrifugation of flocculated clay suspensions. The means of concentrating and drying the size separates should be adjusted to the type of analyses subsequently performed on the size separates.

When exposed to biological incubation tests, size separates should be inoculated because an unknown, but probably large, proportion of the soil microbial biomass is killed during ultrasonic soil dispersion. Excessive concentrations of salts added to flocculate clay and residual contents of microbial inhibitors may also interfere with the outcome of biologically founded tests. Temperatures used to dry size separates should be kept below 40°C in order to reduce changes in biomolecules that might affect subsequent decomposability of the SOM.

A detailed description of laboratory procedures for isolation of primary organomineral size separates is given by Christensen (1985), and by Genrich and Bremner (1974), Watson and Parsons (1974a), Hinds and Lowe (1980a), and Cheshire and Mundie (1981).

III. Soil Organic Matter (SOM) in Density Separates

Whole soils as well as particle size separates have been subject to density fractionation. Usually light-fraction SOM is isolated from whole soils by relatively mild dispersion procedures, whereas isolation of light fractions from size separates most frequently follows successive and more thorough dispersions. The nature and significance of light fractions from whole soils and from size separates are likely to differ, and properties of light fractions from the two sources are therefore treated separately.

A. Light Fractions from Whole Soils

The light fractions of whole soils are generally considered a transitory SOM pool dominated by decomposing plant and animal residues. Charcoal found in light fractions (Greenland and Ford, 1964; Spycher et al.,

1983) may, however, significantly influence results, particularly in soils where natural or anthropogenic burning of vegetation has occurred regularly (Skjemstad et al., 1990). Similarly, relatively inert SOM may be included in light fractions from cultivated peat soils (Oades, 1972; Scheffer, 1977).

Light fractions are liable to include microbial decomposers and microfauna although Ahmed and Oades (1984) demonstrated that the proportion of whole soil biomass present in light fractions of two arable soils was insignificant. Arthopod fecal pellets may also add significantly to light fractions of sand soils (Leuschner et al., 1981).

The amount of light-fraction SOM is likely to be affected by land use, vegetation type, and other factors influencing the balance between litter production and decomposition (e.g., climate, soil characteristics, faunal activity). Accumulation of light-fraction SOM is envisaged to be favored in cold and dry climates, in soils low in pH, and in continuously vegetated soils with a high return of plant litter such as forests and permanent grasslands (Garwood et al., 1972; Whitehead et al., 1975).

Accordingly, the proportion of whole-soil SOM recovered in light fractions varies widely among soils. Light fractions ($<2.0\,g/cm^3$) isolated by Greenland and Ford (1964) from a range of Australian soils comprised 0.03 to 8.2% of soil dry weight and accounted for 1 to 85% and 11 to 100% of the whole soil carbon and nitrogen, respectively. This variation has been confirmed in later studies (Greenland, 1965a; Ford et al., 1969; Oades, 1972).

Due to the dominant influence of plant-derived materials, light fractions show seasonal and spatial variability. Spycher et al. (1983) found large seasonal variation in light fractions ($<1.65\,g/cm^3$) from the 0 to 5 cm horizon of a Douglas-fir stand (Figure 2), with changes in the deeper horizons being less important. Light-fraction SOM in the 0 to 3 cm horizon accounted for 53% of whole soil SOM in that layer (Figure 3), whereas light fractions in the horizons below made up 25 to 35%. These results imply not only that time of sampling and sampling depths are critical to the outcome of density fractionations, but also that light fractions can be a dynamic property of potential use in SOM turnover studies of forest ecosystems. Seasonal variations in light fractions of arable soils have been found to be much lower (Ford and Greenland, 1968; Adams, 1980).

The proportions of whole soil weight and SOM content in light fractions are very sensitive to the density of fractionation liquids (Richter et al., 1975; Ladd and Amato, 1980) as illustrated in Table 3 (Skjemstad and Dalal, 1987). A density of $1.6\,g/cm^3$ produced light-fraction yields that were 4 to 50 times lower than yields obtained with a density of $2.0\,g/cm^3$. A density of $1.6\,g/cm^3$ allocated 1 to 9% of the whole soil carbon to the light fraction, while the proportion at $2.0\,g/cm^3$ was 5 to 31%. Dry weight and proportion of whole soil nitrogen in light fractions

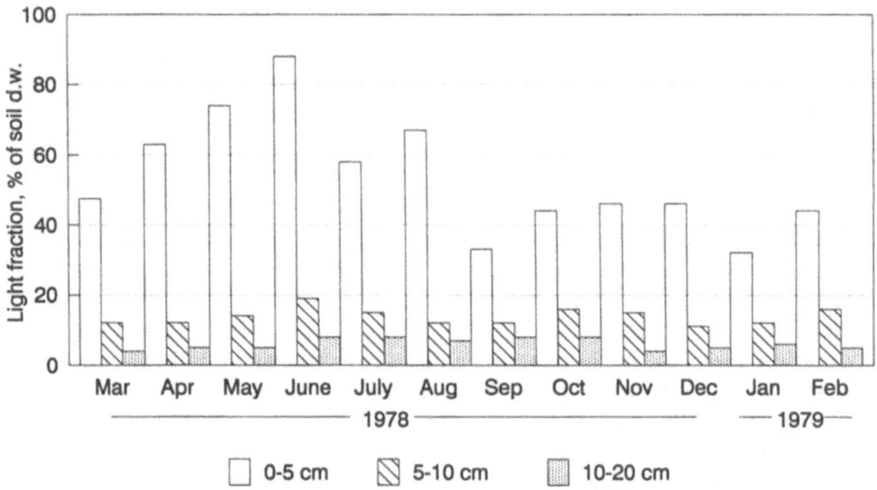

Figure 2. Seasonal variation in light fraction (<1.65 g/cm³) in different soil depths of a 70-year-old Douglas fir stand in Oregon. Samples were passed a 0.85 mm mesh before density fractionation (based on data from Spycher et al., 1983).

Table 3. Distribution of dry weight and carbon among density fractions from two South Queensland subtropical virgin soils (0–10 cm) under Acacia (B_0) or native grass (W_0) and their cultivated counterparts after 35 years of cereal cropping (B_{35} and W_{35}). W site has pH 8.1, 1.6% C, and 72% clay, while B site has pH 7.4, 2.2% C, and 49% clay. Data from Skjemstad and Dalal (1987)

Density (g/cm³)	Distribution of dry weight (% of whole soil)				Distribution of carbon (% of whole soil C)				% loss of carbon due to cultivation	
	B_0	B_{35}	W_0	W_{35}	B_0	B_{35}	W_0	W_{35}	B	W
<1.6	0.58	0.07	0.16	0.04	9.1	1.1	2.6	0.9	95	80
1.6–2.0	3.2	0.2	2.1	1.9	21.7	3.4	12.7	8.1	94	62
2.0–2.4	75.3	79.3	91.2	91.0	62.2	77.5	83.1	89.2	52	37
>2.4	20.9	20.5	6.5	7.1	7.0	18.0	1.6	1.8	0	33
Whole soil	100	100	100	100	100	100	100	100	61	41

of two arable soils examined by Ladd and Amato (1980) increased 5 to 16 and 3 to 9 times, respectively, when fractionation liquid density was increased from 1.59 to 2.06 g/cm³ (see also Section II.C.2).

Table 4 summarizes data on light fractions from forested sites located at different latitudes. Although separation densities differed between studies, the proportion of carbon and nitrogen in light fractions tends to increase with increasing latitude. This trend may reflect the combined

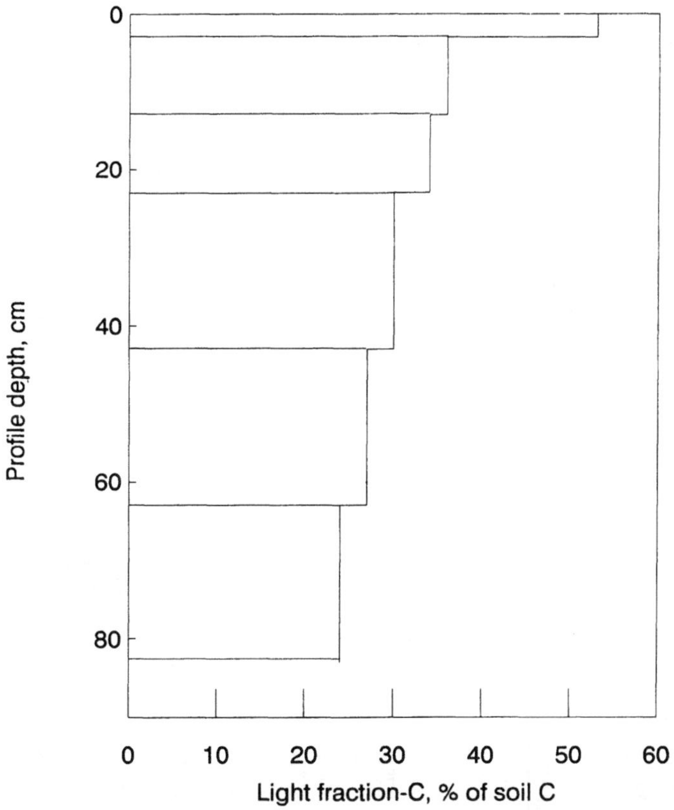

Figure 3. Depth distribution of light fraction (<1.65 g/cm^3) in the forest soil beneath a 70-year-old Douglas fir stand in Oregon. Samples were passed a 0.85 mm mesh before fractionation (based on data from Spycher et al., 1983).

effect of differences in vegetation types and in climatic conditions. In contrast, Molloy et al. (1977) concluded that the heavy rather than the light fraction more accurately reflected the climatic variables throughout a climosequence of New Zealand tussock grasslands, and Dalal and Mayer (1986d) found equilibrium levels of carbon in heavy fractions from virgin and cultivated soils to correlate well with annual precipitation.

Cultivation of virgin sites greatly reduces light fractions (Dalal and Mayer, 1986d, 1987; Skjemstad et al., 1986; Skjemstad and Dalal, 1987), losses of carbon being up to 11 times greater from light than from heavy fractions. Thus, in contrast to permanently vegetated sites (Table 4), arable soils tend to have less significant light fractions (Table 3, Figure 4). A temperate grazed sward was found to maintain twice as much macro-OM than soil under arable tillage (Garwood et al., 1972), and Whitehead et al. (1975) found light fraction (<2.06 g/cm^3) yields of 0.4, 1.5, and

Table 4. Dry weight of light fractions ($<1.65\,g/cm^3$ for references 1 and 3; $<2.0\,g/cm^3$ for reference 2) isolated from the upper 10 to 15 cm layer of forested sites, and proportions of whole soil carbon and nitrogen recovered in light fractions

Reference	Site	Latitude	Vegetation	Light fraction: Dry weight	Carbon	Nitrogen
				(% of whole soil)		
1	Monte Verde, Costa Rica	10° N	Tropical wet forest	12.0	25	24
1	La Selva, Costa Rica	10° N	Tropical wet forest	0.8	5	3
1	Turrialba, Costa Rica	10° N	Tropical wet forest	2.1	9	6
2	Langlands-Logie, Qld., Australia	27° S	Acacia	2.8	26	15
2	Cecilvale, Qld., Australia	27° S	Eucalyptus	2.2	21	10
2	Riverview, Qld., Australia	27° S	Eucalyptus	2.4	29	14
2	Thallon, Qld., Australia	29° S	Eucalyptus	1.8	17	8
3	Oak Ridge, TN, USA	36° N	Liriodendron	8.9	21	n.d.
1	Waldo Lake, OR, USA	43° N	Hemlock	11.1	47	27
1	H.J. Andrews, OR, USA	44° N	Conifer	7.3	28	26
4	McKenzie Bridge, OR, USA	44° N	Douglas fir	11.5	45	30
1	Cascade Head, OR, USA	45° N	Conifer	9.3	24	14
			Alder	14.7	31	26
1	Wind River, WA, USA	46° N	Douglas fir	6.4	36	24
			Alder/Douglas fir	13.6	42	40

References: 1, Sollins et al. (1984); 2, Dalal and Mayer (1986a, 1986d, 1987); 3, Malone and Swartout (1969); 4, Spycher et al. (1983)

n.d., not determined

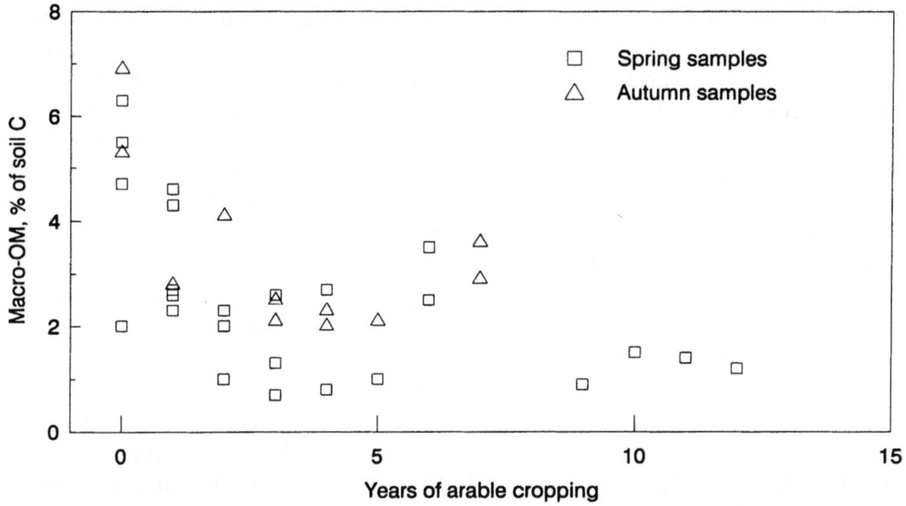

Figure 4. Macro-OM content in 0 to 12 cm of Northern Ireland soils (acid brown earths derived from mica-schist and Silurian parent material) in relation to years of arable cropping. Mean soil pH and carbon content was 6.1 and 4.2%, respectively. Macro-OM was material retained on a 0.25 mm mesh and isolated by repeated flotation in water (based on data from Adams, 1980).

5.4% of soil weight under arable cultivation, 8 year-old grass sward and permanent pasture, respectively. Janzen (1987) found light fractions (<1.6 g/cm^3) to account for <0.4% of the soil weight and <4% of the nitrogen in Canadian crop rotations (fallow, wheat, and forage crops). For a range of arable German coarse sand soils, light fractions (<1.8 g/cm^3) made up 0.1 to 0.4% of whole soil weight and 7 to 12% of soil carbon (Leuschner et al., 1981). Australian wheat/pasture/fallow rotations separated at a density of 2.0 g/cm^3 showed light fractions encompassing 3 to 26% of whole-soil nitrogen (Ford and Greenland, 1968), the light fractions of these soils accounting for 0.3 to 3.7% of whole soil weight (Ford et al., 1969).

These results suggest that noncomplexed SOM in long-term arable soils may not interfere significantly with the distribution of organomineral SOM among particle size separates, whereas removal of noncomplexed SOM may be essential before size fractionation of soils under continuous vegetation.

Light fractions isolated by Greenland and Ford (1964) showed carbon and nitrogen concentrations of 20 to 32% and 1.1 to 2.0%, respectively. Similar values were found by Ford et al. (1969), Molloy and Speir (1977), and Sollins et al. (1984), whereas Dalal and Mayer (1986d, 1987) found considerably lower concentrations (7 to 21% C, 0.3 to 1.1% N). Compared with concentrations in whole soils, light fractions of Greenland and

Ford (1964) were 10 to 30 times higher in carbon and nitrogen, whereas more recent studies have demonstrated lower enrichments. Sollins et al. (1984) and Dalal and Mayer (1986d, 1987) found light fractions with carbon contents 2 to 7 and 8 to 12 times higher than whole soils, respectively. Corresponding nitrogen enrichments were 1.5 to 4 and 3 to 8.

SOM in light fractions is usually higher in C:N ratio than in whole soils and heavy fractions, reflecting the dominant influence of plant litter on this SOM pool. The C:N ratios of light fractions isolated by Greenland and Ford (1964) ranged from 12 to 30, and despite differences in methodology, similar light fraction C:N ratios have been found for New Zealand tussock grasslands (Molloy and Speir, 1977; Molloy et al., 1977), arable coarse sand soils in Germany (Leuschner et al., 1981), grassland/arable soils of Northern Ireland (Adams, 1980), virgin subtropical grassland, Acacia and Eucalyptus soils of Queensland, Australia (Dalal and Mayer, 1987), and for Costa Rican tropical wet forests (Sollins et al., 1984). Although the C:N ratios of light fractions isolated from alder and conifer stands in the U.S. Pacific Northwest averaged 41 and encompassed values as high as 55 (Spycher et al., 1983; Sollins et al., 1984), the ranges of C:N ratios in light fractions from different ecosystems in different geographical regions appear relatively narrow and surprisingly similar. Moreover, Dalal and Mayer (1987) found that the C:N ratio of light fractions was not significantly changed by long-term cultivation. This similarity in C:N ratios probably reflects that most studies have discarded fresh and easily recognizable plant residues of higher and more variable C:N ratios from the samples before isolation of light fractions. Therefore, only litter material in an intermediate state of decomposition is included in light fractions. For example, Molloy et al. (1977) isolated light fractions from New Zealand tussock grasslands with C:N ratios of 18 to 38 (mean 26), whereas the C:N ratios of various plant parts were 44 to 93 (green leaves), 46 to 335 (standing dead leaves), 40 to 166 (litter), and 57 to 113 (roots).

The more advanced state of decomposition of plant residues in light fractions is also reflected by a relatively high ash content, typically ranging from 20 to 50% of fraction dry weight (Greenland and Ford, 1964; Malone and Swartout, 1969; Molloy and Speir, 1977; Adams, 1980; Leuschner et al., 1981; Spycher et al., 1983). Phytoliths (Greenland and Ford, 1964), pumice particles (Spycher et al., 1983), amorphous oxides (Ford et al., 1969), and clay minerals (Spycher and Young, 1979; Turchenek and Oades, 1979; Leuschner et al., 1981) have all been found to contribute significantly to the ash content of light fractions. For coarse sand soils, Leuschner et al. (1981) found that most of the clay and silt particles were in the light fractions.

Light fractions are enriched in carbohydrates relative to whole soils and heavy fractions (Oades, 1972; Whitehead et al., 1975; Molloy et al., 1977; Murayama et al., 1979; Dalal and Henry, 1988). From 3 to 26% of the light-fraction carbon may be present in carbohydrates. In contrast to heavy fractions, the light fractions may show considerable variation in sugar composition. The proportion of whole soil carbohydrates present in light fractions may account for 8 to 45%. While Molloy et al. (1977) considered that plant-derived sugars dominated the light-fraction carbohydrate pool and that microbially derived sugars were abundant in the heavy fraction, Murayama et al. (1979) concluded from an incubation study with ^{14}C-labeled ryegrass material that fractionation by density did not satisfactorily distinguish carbohydrates from plant and microbial sources. Dalal and Henry (1988) were also led to the conclusion that no clear distinction can be made between light and heavy fractions with regard to the origin of carbohydrates. They found that cultivation reduced the carbohydrate content of light fractions, but that the loss of carbohydrates paralleled that of total carbon, demonstrating that the carbohydrate pool was no more liable to decomposition than the total SOM pool.

Greenland and Ford (1964) found the proportion of SOM in light fractions to decrease as soil clay content increased. Similarly Richter et al. (1975) found a highly significant correlation between clay content and SOM in heavy fractions, and Leuschner et al. (1981) demonstrated that the proportion of SOM in heavy fractions was directly related to the clay and silt content of sandy soils. Thus, it appears that light fractions are quantitatively more important in sandy than in clayey soils, although differences in the dispersivity of soils with different clay contents should be considered too.

Studies on the turnover of light-fraction SOM from whole soils reveal conflicting results. Sollins et al. (1984) found that the proportion of nitrogen mineralized from light fractions ($<1.6\,g/cm^3$) of forested sites correlated well with the nitrogen content of that fraction, whereas nitrogen mineralization from heavy fractions and whole soils showed no such correlation. Janzen (1987) observed that the amount of mineral nitrogen released from arable soils was highly correlated with light-fraction ($<1.6\,g/cm^3$) nitrogen content. Light fractions ($<2.0\,g/cm^3$) isolated by Ford and Greenland (1968) contained 3 to 26% of whole soil nitrogen but accounted for 25 to 60% of the mineral nitrogen released upon incubation of arable soils. Dalal and Mayer (1987) stated that rates and proportions of light-fraction ($<2.0\,g/cm^3$) carbon and nitrogen lost on cultivation of virgin soils generally were much higher than from heavy fractions. Adams (1980) found that arable soils incubated with macro-OM produced more CO_2 but released less mineral nitrogen than unamended soils. It was concluded that macro-OM was neither an immediate nor a major source of mineral nitrogen in these soils. Although experimental procedures differ, it has been demonstrated (Bernard-Reversat, 1981; Sollins et al., 1984) that the proportion of nitrogen mineralized from light fractions ($<1.6\,g/cm^3$) decreases with increasing C:N ratio and that more nitrogen can be mineralized from heavy (low C:N) than from light (high C:N) fractions. Greenland and Ford (1964) also noted that light fractions from a permanent grassland and with a C:N ratio of 12 released more mineral nitrogen during incubation than light fractions from soil under continuous wheat cultivation and with a C:N ratio of 23. It was stated that immobilization of mineral nitrogen could be expected during decomposition of light fractions with high C:N ratios.

The different observations on the turnover of light-fraction SOM are not easily reconciled, and part of the discrepancy may be ascribed to differences in procedures used for isolation of light fractions and for incubation tests. The classical concept of C:N ratio supports a higher nitrogen mineralization rate at lower C:N ratios (heavy fractions and whole soils) than at higher C:N ratios (light fractions), whereas mineralization of carbon would increase with increasing C:N ratio. Accordingly, light fractions could be an important pool of carbon available for immobilization of mineral nitrogen in soils, and light-fractions dynamics may be critical in reducing losses of excess mineral nitrogen from intensively managed arable soils. The active role of light-fraction material in nitrogen immobilization in supported by the results of Ladd et al. (1977c) showing that destruction of the soil microbial biomass in calcareous soils incubated with straw was accompanied by a significant decrease in labeled and native nitrogen of the light fraction ($<1.59\,g/cm^3$). Ladd et al. (1977a) found that about 20% of the immobilized nitrogen was in this fraction.

Skjemstad et al. (1990) studied the carbon turnover of light fractions ($<1.6\,g/cm^3$) isolated from an Australian native rainforest and from nearby sites where rainforest (C_3 plants) had been replaced by pasture (C_4 plants). After 35 and 83 years under pasture, about 60% and 20% of the light-fraction SOM was derived from the previous rainforest, respectively. Although charcoal was found to be present in light fractions, such apparent longevity of light-fraction material is not easily compatible with the concept that light fractions are dominated by partly decomposed plant remains with a relatively high turnover rate.

The proportion of SOM in light fractions has been claimed to provide a better and earlier indication of consequences of different soil managements than the levels of SOM in whole soils, but it appears that not all available information supports this concept. The role of light-fraction SOM in the turnover of carbon and of nutrients may well be different. The exclusion of recently shed plant residues may cause light-fraction turnover to be blurred by simultaneous loss and gain of light-fraction SOM. For virgin soil rich in light-fraction SOM, gains from coarse plant residues initially present may make up for subsequent management-induced losses in light-fraction SOM.

The bulk of evidence does, however, suggest that light fractions from whole soils may form a distinct and functional pool of significance to the SOM turnover in virgin and in recently cultivated soils, but also that much remains to be learned about the nature and role of this SOM pool. For long-term arable soils growing annuals, light fractions are probably less significant. Where perennial crops are grown regularly or when inputs of organic manures (such as, plant residues, animal manure, composts, sludge) are significant, the nature and seasonal variation in light-fraction SOM may be important to carbon turnover and nutrient cycling.

Finally, procedures for isolation of light fractions from whole soils, including determination of critical densities, deserve much more attention. More recent studies favor a critical density of 1.6 g/cm^3 (Ladd and Amato, 1980; Spycher et al., 1983; Ahmed and Oades, 1984; Sollins et al., 1984) rather than 2.0 g/cm^3 in order to obtain light fractions with reduced contents of organomineral complexes and ash, and with higher contents of noncomplexed macro-OM.

B. Densimetric Analyses of Size Separates

Density fractionation has been applied to primary particle size separates in attempts to delineate SOM pools with different turnover within a given size separate (Ladd et al., 1977a,b,c; Amato and Ladd, 1980; Ladd and Amato, 1980), to elucidate mechanisms responsible for the association between organic and inorganic soil constituents (Turchenek and Oades, 1974, 1978, 1979), and to characterize the SOM in size separates by solid-state ^{13}C-nuclear magnetic resonance (NMR) (Oades et al., 1987; Baldock et al., 1990). In most studies, however, the use of densimetric analysis has been limited to the sand size separates in order to isolate noncomplexed macro-OM.

Ladd et al. (1977a) found that sand size (50 to 2000 μm) light fractions (<1.59 g/cm^3) accounted for 0.3 to 1.1% and 3 to 7% of arable soil weight and nitrogen content, respectively. Ahmed and Oades (1984) found similar contents of nitrogen in light and heavy (1.6 g/cm^3) subfractions of sand, 5% of whole soil nitrogen being present in each fraction. For whole soil carbon, the light and heavy sand size fractions accounted for 11 and 4%, respectively. The light fraction (<2.0 g/cm^3) of sand from a range of Canadian A_h-horizons was found to make up 10 to 30% of whole soil carbon (McKeague, 1971). While 10% of the sand material was in the light fraction and had a carbon concentration of about 30%, the heavy sand fraction accounted for 90% of the weight and had a carbon content of 0.3%. Microscopic examination revealed that light fractions included organic fragments and amorphous brown composites, but also clean quartz grains.

Zhang et al. (1988) examined the amount of floatable macro-OM in sand size separates of 12 pairs of native and cultivated soils. The floatable SOM in sand accounted for 13 to 19% of the carbon in native soils (prairie and deciduous forest) and 8 to 11% in sites cultivated for 120 years. The C:N ratios indicated that floatable macro-OM was compositionally distinct from SOM associated more firmly with the sand separate. Native soils appeared to have more organic matter in the sand than cultivated counterparts.

The weight of floatable light-fraction material from sand size separates of a native prairie in Missouri, USA, was 0.3% for 25 to 50 μm, 4.6% for 50 to 200 μm, and 1.2% for 200 to 2000 μm, the corresponding values for a cultivated counterpart (Sanborn Field) being 0.1 to 0.3%, 0.2 to 1.2%, and 0.1 to 0.4% (Balesdent et al., 1988). The floatable SOM accounted for 20% of whole soil carbon, the heavy sand size separates holding about 6%. Similar losses of SOM due to cultivation were observed for light and heavy sand fractions. Similarly, SOM in sand size separates of three canadian grassland soils of different texture accounted for 20 to 37% of whole soil carbon (Tiessen and Stewart, 1983). However, the SOM floated off from the sand (50 to 2000 μm) was found to be very labile and to account for a significant part of the cultivation-induced loss of whole-soil SOM. Four years of cultivation reduced sand size floatable SOM with 43%, the loss after 65 to 90 years being 67 to 76%. Floatable SOM from sand apparently reached a new equilibrium level after 60 years of cultivation. The loss of sand size SOM was ascribed to physical breakdown and transfer of macro-OM to finer size separates, and to microbial decomposition.

The light fraction of sand separates appears to hold a significant part of the SOM in permanently vegetated soils. Cultivation generally induces a massive loss of light sand sized SOM. There are indications that SOM isolated along with sand during size fractionation is composed of at least two separate SOM pools. Light fractions usually encompass the major part of SOM in sand separates and are mainly made up of partly decomposed, particulate plant and animal remains. The SOM in heavy sand size separates may be present as organic coatings on the mineral quartz grains as observed in water-repellant sands (Emerson et al., 1986). Zhang et al. (1988) demonstrated that the SOM associated with the heavy sand fraction had very high C:N ratios.

Altogether the available information strongly suggests that light-fractions SOM isolated from sand size separates and from whole soils behave similarly during SOM turnover, and may be identical SOM pools.

Spycher and Young (1977) subjected clay size separates (<2 μm) to density fractionation using liquid densities ranging from 1.7 to 2.3 g/cm^3. Generally, the carbon and nitrogen contents as well as the C:N ratios decreased gradually with increasing density, and a distinct density for isolation of a light clay fraction encompassing noncomplexed SOM could not be identified. Later studies (Spycher and Young, 1979; Young and Spycher, 1979) including a wider range of soils confirmed the existence of a continuum in organic and mineral properties of clay between density extremes. Since these studies were based on limited soil dispersion (simple shaking), stable microaggregates may have been included in light fractions.

Turchenek and Oades (1974, 1978, 1979) exposed sand, silt, and clay size separates to densimetric analyses using liquid densities ranging from 1.8 to 2.6 g/cm^3. The SOM content and C:N ratios generally decreased with increasing density, the decrease being more gradual in clay and silt than in sand size separates. Within most size separates, the C:N ratios ranged widely between the lightest and heaviest fraction. The SOM of fractions <2.06 g/cm^3 was considered less humified than that of heavier fractions, and SOM in light fractions of sand (20 to 2000 μm) and coarse silt (5 to 20 μm) seemed less aromatic than light fractions from fine silt (2 to 5 μm) and coarse clay (0.4 to 2 μm). Whereas light fractions of the coarser separates included relatively undecomposed materials, light SOM from finer fractions was taken to be microbial products rather than plant materials. The SOM of heavy fractions was considered more aliphatic in nature. This was later supported by solid-state ^{13}C-NMR studies (Oades et al., 1987).

The fractions <2.06 g/cm^3 were rich in SOM, and showed relatively high contents of Fe, Al, Ca, and Mn together with clay minerals (Turchenek and Oades, 1978). Kaolinite, illite, quartz, and feldspars were associated mainly with heavier fractions which were low in SOM.

Ladd et al. (1977a) subjected freeze-dried clay and silt separates from two calcareous soils to densimetric analyses in order to deliniate active and passive pools of nitrogen during immobilization and mineralization processes. The light fraction (<2.06 g/cm^3) of silt (2 to 50 μm) accounted for 7 to 23% of the silt weight but was 10 to 12 times higher in nitrogen content than the heavier subfraction. About 50 to 70% of the silt nitrogen was in the light fraction. Light fractions from coarse (0.2 to 2 μm) and fine clay (<0.2 μm) accounted for only 1 to 8% and 15 to 22% of the nitrogen present in the respective size separates.

The organic ^{15}N of light and heavy subfractions of silt and fine clay changed markedly during periods of immobilization and mineralization (Ladd and Amato, 1980), but changes in complementary light and heavy fractions followed similar trends, and differences between subfractions were not significant. It was concluded that densimetric fractionation at 2.06 g/cm^3 yields light and heavy fractions whose nitrogenous components are similarly available to decomposition. In accordance with previous studies (Turchenek and Oades, 1979; Young and Spycher, 1979), Ladd and Amato (1980) concluded that differences between complementary density fractions may more easily be obtained using densities below 2.0 g/cm^3. For one soil, the proportion of native nitrogen accounted for in the light fraction of silt far exceeded that in the light fraction of whole soil. The energy input differed significantly, however, between procedures used to disperse silt and whole soil samples, and sample pretreatments (e.g., freeze-drying) may have influenced the results of densimetric analyses (Ladd et al., 1977a).

To summarize, densimetric analyses of particle size separates suggest that a major part of the SOM in clay and silt separates may be associated with a smaller part of the inorganic constituents present in these size classes, and that the SOM-rich fractions of clay and silt have specific densities below 2.2 to 2.4 g/cm^3. Except for sand size separates, the available information does not, however, demonstrate the existence of distinct pools of noncomplexed SOM in size separates. Most likely a continuum exists in the composition of clay and silt size organomineral complexes, the ratio between SOM and inorganic components varies continuously. Thus, densimetric analyses of clay and silt demonstrate the heterogeneity associated with organomineral complexes in soil, but do not show distinct and homogeneous pools of SOM with uniform chemical characteristics and biological turnover. Since most densimetric studies have relied on partly disaggregated rather than completely dispersed samples, the inclusion of unknown proportions of microaggregates in the light fractions isolated from supposedly primary particle size separates obviously complicates interpretation of results.

IV. Chemical Characteristics of SOM in Size Separates

A. Distribution of SOM

The degree of soil disaggregation reached during dispersion of samples is decisive for the resulting distribution of SOM between particle size separates. Several studies based on limited soil dispersion have demonstrated significant accumulations of SOM in silt and sand size separates (Kyuma et al., 1969; Healy and Claridge, 1974; Hamblin, 1977; Cameron and Posner, 1979; Young and Spycher, 1979; Tsutsuki and Kuwatsuka, 1989), undispersed aggregates of smaller particles rich in SOM contributing significantly to the SOM content in these separates. For soils high in light-fraction SOM, inclusion of macro-OM may cause a high SOM content in the sand (Angers and Mehuys, 1990; Elustondo et al., 1990).

Gregorich et al. (1988) examined the SOM distribution between size separates after application of step-wise increased levels of ultrasonic dispersion (Figure 5). When the soil was dispersed completely (1500 J/ml soil suspension), clay (<2 μm), silt (2 to 50 μm), and sand (50 to 2000 μm) contained 60, 30, and 5% of the whole soil SOM, respectively, whereas limited dispersion (e.g. 100 J/ml) allocated 28, 54, and 13% to these separates. Changes in the distribution of SOM between size separates resulted from changes in fraction weight and SOM content. Raising the ultrasonic energy input from 100 to 1500 J/ml caused contents

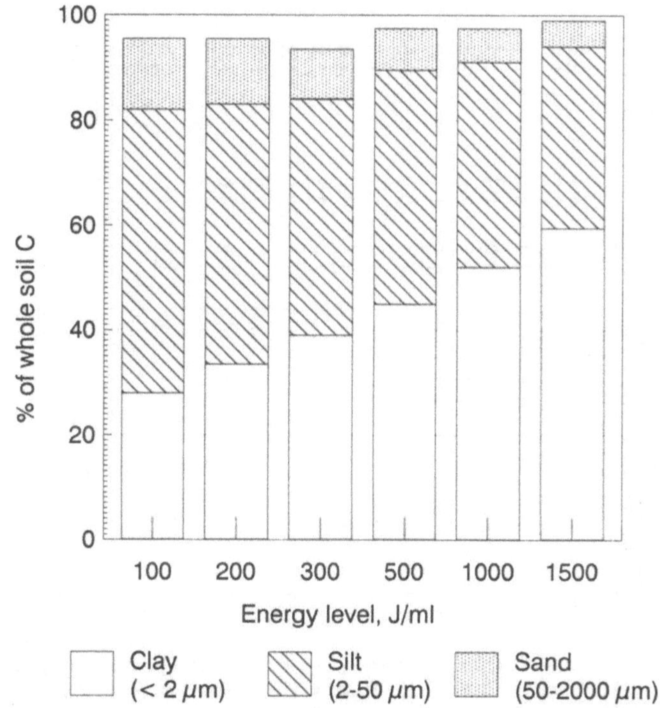

Figure 5. Distribution of whole soil carbon between size separates after dispersion of soil at different ultrasonic energy levels. Soil was from 0 to 15 cm of a cultivated Aquic Eutrochrept (gleyed melanic brunisol). Based on data from Gregorich et al. (1988).

of carbon in sand and silt to decline from 1.0 to 0.5% and 2.3 to 1.4%, respectively, while clay SOM content increased from 3.9 to 6.3%.

To compare concentrations of SOM in clay and silt isolated from different soils, the carbon enrichment ratio (mg C/g separate to mg C/g whole soil) was calculated for each separate (Table 5). The use of enrichment factors excludes the effects of different SOM levels in whole soils. Studies based on poorly dispersed soils were not considered. The highest concentrations of carbon were found in fine silt and in clay size separates. Fine silt appeared 1.3 to 4.0 times higher in carbon than whole soils, whereas carbon enrichments ranged much wider for clay (Table 5). Other studies (Figure 6) have demonstrated even greater SOM enrichments of clay and silt, especially for soils low in these separates (Christensen, 1987b; Leinweber and Reuter, 1988). Excluding soils high in light-fraction SOM, sand size separates usually show SOM enrichment factors <0.1. In Mollisols, carbon concentrations tend to peak in fine silt while in Alfisols the clay size separates show the highest carbon concentrations. Differences between individual soils do not appear to be related to whole soil

Table 5. Ca..on enrichment (%C in separate to %C in whole soil) in different silt and clay size classes isolated by ultrasonic dispersion from predominantly cultivated soils. Separate showing the highest enrichment within a soil is underlined

Reference	USDA Soil Taxonomy	Site (sampling depth)	%C	pH	Carbon enrichment: Silt (μm) 50–20	50–5	50–2	20–5	20–2	5–2	Clay (μm) 2–0.2	<2	<0.2
1	Typic Argiudoll	Iowa, USA (0–25 cm)	2.18	5.1	0.2				1.1		2.2		2.3
3	Typic Cryoboroll	Canada, A-horizon	2.41	n.d.		0.6				2.5	2.2		1.3
4	Typic Haploboroll	Sask., Canada, Ap-horizon	2.50	5.6	0.1			1.1		2.4	2.6		2.1
4	Boralfic Cryoboroll	Sask., Canada, Ap-horizon	4.60	5.0	0.2			1.0		1.6	1.4		1.4
4	Aridic Natriboroll	Sask., Canada, Ap-horizon	2.10	5.3	0.1			1.2		2.9	2.8		2.4
7	Haploboroll	Sask., Canada (0–30 cm)	6.29	6.2	0.3			1.0		1.5	1.1		1.0
7	Calciaquoll	Australia (1–10 cm)	5.35	8.0	0.2			1.2		1.3	0.9		0.5
8	Aquoll	Ontario, Canada (0–15 cm)	3.29	6.3		0.8				1.7	1.1		1.1
11	Typic Argiboroll	Sask., Canada, Ap-horizon	1.71	n.d.		0.6				3.8	2.0		1.6
11	Udic Haploboroll	Sask., Canada, Ap-horizon	3.32	n.d.		0.7				3.9	2.5		2.0
1	Typic Hapludalf	Iowa, USA (0–20 cm)	1.51	5.4	0.2				1.3		2.8		3.8
2	Udollic Ochraqualf	Missouri, USA (0–10 cm)	1.13	6.1	0.1				0.6		2.9		2.6
7	Vertic Rhodoxeralf	Australia (0–10 cm)	1.04	6.0	0.2			0.8		1.9	2.6		3.0
9	Typic Hapludalf	Denmark (0–20 cm)	1.63	6.8	0.1				3.0			5.1	
3	Vertic Cryorthent	Canada, A-horizon	2.44	n.d.		0.9				1.5	1.3		0.6
9	Mollic Fluvaquent	Denmark (0–20 cm)	1.80	6.8	0.1				2.0			3.3	
5	Aquic Entrochrept	Canada (0–15 cm)	2.31	6.8			0.6					2.7	
6	Chromic Pelloxererts	South Australia (0–10 cm)	1.14	8.2			2.1				1.6		1.0
10	Typic Haplorthox	Sao Paulo, Brazil (0–10 cm)	1.41	n.d.	0.7				2.2			1.5	
9	Orthic Haplohumod	Denmark (0–20 cm)	1.73	5.6	0.3				7.0			11.1	

References: 1, Zhang et al. (1988); 2, Balesdent et al. (1988); 3, Tiessen and Stewart (1983); 4, Curtin et al. (1987); 5, Gregorich et al. (1988); 6, Amato and Ladd (1980); 7, Turchenek and Oades (1979); 8, Catroux and Schnitzer (1987); 9, Christensen (1985); 10, Bonde et al. (1992); 11, Anderson et al. (1981)

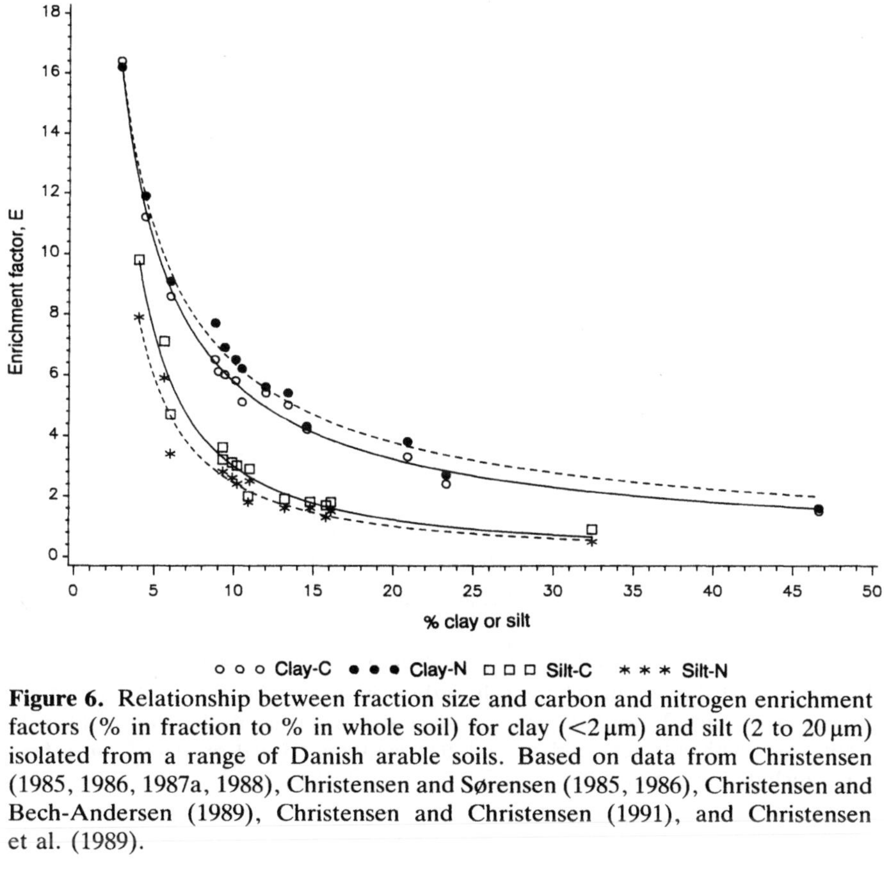

Figure 6. Relationship between fraction size and carbon and nitrogen enrichment factors (% in fraction to % in whole soil) for clay (<2 μm) and silt (2 to 20 μm) isolated from a range of Danish arable soils. Based on data from Christensen (1985, 1986, 1987a, 1988), Christensen and Sørensen (1985, 1986), Christensen and Bech-Andersen (1989), Christensen and Christensen (1991), and Christensen et al. (1989).

pH or carbon content. Since incomplete disaggregation is indicated for some soils, differences observed between soils may reflect both inherent soil-type characteristics and the ease with which different soils are dispersed when exposed to ultrasonic vibrations.

The SOM enrichment of clay and silt has been found to be inversely related to the proportion of these size separates in whole soil (Christensen, 1985, 1987b; Christensen and Sørensen, 1985; Leinweber and Reuter, 1988; Gregorich, 1989; Elustondo et al., 1990), whereas the SOM enrichment of sand does not relate to sand yields. Figure 6 depicts the relationships between SOM enrichment of clay and silt and the yields of these size separates in a number of Danish arable soils, including Orthic Haplohumods, Typic Agrudalfs, Typic Hapludalf, and Mollic Fluvaquents. Enrichments decline following a power function as clay and silt yields increase. Clay shows higher enrichment factors than silt for similar yields of dry weight, and clay size separates tend to be more

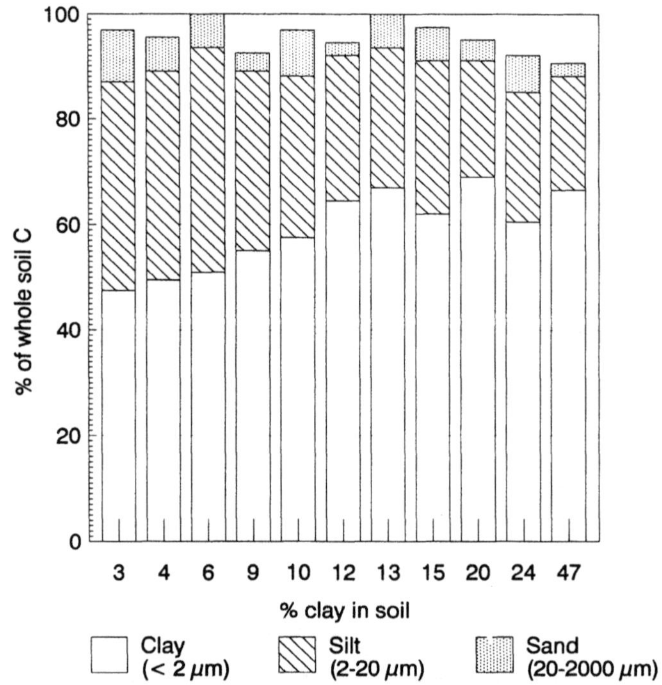

Figure 7. Distribution of whole soil carbon between clay (<2 μm), silt (2 to 20 μm), and sand (20 to 2000 μm) size separates from the A_p-horizon of Danish arable soils with different carbon and clay contents. Based on data from references included in Figure 6.

enriched in nitrogen than in carbon, the opposite being true for silt. The differences between carbon and nitrogen enrichments of clay and silt are reflected in the C:N ratio, the clay-associated SOM having a lower C:N ratio than SOM in silt (see Section IV.B and V.B).

Most of the SOM in arable soils low in macro-OM is found in clay and silt, clay size separates often accounting for more than 50% of whole soil SOM (Anderson et al., 1981; Tiessen and Stewart, 1983; Christensen, 1985; Dalal and Mayer, 1986c; Curtin et al., 1987; Balesdent et al., 1988; Gregorich et al., 1988; Bonde et al., 1992). Figure 7 shows the distribution of whole soil carbon among size separates from a range of Danish Ap-horizons with different clay and carbon content and cultivated for long periods. Between 48 and 69% of soil SOM was in the clay, while silt and sand accounted for 21 to 43% and 2 to 10%, respectively. For soils with <20% of clay, the proportion of whole soil SOM in clay tends to increase with increasing clay content while that in silt decreases. SOM in

Table 6. Ratios of C:N of native and labeled SOM in whole soils, clay (<2 μm), and silt (2–20 μm) separates of soils incubated with ^{15}N-ammonium and ^{14}C-hemicellulose for 5 years (Christensen and Sørensen, 1986)

		C:N ratio of					
		Labeled SOM			Native SOM		
Soil	% Clay	Clay	Silt	Whole soil	Clay	Silt	Whole soil
A. Ochrept	5.7	10.1	11.8	10.4	9.8	15.4	11.2
L. Agrudalf	12.4	8.4	13.0	8.9	8.8	11.5	9.5
H. Fluvaquent	23.2	12.3	12.3	12.4	7.3	13.0	8.9
R. Fluvaquent	47.4	10.3	5.9[a]	9.9	8.5	12.5	9.5
Mean		10.3	12.4	10.4	8.6	13.1	9.8

[a] Not included in mean value

sand was not related to sand yields, and the overall distribution of SOM was not related to whole soil SOM content.

In a number of soils, however, the silt size separates appear to contain the highest proportion of whole soil SOM (Watson and Parsons, 1974a; Hinds and Lowe, 1980b; Zhang et al., 1988; Elustondo et al., 1990). A complete soil dispersion was generally indicated in these studies, and the reasons for the different distributions of SOM in different soils remain unknown. Relationships to contents of clay or silt could not be established.

B. Organic Constituents

The C:N ratio of SOM in size separates declines with decreasing particle size (Chichester, 1969; Watson and Parsons, 1974a; Turchenek and Oades, 1979; Anderson et al., 1981; Tiessen and Stewart, 1983; Christensen, 1985; Zhang et al., 1988; Elustondo et al., 1990). The C:N ratio of clay is generally lower than that of whole soils, while C:N ratios of silt and sand are higher, especially if sand includes significant amounts of macro-OM. Zhang et al. (1988) removed the macro-OM from sand size separates (50 to 2000 μm) and found extremely high C:N ratios (52 to 132) for the SOM left in the heavy sand fraction. For soils incubated for 5 years with ^{15}N-ammonium and ^{14}C-hemicellulose, Christensen and Sørensen (1986) found that the C:N ratio of native SOM in clay was lower and that in silt higher than the whole soil C:N ratio, differences being less manifest for labeled SOM (Table 6). No consistent changes are found in C:N ratios of different size separates following cultivation of native soils (Tiessen et al., 1984b; Dalal and Mayer, 1987; Zhang et al., 1988; Elustondo et al., 1990), and changes in SOM content induced by various agronomic practices are not reflected in the C:N ratios of clay and silt (Adams, 1982; Christensen, 1988). The nitrogen mineralization from particle size separates during incubation correlates with the C:N ratio, the separates with narrow C:N ratios releasing more mineral nitrogen than size separates with wider C:N ratios (Chichester, 1969; Cameron and Posner, 1979; Lowe and Hinds, 1983; Catroux and Schnitzer, 1987). However, the C:N ratio of SOM must be considered a less informative indicator of SOM quality (Fog, 1988).

The C:N ratio of size separates and its relationship to nitrogen mineralization may be influenced by mineral-fixed ammonium. Clay appears to release mineral-fixed ammonium more readily than silt, and clay may contain more mineral-fixed ammonium than the silt and sand size separates (Kowalenko and Ross, 1980). The soils included in Table 6 have been analyzed for mineral-fixed native and labeled ammonium by Jensen et al. (1989). Less than 1% of the labeled nitrogen was present as mineral-fixed ammonium in soils with 6 and 12% clay, the soil with 23% clay having 3% of the labeled nitrogen bound as mineral-fixed ammonium (Table 7). The native mineral-fixed ammonium in whole soils, clay and silt accounted for 3 to 6% of the total nitrogen content. Neglecting the pool of mineral-fixed ammonium in these soils is not critical for results on C:N ratios. In contrast, the proportion of labeled nitrogen present as mineral-fixed ammonium in the soil with 47% clay accounted for 23 to 38%, corresponding values for native nitrogen ranged from 10 to 16%. The high contents of labeled mineral-fixed ammonium in this soil may have inflated the C:N ratio of SOM (Table 6). Similar levels of mineral-fixed ammonium were obtained by Kowalenko and Ross (1980) for a clay loam with 37% clay, whereas Watson and Parsons (1974a) found that

Table 7. Mineral-fixed native and labeled ammonium in whole soil, clay, and silt separates of soils incubated for 5 years with ^{15}N-ammonium and ^{14}C-hemicellulose (Jensen et al., 1989)

Soil	Fraction	Fraction weight (mg/g soil)	Mineral-fixed NH_4^+-N				
			μg/g Fraction	μg/g Whole soil		% Total-N in fraction	
				Native	Labeled	Native	Labeled
A. Ochrept	Clay	57	255	14.4	0.1	3	<1
	Silt	63	72	4.5	0.0	4	0
	Whole soil	1000	19	19.2	0.1	4	<1
L. Agrudalf	Clay	124	312	38.6	0.1	4	<1
	Silt	161	111	17.9	0.0	6	0
	Whole soil	1000	55	55.2	0.1	4	<1
H. Fluvaquent	Clay	232	412	95.5	0.8	6	3
	Silt	132	134	17.6	0.0	4	0
	Whole soil	1000	127	125.6	1.3	6	5
R. Fluvaquent	Clay	474	430	197.2	6.6	12	23
	Silt	321	166	57.3	1.4	16	38
	Whole soil	1000	248	236.2	12.0	10	31

mineral-fixed ammonium in clay and silt from five soils including a calcareous clay with 60% clay, accounted for 1 to 4% of the total nitrogen. Consequently, studies based on organomineral complexes and addressing the nitrogen turnover in clay-rich soils should consider the pool of mineral-fixed ammonium, especially when comparing different size separates within a soil and corresponding size separates isolated from soils of widely different texture and mineralogy.

The SOM in size separates has been characterized by analyses for hydrolyzable amino acids. Watson and Parsons (1974a) found similar proportions of 6N HCl hydrolyzable amino acid-N in clay and silt separates of five soils (38 to 44% of total nitrogen). The distribution of individual amino acids was generally similar for clay and silt, and similar to distributions reported for hydrolysates of whole soils. The suite of amino acids in SOM extracted from clay was almost similar to that of intact clay SOM (Watson and Parsons, 1974b).

Following limited dispersion of a clay loam (Aquoll), Schnitzer and Ivarson (1982) found some variation between size separates in hydrolyzable amino acid-N (28 to 42% of total nitrogen) and amino sugar-N (0.6 to 1.9%); the finer separates tended to be richer in amino acids. A more complete dispersion of the same soil (Catroux and Schnitzer, 1987) showed, however, relatively constant contents of amino acid-N (28 to 32%) and amino sugar-N (4 to 5%) across the size separates. The SOM extracts from size separates were higher in amino acid-N (41 to 52% of total nitrogen) but lower in amino sugar-N (1.4 to 3.0%) than intact SOM. Differences between size separates were insignificant.

Christensen and Bech-Anderson (1989) found that amino acid-N made up 31 to 39% of the total nitrogen in clay and silt, while values for sand were lower (21 to 30%). Amino acid contents in clay and silt were 5 to 10 and 2 to 5 times higher than contents in whole soils, respectively, sand being 10 to 20 times lower. The relative distribution of individual amino acids was almost the same for all separates. Annual straw incorporation increased the concentration of most amino acids and the hydrolyzability of sand SOM but was not reflected in the distribution of amino acids between size separates. Possible differences in the contribution of microbial cell wall constituents to the SOM associated with different size separates were examined by analyzing for diaminopimelic acid, which is confined to the cell wall peptidoglycans of procaryotes. The distribution of diaminopimelic acid (Table 8) suggests a higher microbial content in clay relative to silt and does therefore not lend support to the hypothesis that the reduced decomposability of SOM in silt compared with clay (see Section V.A.) can be attributed to a higher proportion of resistant cell wall materials in silt size separates (McGill et al., 1975; Ladd et al., 1977b; Anderson and Paul, 1984).

The proportion of carbon in hydrolyzable carbohydrates tends to increase when particle size decreases (Anderson et al., 1981; Angers and

Table 8. Distribution of 6 N HCl hydrolyzable diaminopimelic acid (DAPA) in whole soil, clay, silt, and sand size separates from the Ap-horizon of two arable soils (Christensen and Bech-Andersen, 1989)

	μg DAPA/g separate		% Total amino acid-N	
Separate (μm)	Loamy sand	Sandy loam	Loamy sand	Sandy loam
Clay (<2)	170	115	0.6	0.9
Silt (2–20)	61	25	0.4	0.5
Sand (20–2000)	1	1	0.5	0.6
Whole Soil (<2000)	15	20	0.5	0.7
Mean (s.d.)			0.5 (0.1)	0.7 (0.2)

Mehuys, 1990; Cheshire et al., 1990), the SOM associated with clay size separates being enriched in sugars relative to silt and sand (Table 9). Soil carbohydrates are a mixture of polysaccharides derived from plants and microorganisms. Because pentoses (e.g., arabinose and xylose) are not synthesized to any great extent by the microorganisms but are quantitatively important constituents of plant residues, various hexose/pentose ratios have been used to indicate the origin of soil carbohydrates (Turchenek and Oades, 1979; Murayama, 1981, 1984). Typically the ratio is based on contents of galactose + mannose and arabinose + xylose, galactose and mannose being minor constituents of plant polysaccharides. The galactose + mannose to arabinose + xylose ratio is generally higher for clay than for silt and sand size separates (Table 9), indicating a greater microbial input of carbohydrates to clay than to silt and sand. This is in accordance with evidence based on the distribution of diaminopimelic acid (Table 8). Turchenek and Oades (1979) found somewhat higher values, the fine silt (2 to 5 μm) and coarse clay (0.4 to 2 μm) had ratios of 2.5 to 3.5. Finer clay subfractions and separates >5 μm had sugar ratios of 2.2 to 2.4 and 1.7 to 1.9, respectively. Based on samples dispersed by a cation resin, Dalal and Henry (1988) found no consistent trends across soils and size separates, while Ahmed and Oades (1984) found a higher ratio for clay than for silt isolated after limited soil dispersion.

Natural waxes extracted from various size separates with chloroform and hexane have been found to account for 1 to 7.5% of the SOM in size separates (Schnitzer and Schuppli, 1989b), but no consistent trend was observed in the extractability of various size separates. Similar extractions of fine clay (<2 μm) from an Aquoll, a Udic Boroll, and a Typic Boroll showed that 3.4, 2.5, and 8.9% of the SOM was in wax-like material, respectively (Schulten and Schnitzer, 1990). The chemical composition of the extracts appeared very similar.

More SOM can be extracted with alkaline pyrophosphate from fine clay (<0.2 μm) than from any other fraction, the SOM in coarse silt (5 to 50 μm) being least extractable (Anderson et al., 1981; Catroux and

Table 9. Hexose to pentose ratio of SOM and the proportion of total carbon present as hydrolysable carbohydrate-C in sand (2000–20 μm), silt (20–2 μm) and clay (<2 μm) isolated from Ap horizons of different soils

			Carbohydrate-C (mg C/g separate-C)			Galactose + Mannose / Arabinose + Xylose		
Reference	Soil	% Clay	Sand	Silt	Clay	Sand	Silt	Clay
1	Askov sand, Denmark	5.7	28	77	91	0.71	0.92	1.24
	Ll. Valby, Denmark	9.0	40	97	86	1.29	1.39	1.50
	Højer, Denmark	24.3	42	69	90	0.95	0.95	1.47
	Ribe, Denmark	45.9	—	29	101	—	1.34	1.43
2	Insch, Scotland	13.5				0.58	1.31	1.52
	Countesswells, Scotland	11.9				0.61	1.25	1.64
3[a]	Kamouraska, Canada	47	69	51	73	0.61	1.40	1.61

Reference: 1, Cheshire et al. (1990); 2, Cheshire and Mundie (1981, 1990); 3, Angers and Mehuys (1990)

[a] For reference 3., sand was 6000–50 μm and silt 50–2 μm

Schnitzer, 1987). The lower extractability probably reflects a higher stability of SOM associated with silt, although SOM residing in organomineral separates after two successive extractions with alkali has been shown to possess considerable nitrogen mineralization capacity (Catroux and Schnitzer, 1987).

The SOM dissolved during the particle size separation procedure may account for 1 to 11% of whole soil SOM (Chichester, 1969; Cameron and Posner, 1979; Hinds and Lowe, 1980c; Christensen and Sørensen, 1985; Christensen, 1987a; Cheshire and Mundie, 1990). Unless soluble SOM is isolated separately, it will be collected with the finest separate and may add to the high extractability of SOM in fine clays. Furthermore, alkaline extracts of clay separates may show very high ash contents (Schnitzer and Schuppli, 1989a), suggesting that clay size organomineral complexes are included in the extract.

^{13}C-NMR spectroscopy performed on alkaline extracts of a Vertisol has shown that the SOM fraction highest in alkyl material was associated with clay size separates (Skjemstad et al., 1986). Catroux and Schnitzer (1987) also observed that alkaline SOM extracts of clay from an Aquoll were more aliphatic in nature than SOM extracted from any larger separate.

Theng et al. (1986) applied solid-state ^{13}C-NMR to clay from an acid humic silt loam and found little evidence for the occurrence of aromatic structures, the ^{13}C-NMR spectra being indicative of aliphatic SOM consisting essentially of substituted polymethylene chains. Almost two-thirds of the SOM in clay examined by Oades et al. (1987) was found to be alkyl in nature, and an accumulation of long-chain material was indicated. In accordance with Schulten and Schnitzer (1990), this material was considered to include natural waxes strongly associated with the clay minerals. Schnitzer et al. (1988) concluded that at least 50% of the SOM in intact fine clay (<0.2 μm) from a Typic Haplorthod was made up of long-chain aliphatics, and Baldock et al. (1990) similarly observed significant amounts of O-alkyl and alkyl in clay (<2 μm). Appreciable contents of aromatic structures were also seen for native SOM but not for recently formed SOM.

Catroux and Schnitzer (1987) found the aromaticity of SOM extracted from silt (2 to 45 μm) to be higher than that extracted from clay (<2 μm). Using solid-state ^{13}C-NMR on size separates from a Rhodoxeralf, Oades et al. (1987, 1988) confirmed that the proportion of SOM present in aromatic structures generally was greatest in silt (2 to 20 μm). Schnitzer and Schuppli (1989a) studied alkaline extracts of an Aquoll and a Udic Boroll by ^{13}C-NMR spectroscopy and found the aromaticity of medium silts (5 to 20 μm) to be almost twice as high as the aromaticity of coarse clays (0.2 to 2 μm). While the aromaticity of medium silt from the Boroll was 58%, that calculated for the corresponding size separate of the Aquoll was only 30%.

In conclusion, ^{13}C-NMR spectroscopy has revealed substantial differences in the chemical constitution of SOM in clay and in silt. While aromatic structures appear to be concentrated in silt size separates, SOM in clay seems dominated by aliphatics. Recent solid-state ^{13}C-NMR studies (Baldock et al., 1989, 1990) have shown that little aromatic material may be synthesized during decomposition of simple substrates. Alkyl, O-alkyl, acetal, and carboxyl structures were synthesized and accumulated predominantly in the clay size fraction (<2 μm). Evidence from ^{13}C-NMR studies therefore suggests a higher microbially derived SOM input to clay than to silt, a concept supported by the results on the distribution of soil carbohydrates and diaminopimelic acid and by measurements of the natural ^{15}N-enrichment of size separates (Tiessen et al., 1984a; Catroux and Schnitzer, 1987).

C. Other Chemical Features Related to SOM

Curtin et al. (1987) examined the cation exchange capacity (CEC) of size separates isolated from four ultrasonically dispersed Mollisols. The CEC decreased as particle size increased from <2 μm to 50 μm, the particles >50 μm showing negligible CEC. The silt size separate (2 to 20 μm) had substantial CEC indicative of a high potential surface reactivity. Dudas and Pawluk (1970) found that the CEC of coarse (1 to 2 µm), medium (0.2 to 1 µm), and fine clay (<0.2 µm) was reduced by 50, 40, and 11 to 29% respectively when SOM was partly removed by hydrogen peroxide, demonstrating that a substantial proportion of the CEC was due to SOM. Drake and Motto (1982) found the CEC of SOM to be six times higher than the CEC of clay.

Curtin et al. (1987) also observed that the pH values of the 0.2 to 20 µm separates were 0.3 to 0.4 pH units lower than whole soil, fine clay (<0.2 µm), and the >20 µm separates. Titratable acidity peaked in coarse clay (0.2 to 2 µm), but 37% of whole soil titratable acidity was accounted for by silt (2 to 20 µm). Thus, silt may contribute significantly to lime requirements of soils. The SOM as well as sesquioxides contributed to titratable acidity, and acidity was significantly correlated with SOM and with citrate/dithionite/bicarbonate extractable iron and aluminum. The results clearly indicate that separates coarser than clay are of great quantitative importance to the physicochemical reactions of soils.

Using ^{75}Se-labeled sodium selenite, Christensen et al. (1989) examined the effect of SOM on the selenite fixation capacity of clay, silt, and sand size separates from an Alfisol (Typic Agrudalf). Intact clay (<2 µm) as well as silt (2 to 20 µm) showed high fixation capacities, the fixation capacity of sand (20 to 2000 µm) being significantly lower. Hydrogen peroxide treatments reduced the selenite fixation capacity of whole soil, silt, and sand to 38, 17, and 1% respectively of that of intact samples. In contrast, clay maintained a fixation capacity corresponding to about 50%. The fixation capacity of peroxide treated "natural" clay was in accordance with values obtained previously for mined "pure" clay minerals (Hamdy and Gissel-Nielsen, 1977). The content of dithionite/citrate extractable iron and aluminum in size separates was not significantly affected by the peroxide treatments. It was concluded that the fixation capacity of sand and silt was mainly related to the SOM content, while only half the capacity of clay was accounted for by SOM.

The CEC and sorption capacities (e.g., of cations and pesticides) are related to the surface area of particles. Using the ethylene glycol monoethyl ether method (EGME) Broersma and Lavkulich (1980) found the average specific surface area of SOM-free coarse silt (20 to 50 µm), fine silt (2 to 20 µm), coarse clay (0.2 to 2 µm), and fine clay (<0.2 µm) from Ah-horizons to be, respectively, 12, 43, 113, and 170 m^2/g. Similarly, Gregorich et al. (1988) found a specific surface area of 201 m^2/g for clay (<2 µm). Relating the amount of SOM in each size separate to the corresponding SOM-free surface area, Broersma and Lavkulich (1980) observed that the weight of SOM per unit area increased with decreasing specific surface area. Thus, coarse silt had 0.97 µg SOM/cm^2 and fine clay 0.50 µg/cm^2. With the BET equation, Turchenek and Oades (1978) estimated the surface area of intact and SOM-free coarse (0.4 to 2 µm), medium (0.1 to 0.4 μm), and fine clay (<0.1 μm). For SOM-free Rhodoxeralf separates, the surface area was 54, 76, and 149 m^2/g for coarse, medium, and fine clay, respectively, the corresponding values for the Haploboroll being 85, 132, and 211 m^2/g. It could be expected that the SOM would add to the specific surface area of SOM-free separates, but the specific surface area of intact clay separates was found to be considerably lower (28 m^2/g) and to show much less variation between soils and clay size separates. To account for increased surface areas following SOM removal by hydrogen peroxide, Oades (1984) suggested that clays were embedded in a matrix of SOM. The concept of SOM having a very large specific surface area as determined by EGME (e.g., 500 to 800 m^2/g) has been seriously questioned by Chiou et al. (1990). Using the BET method, they measured specific surface areas of less than 1 m^2/g for SOM preparations and soils very high in SOM.

The role of soil minerals in the abiotic polymerization of various phenolic compounds has been reviewed by Wang et al. (1986), and it was found that SOM-free clay and silt isolated from a range of soils may catalyze oxidative polymerization. The catalytic power of silt ranged from 23 to 67% of that of clay size separates, and removal of sesquioxides by citrate/dithionite/bicarbonate reduced the oxidative power of clay and silt by 1 to 45% and 7 to 25%, respectively (Wang et al., 1983). The catalytic polymerization of phenolics differed among various mined "pure" clay minerals and quartz also exhibited some catalytic power (Wang et al., 1978). Since the samples tested were treated repeatedly with hydrogen peroxide to remove SOM, the catalytic abilities of intact primary organomineral separates remain unknown. Mineral particles embedded in SOM may well possess considerably less catalytic power than do SOM-free preparations.

V. Dynamics of SOM in Size Separates

A. Mineralization During Incubation

The decomposability of SOM associated with different size separates has been assessed during incubation under anaerobic (Chichester, 1969, 1970; Cameron and Posner, 1979; Christensen and Christensen, 1991) and aerobic conditions (Cameron and Posner, 1979; Lowe and Hinds, 1983; Catroux and Schnitzer, 1987; Christensen, 1987a; Gregorich et al., 1989). Regardless of aeration status, the proportions of carbon, nitrogen, and sulfur mineralized from various size separates generally increase with decreasing particle size. Chichester (1969) found that the relative quantities of nitrogen mineralized under water-logged conditions varied directly with the nitrogen content of the separates and inversely with the corresponding C:N ratios. The decomposability of SOM in size separates differed, however. While various clay size separates mineralized between 14 and 32% of their nitrogen content during 6 weeks at 35°C, nitrogen mineralized from larger size separates was 5.4 to 7.4%. Similarly, Cameron and Posner (1979) found that the mineralizability of nitrogen was higher from SOM in clay than from SOM in silt and sand size separates. Aerobic and anaerobic incubation produced similar results for clay, whereas mineralization was higher from silt and sand incubated anaerobically.

Lowe and Hinds (1983) examined the nitrogen and sulfur mineralization from silt (2 to 50 μm) and three clay size separates (1 to 2 μm, 0.2 to 1 μm, and 0.2 μm) isolated from five gleysolic soils. The results of Chichester (1969) and Cameron and Posner (1979) were generally confirmed and the behavior of the five soils was found to be rather similar. The mineralizations of nitrogen and of sulfur were highly correlated. For individual separates incubated aerobically at 37.5°C for 8 weeks, the proportions of nitrogen and sulfur mineralized from SOM ranged from 4.5 to 25.4% and from 2.7 to 43.3%, respectively. The proportion of sulfur mineralized from individual size separates was usually markedly greater than that of nitrogen.

The results obtained by Catroux and Schnitzer (1987) also agreed with previous observations on nitrogen mineralization from intact size separates, but it was also demonstrated that the SOM residing in size separates treated with hydrofluoric acid and extracted with strong alkali was readily decomposable (Table 10). Although 75 to 95% of the initial SOM content was extracted and the SOM remaining could be classified as humin, it was by no means biologically inert. The extractions decreased the mineralizability of sand and clay size separates, whereas the decomposability of the SOM left in coarse silt was higher than that of untreated coarse silt SOM.

Christensen (1987a) measured the CO_2 evolution from clay (<2 μm), silt (2 to 20 μm), and sand size (20 to 2000 μm) separates isolated from two field experiments with straw burned or removed. Whole soil and size separates were incubated aerobically at 20°C for 7 weeks (Table 11). The decomposability of SOM decreased in the following order: sand, clay, whole soil, and silt. The decomposition rate of clay SOM was twice that observed for silt, and while straw incorporation increased the decomposition rate of whole soil and sand-associated SOM, minor effects were noted for clay and silt. The relatively high decomposition rate of SOM in sand was due to the presence of macro-OM. The higher mineralization rate of carbon in sand does not conflict with results, demonstrating a low nitrogen mineralization rate of sand. The high CO_2 production from sand indicates a high decomposability of macro-OM which may well be coupled with immobilization of nitrogen. Adams (1980) found that the addition of macro-OM to soil led to an increased carbon and a decreased nitrogen mineralization.

Table 10. Nitrogen mineralization from whole soil and size separates incubated aerobically for 4 weeks at 28° C. Samples were either untreated (intact) or extracted with NaOH. Data from Catroux and Schnitzer (1987)

Separate (μm)	Nitrogen content (mg N/g fraction)			Nitrogen mineralization					
				% of fraction-N			mg N/kg whole soil		
	Intact	Extract 1[a]	Extract 2[b]	Intact	Extract 1[a]	Extract 2[b]	Intact	Extract 1[a]	Extract 2[b]
Sand (2000–45)	1.1	0.6	0.5	3.8	2.3	1.0	12.6	4.1	1.6
Coarse silt (45–10)	2.3	1.8	0.9	2.4	3.8	4.4	15.7	19.5	11.5
Fine silt (10–2)	5.6	3.9	3.3	4.5	2.8	4.3	65.3	28.1	36.1
Coarse clay (2–0.2)	4.5	3.1	2.5	7.8	2.6	4.1	48.1	11.6	14.1
Fine clay (<0.2)	4.3	2.3	n.d.	11.0	1.7	n.d.	6.1	0.5	n.d.
Sum of separates							147.8	63.8	63.3
Whole soil (<2000)	3.4	2.4	1.5	4.7	4.0	4.8	160.5	95.2	72.4

[a] Samples extracted with 0.5 *M* NaOH, before incubation

[b] Samples extracted with 0.5 *M* NaOH, exposed to 0.1 *M* HCl and 48% HF, and then extracted with 0.5 *M* NaOH before incubation

Table 11. Carbon mineralization from samples of whole soil and size separates incubated aerobically for 7 weeks at 20° C. Data from Christensen (1987a)

		Decomposition rate constant (μg CO_2-C/g of sample-C/day)		Respiration loss after 7 weeks at 20° C			
				% of sample-C		mg C/kg whole soil	
Separate (μm)	Straw[a]	CS-soil	SL-soil	CS-soil	SL-soil	CS-soil	SL-soil
Sand (2000–20)	Burned	190	490	0.9	2.4	16	10
	Incorporated	400	1300	2.0	6.5	36	40
Silt (20–2)	Burned	80	230	0.4	1.2	40	38
	Incorporated	90	280	0.5	1.4	43	49
Clay (<2)	Burned	180	350	0.9	1.8	106	130
	Incorporated	170	390	0.9	2.0	109	148
Sum of separates	Burned			0.7	1.6	162	178
	Incorporated			0.8	2.0	188	237
Whole soil (<2000)	Burned	110	280	0.6	1.4	129	164
	Incorporated	170	380	0.9	1.9	205	241

[a] Straw (4 t/ha) incorporated or burned annually for 18 years. CS-soil, coarse sand soil (pH 7.0, 6% clay, 2.4% C); SL-soil, sandy loam soil (pH 7.1, 15% clay, 1.2% C)

For size separates isolated from 1 to 2 mm aggregates and incubated aerobically at 24°C for 3 weeks, Gregorich et al. (1989) found a greater CO_2 evolution from sand than from clay SOM. While 4.9 and 1.5% of the carbon in sand (2000 to 50 μm) and silt (50 to 2 μm) evolved as CO_2, only 0.7% of the clay carbon mineralized during incubation. The ranking of size separates with regard to decomposability differed from that obtained by Christensen (1987a), the silt-associated SOM being twice as labile as that found in clay. However, the silt separate of Gregorich et al. (1989) encompassed the 20 to 50 μm particles which in the study of Christensen (1987a) were included in the sand, and the sum of CO_2-C produced by size separates was only 60% of the CO_2-C evolved from whole soil samples, leaving more than 40% of the potential CO_2 production unaccounted for.

The bioavailability of SOM was examined by Christensen and Christensen (1991) by measuring the potential denitrification (N_2O evolution) from clay (<2 μm) and silt (2 to 20 μm) during anaerobic incubation with excess nitrate. Samples were from a fine sandy loam (Typic Agrudalf) and were used unfrozen or exposed to one freeze/thaw cycle before being incubated. The availability of clay SOM for denitrification was twice that observed for silt-associated SOM, and the N_2O production rates following a lag period increased more steeply for clay than for silt (Figure 8). The availability of SOM in clay and silt after one freeze/thaw cycle was only half that observed for unfrozen samples, the reduced availability of SOM being ascribed to formation of microaggregates in which SOM may be less accessible to the denitrifiers. In contrast, whole soil and macroaggregates showed an increased SOM availability following freeze/thaw.

Chichester (1970) found the sum of nitrogen mineralized from size separates to be similar to the amount mineralized from whole soil. Good agreement between whole soil mineralization and cumulated mineralization from size separates was also demonstrated for nitrogen by Catroux and Schnitzer (1987) and for carbon by Christensen (1987a) (Tables 10, 11). Other studies have, however, failed to reach a balance between whole soil and the sum of the separates, the sum of contributions from size separates being lower (Gregorich et al., 1989) or higher (Christensen and Christensen, 1991) than that of whole soils.

The total mineralizations of SOM from various size separates during incubations have been compiled in Table 12. In general, clay accounts for more than 60% of the mineralizations, while some 30% is made up of mineralizations from silt size separates. Sand generally contributes less than 10%. In two studies, however, silt shows the largest contribution to the total mineralization, but the reason for the different behavior of these soils is not known. While the study of Gregorich et al. (1989) reached a relatively low recovery of whole soil mineralization, limited dispersion may have caused clay-containing microaggregates to be included in silt size separates examined by Catroux and Schnitzer (1987).

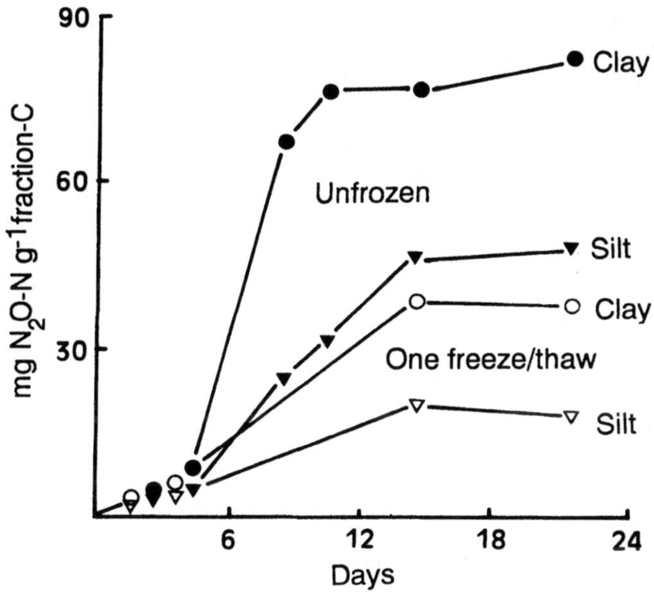

Figure 8. Bioavailability of SOM in clay (<2 µm) and silt (2 to 20 µm) measured by N_2O evolution from samples incubated anaerobically with excess nitrate at 21°C. Samples were incubated directly (unfrozen, closed symbols) or exposed to one freeze/thaw cycle (open symbols) before incubation. Data from Christensen and Christensen (1991).

In summary, the potential bioavailability of SOM increases as particle size decreases, and the mineralization rate of clay SOM is generally twice that observed for silt. Mineralization rates for sand-associated SOM appear to be high for carbon but low for nitrogen, and rates depend on the content of macro-OM in the sand. Results from studies where whole soil mineralization was similar to the sum of contributions from size separates do not indicate the presence of a massive pool of SOM physically protected within aggregates but released and made available during ultrasonic dispersion of soils. Generally, clay accounts for most of the mineralization in soil (about 60%), but the contribution from silt is also substantial (about 30%). Sand may contribute 10%, but the presence of significant contents of light-fraction SOM may cause high variability in the mineralization of carbon from sand size separates.

A greater mineralization rate from clay than from silt-associated SOM may result from differences in the bioavailability of the particle-bound SOM. However, different types of organomineral complexes may differently affect the microenvironment created in the immediate vicinity of particle surfaces, layered clay-type minerals probably creating a more favorable environment for the decomposer organisms than relatively inert

Table 12. Contribution of various size separates to total mineralization of nitrogen (N), sulfur (S), and carbon (C) during incubation, calculated as percentage of recovered mineralization from all size separates

Separate (μm)	Chichester (1969) N	Cameron and Posner (1979) N	Lowe and Hinds (1983)[a] N	Lowe and Hinds (1983)[a] S	Catroux and Schnitzer (1987) N	Christensen (1987a)[b] C	Gregorich et al. (1989) C
Sand							
2000–200		3.2					
2000–50	4.0				8.5		20.5
2000–20						13	
200–50		3.7					
Sum	4.0	6.9			8.5	13	20.5
Silt							
50–20	1.4						
50–10					10.6		
50–4		7.5					
50–2			37.5	34.8			44.7
20–10	3.0						
20–2						23	
10–2	13.6				44.2		
4–2		15.0					
Sum	18.0	22.5	37.5	34.8	54.8	23	44.7
Clay							
<2						65	34.7
2–1	12.4		17.5	21.1			
2–0.2		47.4			32.5		
1–0.2	34.7		34.6	35.2			
<0.2	30.9	20.1	10.3	8.8	4.1		
Sum	78.0	67.5	62.5	65.2	36.7	65	34.7

[a] Mean of five soils
[b] Mean of two soils

minerals such as quartz. Thus, pure clay minerals have been shown to enhance yields of biomass and production of microbial metabolites (Filip, 1977) and to have greater catalytic potentials (Wang et al., 1983). Differences between size separates in CEC, pH etc. have also been demonstrated (Curtin et al., 1988).

B. Stabilization of Decomposition Products

Detailed reviews on the interaction between soil inorganics, microorganisms, metabolic processes, and SOM components can be found in Huang and Schnitzer (1986), Stotzky (1986), Oades (1989), and Coleman et al. (1989). This section considers the distribution of decomposition products and residual substrates among size separates from soils incubated with isotope-labeled materials.

Examining a silt loam where ^{15}N-ammonium sulfate had been incorporated into SOM after successive cycles of cropping and crop residue return, Chichester (1970) was probably the first to observe substantial differences in the distribution pattern of native and labeled SOM. While the proportion of labeled nitrogen in clay (<2 μm) was 6.7 times that in silt (2 to 53 μm), the proportion of native nitrogen was only 4.2 times higher, indicating a relative enrichment of newly formed SOM in the clay.

Ladd et al. (1977a,b,c) incubated two calcareous soils with ^{15}N-nitrate and glucose or wheat straw and followed in detail the changes in the distribution of labeled SOM formed during the incubation. In soils amended with glucose, the labeled nitrate was immobilized within 2 days. At this early stage of decomposition, the proportion of labeled SOM found in clay exceeded that of native SOM, the opposite being true for silt (2 to 50 μm). After 16 days the proportion of labeled SOM recovered in silt had increased and that in clay had decreased, but the distribution patterns of labeled and native SOM were still different after 160 days. It was inferred that the silt would accumulate an increasing proportion of the more stable SOM as decomposition progressed. Lower decomposition rates of silt-associated SOM relative to clay and transfer to silt of stabilized decomposition products from other size separates were held responsible. The transfer could be in relatively resistant cell wall materials formed by successive microbial populations. Ladd et al. (1977c) found that destruction of the microbial biomass was accompanied by a significant decrease in silt SOM, the proportion of SOM in the biomass being greater for labeled than for native SOM.

Later Amato and Ladd (1980) incubated soil with double-labeled medic residues for 130 days, and the results essentially confirmed previous observations. It was concluded that SOM formed during the decomposition of added substrates becomes distributed among most size separates, but that the newly formed SOM in clays is more readily transferred or decomposed. In the long term, substrate-derived SOM will continue to decline disproportionately rapidly from clay and from microbial biomass, and silt will tend to accumulate higher proportions of the more stable residual SOM.

Cheshire and Mundie (1981) examined the distribution of native and labeled carbohydrates among size separates from soils incubated with ^{14}C-labeled glucose (28 days), ryegrass (1 year), and straw (4 years). After adding glucose, the highest specific activities of the sugars were in the clay (<2 μm). Clay was relatively enriched and silt (2 to 20 μm) depleted in labeled carbohydrates. Soils incubated with plant materials had substantial proportions of labeled glucose, xylose, and arabinose (sugars dominating plant materials) in the sand separates (20 to 2000 μm).

However, sugars of microbial origin (mannose, rhamnose, and fucose) showed greatest concentrations in the clay and their distribution resembled that found after the addition of glucose. Thus, if the plant-related sugars are excluded, the distribution of labeled sugars was relatively similar for all incubations.

A preferential accumulation of labeled SOM in clay was also reported for soil incubated with ^{13}C-glucose for 5 weeks (Baldock et al., 1990). The residual labeled carbon was in predominantly alkyl and O-alkyl structures as seen by ^{13}C-NMR, and significant amounts of acetals and carboxyls were present in clay. Little aromatic or phenolic SOM was synthesized during incubation.

To expand the time perspective, Christensen and Sørensen (1985) fractionated soils differing in clay contents and incubated for 5 to 6 years with ^{14}C-glucose, -hemicellulose and -straw. The hemicellulose incubations included ^{15}N-ammonium-sulfate, and a soil incubated for 18 years with ^{14}C-straw was also fractionated. After 5 to 6 years, carbon enrichments (%C in fraction to %C in whole soil) of clay were higher for labeled than for native SOM, silt enrichments being lower (Figure 9). A similar pattern was found for nitrogen (Christensen and Sørensen, 1986), and in accordance with data included in Figure 6, clay enrichment factors were higher for nitrogen than for carbon. The distribution pattern was not related to kind of organic amendment. The highest specific activity was in clay, and in accordance with studies based on short-term incubations, the proportion of SOM in clay was greater for labeled than for native SOM. The opposite trend was seen for silt. In contrast, similar specific activities were found for SOM in clay and silt from the 18-year incubation. Moreover, both clay and silt showed similar proportions of labeled and native SOM (Figure 10).

Thus, the distribution pattern of SOM formed during decomposition of a variety of organic substrates appears similar but remains different from that of the bulk of native SOM for up to several years. When a pulse of readily available substrate is added, the characteristic distribution of substrate-derived SOM is established within a few days. For complex substrates the distribution pattern takes longer to become manifest because of interference from unprocessed substrate. Once established, however, the distribution pattern of substrate-derived SOM can be traced for extensive periods, indicating that the turnover and possible transfer of substrate-derived SOM between size separates are slow and the mechanisms operative over time spans up to decades.

Studying ten soils with clay contents ranging from 17 to 37% and incubated with ^{14}C-glucose for 90 days, Gregorich (1989) found that clay-enrichment factors for native and labeled carbon showed a close negative relationship to soil clay content. But the slope of the linear regression was significantly greater for labeled than for native SOM. Thus, clay from more sandy soils became relatively more enriched in new SOM than clay

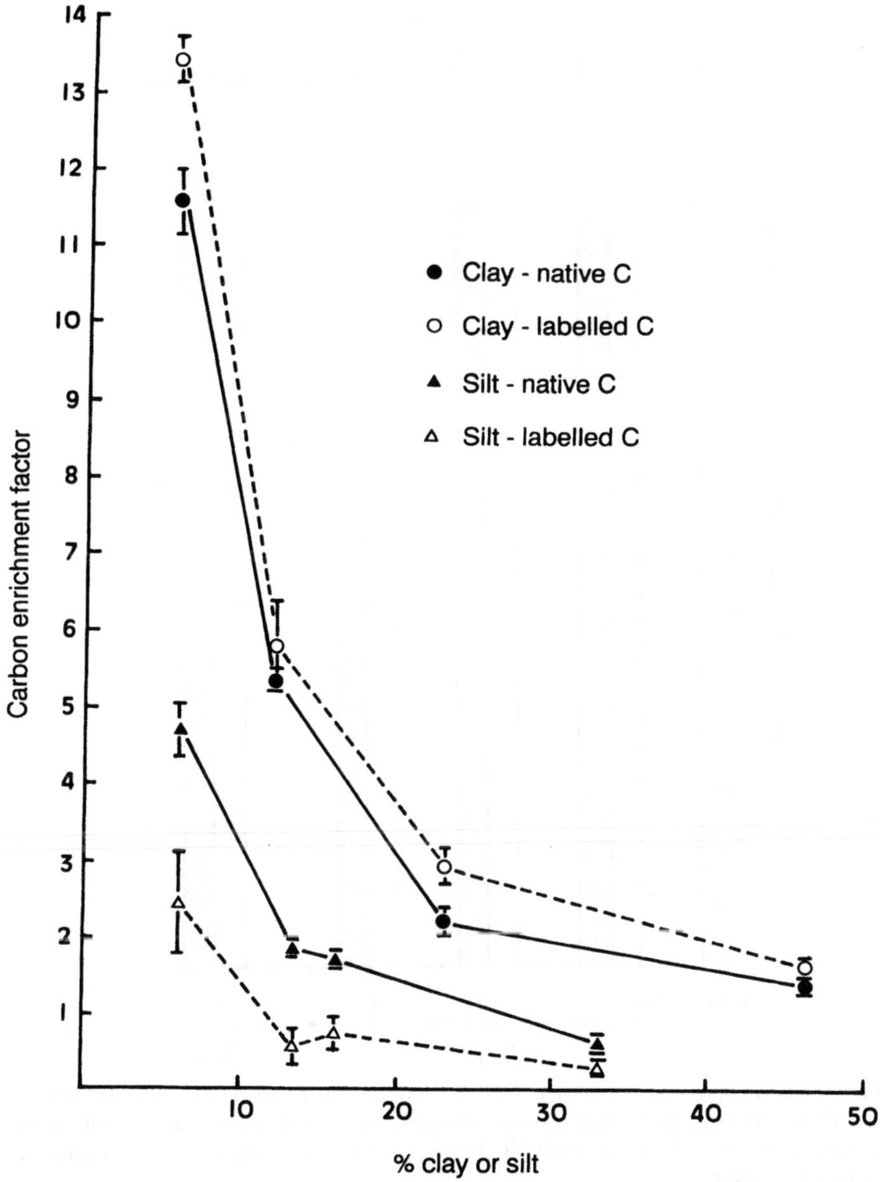

Figure 9. Relationship between fraction size and carbon enrichment factors (%C in fraction to %C in whole soil) for labeled and native SOM in clay (<2 μm) and silt (2 to 20 μm) isolated from soils incubated for 5 to 6 years with ^{14}C-labeled glucose, hemicellulose and barley straw. Bars indicate ±1 standard deviation (from Christensen and Sørensen, 1985).

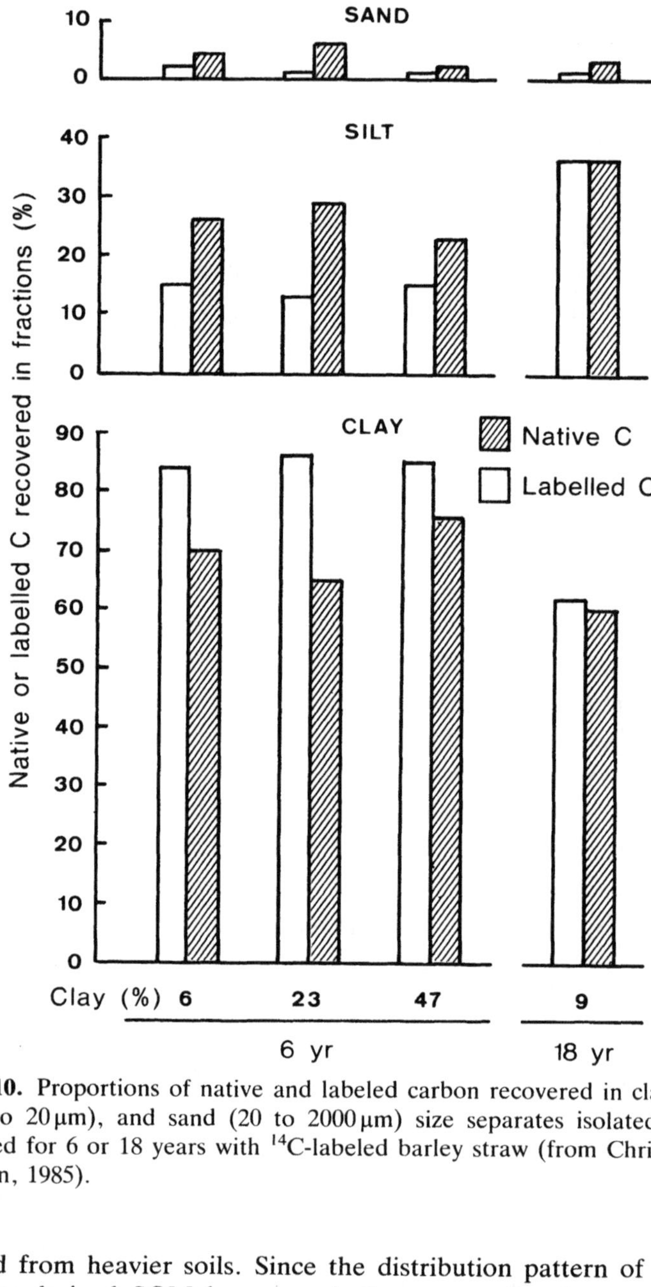

Figure 10. Proportions of native and labeled carbon recovered in clay (<2 μm), silt (2 to 20 μm), and sand (20 to 2000 μm) size separates isolated from soils incubated for 6 or 18 years with ^{14}C-labeled barley straw (from Christensen and Sørensen, 1985).

isolated from heavier soils. Since the distribution pattern of native and substrate-derived SOM becomes similar eventually, the results suggest that clay SOM formed during the initial and rapid decomposition phase is less resistant to further turnover in sandy soils than in more clayey soils. Enrichment factors for native SOM in silt (2 to 50 μm) were also linearly correlated with silt yields, whereas enrichment factors for labeled SOM were not. A similar trend appears for silt data in Figure 9.

Sørensen (1975, 1981, 1983) found that the amount of labeled SOM retained in soil was proportional to the soil clay + silt content. The distribution pattern was established within the initial 10 days of incubation when biological activity was most intense. Subsequent decay rates for labeled SOM were, however, largely independent of clay + silt content. Dalal and Mayer (1986b) demonstrated that the rate of SOM loss upon cultivation of virgin soils was inversely related to soil clay content. It appears that the amount of clay in soil is more decisive for the SOM level established than for the subsequent decomposition rates. The higher SOM enrichment of clay and silt (Figure 6), and the reduced resistance against subsequent turnover of newly formed clay SOM in sandy soils (Gregorich, 1989), suggests that these soils may be less efficient in stabilizing new SOM than soils more rich in clay. This concept agrees with results of Sørensen (1967, 1972) demonstrating that the addition of various mined clay minerals to a sandy soil caused a higher retention of labeled microbial products. Field studies with labeled plant residues have also indicated that retention of labeled SOM tends to increase with increasing soil clay content (Jenkinson, 1977; Ladd et al., 1985; Amato et al., 1987).

Individual soil particles are not envisaged to have a finite capacity for binding SOM, but it is hypothesized that the strength with which recently formed SOM initially becomes attached to organomineral complexes will decline as the distance between the mineral surface and the new SOM increases, and that SOM associated with soil minerals impedes stabilization of new SOM on those portions of the mineral surfaces which are already covered in SOM. Although it has been shown that sorption of hydrophobic ring compounds on mineral surfaces is enhanced when minerals are coated with a layer of humic substances (Murphy et al., 1990) and that CEC is positively related to SOM, clay and silt particles already embedded in a matrix of stabilized SOM probably exhibit surfaces that are less capable of binding new SOM strongly enough for it to resist further biological attack. Indeed, adsorption and other reversible binding mechanisms may increase the decomposability of various organic compounds.

Organomineral complexes highly enriched in chemically less reactive SOM are probably also less attractive for microbial adhesion and colony formation. This could result from a reduced ability of the organomineral complex to favorably modify the physicochemical environment of the microorganisms. Such particles may therefore receive reduced inputs of microbial metabolites formed during turnover of substrate-derived SOM. Most of the microbial biomass has been found to be associated with clay and silt (Amato and Ladd, 1980) or with discrete organic particles (Kanazawa and Filip, 1986), and the levels of microbial biomass in soils have been found to increase with increased soil clay content (Ladd et al., 1985; Gregorich et al., 1991). Moreover, the loss of microbial biomass during soil storage was found to be more rapid in sandy than in clayey soil (Anderson and Domsch, 1989).

Substrate-derived SOM may be stabilized against further decay by being incorporated into the soil microbial biomass, by developing a biochemically recalcitrant nature through covalent abiotic or enzymatic reactions with other organic molecules, and by becoming firmly associated with surfaces of organomineral complexes through the action of amorphous sesquioxides or reactive moieties of SOM molecules already present on particle surfaces. Intercalation into clays is also occurring (Lagaly, 1984; Theng et al., 1986; Schnitzer et al., 1988), although the quantitative significance of this process is not clear for natural organomineral complexes.

Segregation of SOM in micropores that are inaccessible to microorganisms has also been suggested as a major stabilizing mechanism in soil (Van Veen and Paul, 1981; Van Veen et al., 1985; Van Veen and Kuikman, 1990) as has a spatial separation in the soil matrix of available SOM and microorganisms. However, the accordance between whole soil mineralization of SOM and the cumulated mineralization from corresponding size separates (Chichester, 1970; Catroux and Schnitzer, 1987; Christensen, 1987a) does not indicate the existence of a large pool of SOM stabilized by microscale physical barriers. Van der Linden et al. (1987) also considered that physical protection of SOM in sensu Van Veen and Paul (1981) was likely to be less important for SOM stabilization in the long-term arable soils of NW Europe. In contrast, disintegration of macroaggregates has been demonstrated to release readily decomposable SOM (Elliott, 1986; Seech and Beauchamp, 1988).

C. Impact of Cultivation and Agronomy

It is well documented that cultivation of native soils and permanent pastures causes substantial reductions in levels of SOM (Dalal and Mayer, 1986a,b; Schlesinger, 1986). Cultivation effects on SOM levels are through complex interactions of the physical, chemical, and biological soil processes, leading to increased decomposition rates and redistribution of SOM. Reduced inputs of plant residues and increased soil disturbance usually follow cultivation, but the exact nature of the changes induced by cultivation depend on the particular agronomic practices adopted and on properties of the virgin soil.

The effect of cultivation and agronomy on the amount and composition of light-fraction SOM (noncomplexed or macro-OM) has been addressed in Section III. Generally, light-fraction SOM is much reduced upon cultivation, and agronomic practices with little return of organic materials to the soil provide little light SOM. In soils cultivated for long periods,

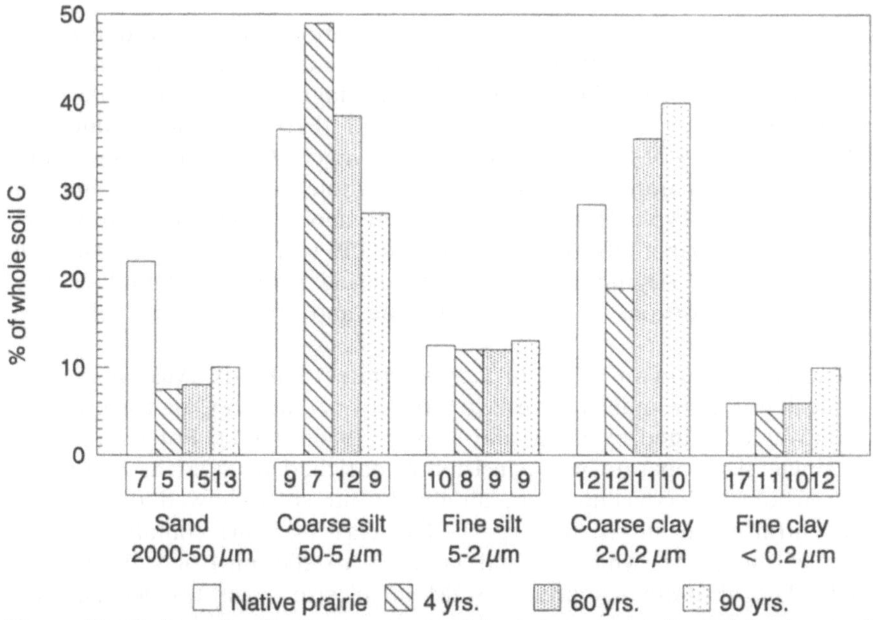

Figure 11. Carbon distribution among particle size separates from A-horizons of Typic Cryoborolls under native prairie or cultivated for 4, 60, or 90 years. Boxed values are 0/00 ^{15}N enrichment relative to air. Based on data from Tiessen and Stewart (1983) and Tiessen et al. (1984a, 1984b).

light fractions or macro-OM in sand usually account for <10% for whole soil SOM.

Tiessen and Stewart (1983) and Tiessen et al. (1983) examined the effects of cultivation on the SOM distribution among size separates from Canadian prairie soils of different textures (Cryoborolls). Floatable macro-OM in sand was readily lost upon cultivation (Figure 11) mainly due to enhanced physical disintegration and subsequent accumulation in finer separates, predominantly in coarse silt. Cultivation decreased the SOM concentration in all size separates. Fine clay and sand showed the greatest decline with much smaller reductions in fine silt and coarse clay.

The proportion of whole soil SOM present in coarse clay increased with time of cultivation. SOM in coarse silt declined after 4 years of cultivation (Figure 11). The proportion residing in fine silt remained almost constant. Losses of SOM from a sandy loam followed a similar pattern, while a heavy clay soil (65% clay) showed greater stability of silt and clay SOM. Losses from sand + macro-OM made up 40% of whole soil SOM loss due to cultivation. More than 50% of the SOM was in the coarse clay, and cultivation caused only a minor decrease in this SOM. The proportion of SOM in coarse silt increased somewhat. Thus, the redistribution of SOM between size separates was generally characterized by a shift towards fine silt and coarse-clay-associated SOM. Cultivation not only reduced the overall SOM level in these Mollisols but caused a disproportionately large reduction in the SOM pools showing high nutrient mineralization potentials. Zhang et al. (1988) also found that cultivation induced a relative shift of SOM from coarser (sand + macro-OM and coarse silt) to the finer size separates. The proportion of whole soil SOM in the fine and coarse clay of Typic Hapludalfs increased significantly upon cultivation while that in coarse (50 to 20 µm) and fine (10 to 2 µm) silt decreased. In contrast, Typic Argiudolls showed increasing proportions in fine clay only; the proportions in coarse clay, fine and coarse silt all remained nearly constant.

The distribution of SOM in size separates from six Australian subtropical Vertisols under native vegetation or cereal cropping for up to 70 years was examined by Dalal and Mayer (1986c, 1987). For virgin soils, no single size separate was consistently enriched in SOM as compared with whole soil concentration, although four soils showed the highest concentration in silt (2 to 20 µm). The clay (<2 µm) contained about 50% of the SOM, while silt and sand (20 to 2000 µm) had about 25% each. The absolute amount of SOM in all size separates decline with period of cultivation. However, the relative proportion of carbon in sand declined rapidly from 26 to 12% while that in clay increased from 48 to 61%. The proportion of SOM in silt showed no consistent trend with time of cultivation. The proportion in clay generally increased exponentially, the rate of increase being inversely related to soil clay content. In contrast to Tiessen and Stewart (1983), the rapid initial decline in sand SOM apparently occurred without a concomitant increase in silt and clay SOM. However, results are not directly comparable because the sand size separate (2000 to 20 µm) of Dalal and Mayer included much material which in the study of Tiessen et al. was assigned to coarse silt (50 to 5 µm). Moreover, soil fractionation methods were very different.

Nevertheless, the major changes in SOM distribution upon cultivation of subtropical Vertisols were consistent with changes observed on Mollisols of colder temperature regimes. Initially, soils release a pulse of carbon and nitrogen as the general SOM level declines rapidly. As cultivation proceeds and the rate of SOM loss decreases exponentially, residual SOM accumulates in clay size separates with relatively high nitrogen mineralization potentials. The residual SOM in long-term arable soils may, therefore, show higher nitrogen mineralization potentials than SOM in virgin soils and permanent pastures. In contrast to nitrogen, carbon mineralization may be reduced when the amount of macro-OM in sand and coarse silt (>20 µm) is rapidly depleted during the initial cultivation phase. Thus, Adams (1980) concluded that macro-OM was an important reservoir of biologically available carbon but not a short-term source of available nitrogen. Comparing carbon and nitrogen turnover in adjacent grassland and cropland (Typic Haploboroll), Schimel (1986)

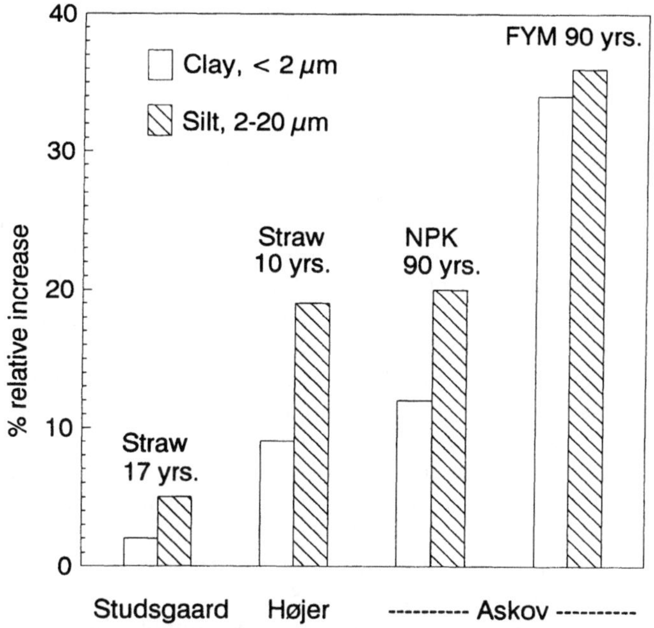

Figure 12. Relative increase in carbon content of clay and silt following straw incorporation (reference: straw removed) and use of mineral fertilizer (NPK) and animal manure (FYM) (reference: unmanured). Studsgaard, Højer, and Askov soils contained 6, 24, and 9% of clay, respectively (from Christensen 1986, 1988).

found higher carbon and lower nitrogen mineralization rates from grasslands than from arable soils. Cultivation increased the turnover of nitrogen but decreased that of carbon, and it was suggested that microorganisms in grasslands are nitrogen limited while those in cultivated soils are limited by the availability of carbon.

The relative increases in clay and silt SOM after annual incorporation of straw and after repeated additions of equivalent amounts of nutrients in mineral fertilizer (NPK) and animal manure (FYM) are shown in Figure 12. Following straw incorporation or use of NPK, the relative increase in silt SOM (2 to 20 µm) was generally twice that observed for clay (<2 µm). Mineral fertilizers influence SOM levels by increasing crop productivity, which causes a greater return of plant residues to soil. The effect of NPK on SOM may thus be similar to that of straw. The FYM treatment caused similar relative increases in clay and silt SOM, probably because animal manure provides an additional input of biologically processed material. Elustondo et al. (1990) found that compared with maize, the carbon contents of sand (2000 to 50 µm) and silt (50 to 2 µm) were 61 and 15% higher, respectively, in Quebec soils under >5 years of continuous lucerne cropping, while the cropping system had no influence of the SOM content of clay.

The distributions of SOM among size separates following different agronomic practices have qualitative bearing on SOM turnover which cannot be read directly from changes in whole soil SOM content. When plant residues decompose in soil, microbial decomposition products and residual substrate components are distributed among size separates. Silt and clay in particular are thereby enriched in newly formed SOM (Section V.B). However, the proportion associated with silt increases with time, and this SOM is more stable than that found in clay (Section V.A). Agronomic practices resulting in a sustained, greater return of plant residues to the soil will therefore cause silt to accumulate a relatively large part of the SOM derived from decomposition of plant residues, and this SOM will show a relatively high stability against further microbial attack. Long-term use of animal manure, composed of partly decomposed plant structures and microbial residues, will also add to the less stable SOM found in clay. Therefore, for a given soil where similar contents of SOM have been reached either through return of plant residues or by use of animal manure, the FYM-treated soil will contain a larger proportion of more mineralizable SOM.

The differential effects of agronomy on SOM contents in size separates are reflected in the SOM enrichment factors. For a range of German long-term field experiments on different soils, Leinweber (1988) found that SOM enrichment factors (%SOM in separate to %SOM in whole soil) of clay ($<2\,\mu m$) were higher for unmanured and mineral-fertilizer-treated plots than for plots receiving organic manures. Slightly lower values were found for silt (6.3 to $20\,\mu m$). A similar trend was observed by Christensen (1988), and Elustondo et al. (1990) showed that SOM enrichment in clay tended to be higher after continuous maize cropping relative to lucerne. It may be speculated that management-induced changes in the SOM enrichment of size separates within a given soil would reflect changes in fundamental soil properties related to SOM turnover, the decomposability of SOM in different size separates being different (see Section V.A). The use of enrichment factors as indicators of the qualitative consequences of management-induced changes in whole SOM level deserves further examination.

D. Long-term SOM Turnover Studies Involving Isotopes

The turnover of plant-derived SOM in whole soils (Jenkinson, 1977; Sørensen, 1987; Voroney et al., 1989) and in particle size separates (Christensen and Sørensen, 1985; Cheshire et al., 1990) has been followed over 10 to 20 year periods by using experimentally introduced isotopes. In recent years studies on the long-term turnover of SOM in organomineral separates have been based on differences in the natural abundance of isotopes (^{13}C, ^{14}C, ^{15}N). Radiocarbon dating relies on the time-dependent decay of ^{14}C in plant residues which at the time of death were in isotopic equilibrium with $^{14}CO_2$ in the atmosphere. The use of natural ^{15}N abundance draws upon the microbial discrimination against ^{15}N, the catabolic processes favoring an elimination of ^{14}N. Successive microbial transformations of nitrogen-containing materials will therefore cause a progressive ^{15}N enrichment of the SOM left behind, and microbial residues will tend to be enriched in ^{15}N. The feasibility of ^{13}C in SOM studies originates in the different discrimination against $^{13}CO_2$ in plants with C_3 (Calvin) and C_4 (Hatch-Slack) photosynthetic pathways. The $^{12}C:^{13}C$ ratio of SOM will reflect that of the plant residues entering the soil as isotopic discrimination during decomposition is comparatively small (Dzurec et al., 1985). Assuming SOM to have reached isotopic equilibrium, a replacement of the vegetation with another having a different $^{12}C:^{13}C$ ratio will gradually change the isotopic composition of SOM and thereby allow a quantification of the turnover of the original SOM.

Anderson and Paul (1984) applied ^{14}C-dating to Ap-horizons of three Borolls. Coarse clay (0.2 to 2 μm) and fine silt (2 to 5 μm) were found to contain the oldest and thus most stable SOM, but for one soil the coarse silt (5 to 50 μm) appeared to encompass the oldest SOM. SOM in fine clay (<0.2 μm) was invariably the most enriched in ^{14}C and was thus taken to have the fastest rate of turnover. However, fine clay was isolated by evaporating the suspension left after removal of larger size particles (Anderson et al., 1981). In this procedure fine clay accumulates SOM that has come into solution during the fractionation procedure. Soluble SOM may originate from all size separates, but will probably be dominated by contributions from recently added substrates and from cell contents of microbial biomass killed during sample pretreatments and soil dispersion. Thus, soluble SOM is likely to be "very young," and the ^{14}C dates obtained may not be representative of those for particle-associated SOM. Scharpenseel et al. (1986) observed that reversing the order of collecting clay size separates made fine clay the oldest separate.

Using the fractionation method of Anderson et al. (1981), Tiessen et al. (1984a) examined the natural abundance of ^{15}N in size separates from native and cultivated Cryoborolls. Fine silt and coarse clay had intermediate ^{15}N enrichments and showed little change in ^{15}N abundance even with prolonged cultivation (Figure 11), indicating that these separates were of mixed plant and microbial origin and relatively stable. The presence of less processed plant material in sand was indicated by a relatively low ^{15}N enrichment, while the high ^{15}N enrichment in fine clay indicated an accumulation of more processed SOM, probably of microbial origin. Both sand and fine clay showed significant changes in ^{15}N abundance upon cultivation. In soils under native prairie, the fine clay was high in ^{15}N, and this SOM was taken to be dominated by residual material from microbially mediated processes. Again, the inclusion of soluble SOM may have influenced the result obtained for fine clay. The SOM transformations during cultivation of Mollisols in relation to organomineral size separates have been discussed in detail by Tiessen et al. (1984b).

Cerri et al. (1985) examined a Haplorthox under native forest (C_3) and two adjacent clearcut sites planted with sugar cane (C_4) for 12 and 50 years. Size separates from 0 to 20 cm were isolated by the ball-assisted shake method (see Section II.B). Total as well as forest-derived SOM decreased rapidly in all separates during the first 12 years of sugarcane cropping. About 60% of the SOM left in the <50 µm separate after 50 years was derived from the previous forest. The fine sand (50 to 200 µm) contained less of the whole soil SOM, but turnover times of the two >50 µm separates were both about 40 years.

Deforestation greatly reduced the amount of SOM in sand but 38% of the SOM remaining after 50 years of cropping was of forest origin, indicating that part of the macro-OM was very stable. Skjemstad et al. (1990) observed a high retention of forest-derived light SOM (<1.6 g/cm^3) in pastures planted after clearing of rainforest. The stability of light SOM was partly ascribed to the presence of charcoal low in ^{13}C content, and originating in prehistoric burning of vegetation. The site planted to sugarcane for 12 years was cleared by accidental burning of the rainforest (Vitorello et al., 1989), but charcoal did not appear to inflate the results of Cerri et al. (1985). Soil carbon content was, however, more variable on this site. Balesdent et al. (1988) also observed the presence of old "contaminating" carbon in paririe soils cultivated for 97 years. The presence of charcoal and other biologically inert carbonaceous material may restrict the use of natural ^{13}C abundance in SOM turnover studies, and Skjemstad et al. (1990) recommended that light-fraction SOM should be removed before analyzing for ^{13}C.

The sites of Cerri et al. (1985) were studied in greater detail by Vitorello et al. (1989). Samples from 10 to 20 cm were freed of recognizable plant material and extracted with sodium pyrophosphate. Thereby, 33 to 37% of the SOM was removed. In accordance with Cerri et al. (1985), SOM in sand was greatly reduced upon cultivation, whereas concentrations of SOM in fine clay (<0.1 µm) and coarse clay + silt (0.1 to 50 µm) remained constant. The proportion of total SOM in coarse clay + silt increased, however, from 38 to 55% during cultivation. Most changes in the SOM distribution between size separates took place during the initial 12 years after deforestation.

The proportion of SOM derived from sugarcane decreased progressively with particle size, and after 12 and 50 years the 0.1 to 50 µm separate contained 89 and 69% forest-derived SOM, respectively. The corresponding values for fine clay were similar, suggesting that the SOM in fine clay did not differ qualitatively from that of the coarse clay + silt. This is

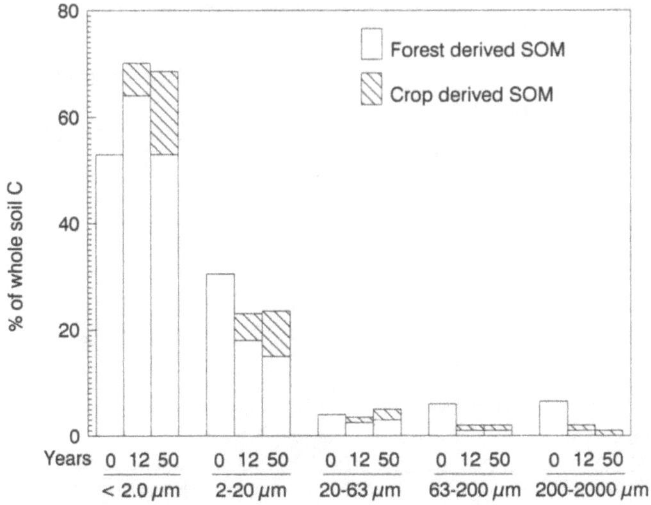

Figure 13. Distribution of SOM between size separates from the 0 to 10 cm of a Haplorthox under native forest (0) or cultivated with sugar cane for 12 and 50 years after deforestation. Based on data from Bonde et al. (1992).

in contrast to results of Tiessen et al. (1984a) and Anderson and Paul (1984) who found a faster turnover of fine clay SOM than of SOM in coarse clay and fine silt. Alkali-extractable SOM was similar to whole soil SOM in isotopic composition, and does not therefore appear to represent a qualitatively distinct SOM pool. While Vitorello et al. (1989) relied on alkali-extracted soil from 10 to 20 cm depth and chemical/mechanical dispersion, Bonde et al. (1992) applied ultrasonic dispersion in water to intact soil from 0 to 10 cm. Fine particles were separated into clay (<2 μm), silt (2 to 20 μm), and fine sand 1 (20 to 63 μm). Silt had the highest concentration of carbon (31 to 93 mg C/g separate), but clay contained the highest proportion of whole soil SOM (Figure 13). The proportion of SOM held in silt, fine sand 2 (63 to 200 μm), and coarse sand (200 to 2000 μm) decreased after forestation, while that in clay increased. Turnover times for SOM in clay, silt, and sand size separates were 59, 6, and 4 years. Thus, in contrast to other soils, silt and sand SOM exhibited similar decomposability.

Studying the natural ^{13}C abundance in size separates from the 0 to 30 cm of a temperate Hapludalf planted with maize (C_4) for 23 years after pine forest (C_3) clearing, Balesdent et al. (1987) found that SOM in silt (2 to 20 μm) contained the lowest proportion of maize-derived SOM (12%) and thus had the slowest turnover. Coarse sand (200 to 2000 μm) had 61% of its SOM from maize, the turnover of fine sand separates (20 to 50 and 50 to 200 μm) being much slower. Coarse clay (0.2 to 2 μm)

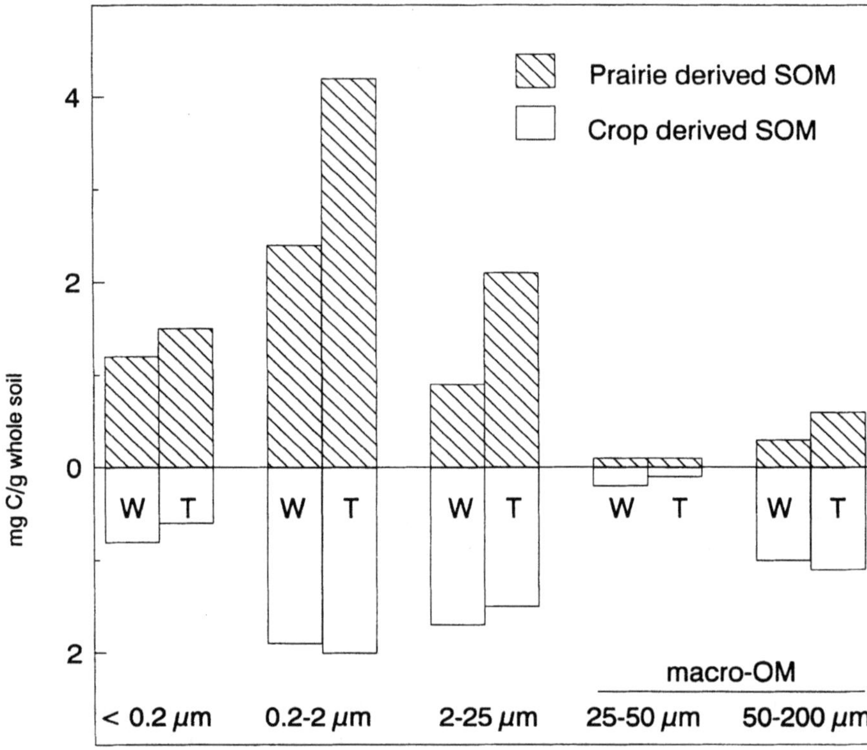

Figure 14. Distribution of prairie- and crop-derived SOM between size separates from Sanborn Field (Udollic Ochraqualf) plots under timothy (T) and wheat (W) for 97 years. Based on data from Balesdent et al. (1988).

and fine clay (<0.2 μm) had 18 and 26% of the SOM coming from maize, indicating that SOM in clay turned over faster than SOM in silt. These results thus conform to the general ranking of size separates with regard to SOM stability.

The SOM associated with clay from a Udollic Ochraqualf prairie (C_4) soil cultivated with timothy or wheat (C_3) for 97 years was found to be most persistent (Balesdent et al., 1988). The major loss of SOM occurred during the initial 27 years and was from macro-OM in separates >25 μm and SOM in fine silt (2 to 25 μm). After 27 years of cultivation, ^{13}C values for fine clay (<0.2 μm) remained constant and higher than those of the larger size separates, suggesting little further turnover of prairie-derived SOM in this separate, or alternatively that isotopic discrimination became significant. Figure 14 shows the distribution of SOM under timothy and wheat after 97 years of cultivation. Timothy showed a greater stabilizing effect on prairie-derived SOM than wheat, this effect being most pronounced in coarse clay and fine silt. In contrast, the stabilization of SOM derived from crop residues differed less between crops, suggesting that the higher level of SOM under timothy was due to a greater protection of old SOM rather than a greater retention of crop-residue-derived SOM. Wheat-derived SOM made up 43 and 45% of the SOM in fine and coarse clay, respectively, while 67% of the SOM in fine silt was from wheat residues.

The major replacement of prairie SOM by crop-derived SOM took place during the initial 27-year period of cultivation. Since the prairie SOM left after this period was relatively stable especially in the clay size separates, subsequent SOM turnover mainly involved SOM of crop origin. In accordance with Vitorello et al. (1989) and Skjemstad et al. (1990), macro-OM fractions were found to include highly stable materials (Figure 14), 36% of the macro-OM in the 50 to 200 μm separate being of prairie origin.

Martin et al. (1990) studied the ^{13}C abundance of a ferrasol (FAO) in the humid tropics whose vegetation had shifted from grass savanna (C_4 species) to a dense woodland (C_3 species) due to protection against fires. After 25 years, 29% of the savanna-derived SOM remained (Table 13). Losses of SOM from the >50 μm separates were much greater (89 to 97%) than losses from clay, and fine and coarse silt (43 to 48%). Savanna-derived SOM accounted for 28, 56, and 70% of the total SOM in coarse silt, fine silt, and clay, respectively, indicating a higher protection of residual savanna SOM in finer size separates.

Natural ^{13}C abundance studies involving more than one sampling after change-over of vegetation (Balesdent et al., 1988; Vitorello et al., 1989; Bonde et al., 1992) clearly demonstrate that individual size separates regardless of size class limits encompass pools of SOM that differ in turnover rates.

Studies based on natural abundance of isotopes have clearly demonstrated a significant potential in evaluating the turnover of SOM in whole soils and size separates. ^{13}C in particular appears to be a very powerful tool. The significance of isotopic discrimination during long-term turnover, of carbon conserved in charcoal and other biologically inert structures, and of variability in the $^{12}C:^{13}C$ ratio of different plant parts and chemical components should, however, be considered as these phenomena may affect results related to SOM turnover.

VI. The Use of Size Separation in Applied Research

Physical fractionation of soil has been adopted in applied and environmental research, including soil fixation of heavy metals and herbicides, soil CEC, surveys of phosphorous and cation reserves in soil (Kaila, 1967; Kaila and Ryti, 1968; Syers et al., 1969; Somasiri et al., 1971; Christenson and Doll, 1973; Tiessen et al., 1983; Zhu et al., 1983), and peat turnover

Table 13. Fraction yield, carbon content, and ^{13}C abundance in size separates from 0 to 10 cm soil under grass savanna (C_4) and under woodland (C_3) evolved on former savanna plots (humid tropics, Ivory Coast). Savanna-derived SOM remaining 25 years after afforestation is given as proportion of total SOM in size separates and as % of original content. Data from Martin et al. (1990)

	Fraction yield (% of soil d.w.)		mg C/g whole soil		Delta ^{13}C (0/00)		Savanna SOM left after 25 years	
Separate (μm)	Grass savanna	Woodland	Grass savanna	Woodland	Grass savanna	Woodland	% of original content	% of SOM present
Sand								
Coarse (2000–250)	44.6	37.3	0.7	0.5	−12.7	−27.0	3	4
Medium (250–100)	32.4	26.1	0.4	0.5	−12.9	−25.5	11	15
Fine (100–50)	9.9	9.8	0.8	0.6	−12.9	−26.3	8	10
Silt								
Coarse (50–20)	5.3	7.6	0.9	1.5	−13.1	−23.8	52	28
Fine (20–2)	5.8	14.1	3.1	2.9	−12.9	−19.4	57	56
Clay + water solubles (<2)	2.0	5.0	2.0	1.4	−14.1	−18.1	53	70
Whole soil	100.0	100.0	7.9	7.1	−12.8	−22.6	29	33

(Williams, 1983; Williams et al., 1987). Kaila (1967) and Kaila and Ryti (1968) demonstrated that clay was 3 to 6 times higher in exchangeable calcium, magnesium, and potassium than silt size separates (2 to 20 μm). Fixation of added potassium was, however, only 20% lower in silt than in clay, and Kaila (1967) concluded that silt played an important role in the release and fixation of potassium in mineral soils. The redistribution of water-soluble cations during soil fractionation may, however, inflate results on readily exchangeable soil constituents. Thus, the total and DTPA extractable zinc was found to be related to size of particles (Khan, 1979; Armour et al., 1990), whereas zinc extracted by a weak $CaCl_2$ solution showed an inconsistent trend across size separates.

Studies on the sorption of herbicides by intact size separates from Mollisols demonstrated that both clay and silt (2 to 20 μm) may account for significant parts of whole-soil sorption capacity (Nkedi-Kizza et al., 1983; Huang et al., 1984), and that contents of SOM and amorphous oxides are important for sorption capacities of size separates.

Fractionation according to particle size has also been used to evaluated the distribution and adsorption of polycyclic aromatic hydrocarbons in river sediments (Evans et al., 1990), and Dong et al. (1983) concluded that particle size distribution and phosphorous content in size separates were important parameters for evaluating the sources and impact of sediment and phosphorous input into waterways. Förstner (1985) discusses various aspects of size fractionation in relation to sources, chemical forms, and reactivities of heavy metals in sediments.

Essington and Mattigood (1990) examined element concentrations in size separates from sludge-amended soil. Concentrations of elements that accumulated in the soil increased in all size separates, but maximum concentrations were found in clay size separates (<2 μm).

The distribution of radionuclides in size separates has been examined for soils contaminated by low-level discharges from the Sellafield Plant (Livens and Baxter, 1988) and for soils from Upper Austria exposed to atmospheric fallout following the Tjernobyl disaster (Gerzabek and Ullah, 1988). Livens and Baxter (1988) found that concentrations in clay were 1.5 to 40 times higher than in bulk soil, and that clay enrichments were greatest in soils low in fine particles. Although SOM was found to have some effect on the distribution of radionuclides, no evidence was found for a simple relationship between SOM and radionuclide activity. The study of Gerzabek and Ullah (1988) showed that 90% of the radioactivity was associated with clay while 10% was in the silt (2 to 63 μm).

Binding of heavy metals, radionuclides, herbicides, and other pollutants to airborne or waterborne particles significantly influences their transport, deposition, reactivity, and potential toxicity. Thus, information on the distribution of pollutants among size separates is essential for determining their fate in various ecosystems and in assessing potential risks related to plant uptake and exposure to humans. Relative to bulk soil, eroded material will be enriched in SOM, nutrients, and pollutants as fine particles are more efficiently moved by wind and water and, thus, more susceptible to long-range transport. The material left behind will be coarser and depleted in pollutants compared with the noneroded soil. The accumulation of pollutants in the clay and silt size separates (particles <20 μm) suggests that treatment of polluted soils and sediments may be confined to the smaller size fractions. For coarse-textured soils and sediments, the isolation of clay and fine silt may greatly reduce the soil volumes to be processed, thereby reducing the costs of the treatment and the amount of waste to be deposited.

Fractionation of soil according to particle size has also been used in the study of aggregates and aggregation. Christensen (1986) found that straw incorporation had no effect on the particle size composition of macroaggregates (>250 μm) obtained by dry sieving. After correcting for sand particles that accumulated in aggregate fractions during the dry sieving process, aggregates appeared to be enriched in clay and silt relative to whole soil. Due to the detachment of sand grains from aggregate surface, it was concluded that aggregates and whole soil were similar in textural composition. Braunack and Dexter (1989) reached a similar conclusion in their review of aggregate properties, whereas Dormaar (1983) found an increase in sand content as wet-stable aggregate size decreased. The SOM content varied among dry aggregate classes, but their SOM content was closely correlated with the content of clay and silt (Christensen, 1986). Differences between the SOM content of clay from whole soil and aggregates were small. This was also true for silt size separates. Thus, dry aggregates were simply random agglomerates of the mineral particles that make up the soil, but whether this holds for wet-stable aggregates too is not known.

Gregorich et al. (1989) studied the decomposability of SOM associated with clay, silt, and sand from 1 to 2 mm wet-stable aggregates after step-wise increased levels of ultrasonic dispersion. The proportion of potentially mineralizable carbon in sand increased with increased dispersion, but the mineralization rate remained constant. In contrast, the proportion of potentially mineralizable clay-associated carbon decreased and its decomposition rate increased abruptly upon microaggregate breakdown. It was suggested that breakdown of sand size microaggregates released recalcitrant SOM which accumulated in the clay size separate.

Using physical fractionation methods, Oades and Waters (1991) recently examined the concept of aggregate hierarchy in soils and its implications for soil management.

VII. Feasibility of Physical Soil Fractionation in SOM Studies

During the past 20 years, ample evidence has accumulated to demonstrate that fractionation of soil according to particle size provides a significant tool in the study of SOM distribution and dynamics. The procedures used for isolation of size separates are, however, crucial to the applicability of the method. It is essential to reach an adequate soil dispersion since limited dispersion procedures cause microaggregates of smaller size particles be included in silt and sand size separates. Chemical dispersants may cause excessive solubilization of SOM. The feasibility of dispersing soil in water by ultrasonic vibrations has been well documented, the outcome depending on instrument features, experimental procedures, and soil characteristics. The complete experimental setup should be tested as a whole, and the particle size distribution obtained for representative soil types should be calibrated against a conventional textural analysis.

For adequately dispersed soils, particles <5 μm appear to be more enriched in SOM than any other fraction. The clay size separates usually account for >50% of the whole soil SOM, while silt contains 20 to 40%. SOM in clay has a lower C:N ratio, a higher content of diaminopimelic acid, microbially derived sugars, and aliphatic SOM components than SOM in silt size separates. Carbon, nitrogen, and sulfur are more readily mineralized from clay than from silt. Studies based on soils incubated with isotope-labeled organic substrates indicate that clay-bound SOM may be important in medium-term SOM cycling while silt-associated SOM may be subject to longer term cycling. Studies on the effect of cultivation and agronomic practices have shown that the various organomineral separates are affected differently when the levels of whole soil SOM change, the SOM associated with silt generally showing the greatest stability. Studies based on the natural abundance of ^{13}C have shown great potential in elucidating the long-term aspects of SOM turnover in size separates. Although the exact mechanisms responsible for the greater stability of silt than of clay-bound SOM are not fully explained, evidence obtained by solid-state ^{13}C-NMR clearly indicates differences in the chemical constituents of the SOM in various size separates. ^{13}C-NMR studies and evidence from conventional chemical analyses thus suggest a higher microbially derived input to clay than to silt size particles. The stabilization and subsequent turnover of SOM in different size separates are the result of physical as well as chemical processes which occur concurrently and affect organic substrates in all stages of decay.

Conceptual and mathematical models of SOM turnover show an increasing awareness of the importance of soil texture (Van Veen et al., 1985; Jenkinson et al., 1987; Parton et al., 1987; Verberne et al., 1990), and it is envisaged that physical soil fractionation may contribute in reducing the discrepancies between the capabilities of SOM fractionation techniques and the requirements of biologically founded and mathematically formulated SOM-turnover models. For this development to proceed, it is, however, necessary to elaborate a closer coupling between research on primary particle size separates and secondary organomineral complexes (micro- and macroaggregates). Early concepts along this line were outlined by Emerson (1959) and later presentations are given by Tisdall and Oades (1982), Elliott (1986), and Oades and Waters (1991). Recently, Van Veen and Kuikman (1990) have discussed soil architecture in relation to microbial SOM turnover. There seems to much to be gained by a closer coupling between soils physics and SOM research. Evidently, research on SOM should be based on a "three-dimensional" soil concept.

Although the concept of noncomplexed OM as a discrete pool of relevance to SOM turnover appears attractive theoretically, the potential of density fractionation is less clarified than that of size separation. The methods applied in density fractionation are poorly described and little critical testing of methods has been performed. Available information does, however, indicate that noncomplexed or macro-OM forms a significant pool to SOM turnover when virgin soils are brought under cultivation. ^{13}C abundance studies do indicate that inert SOM may be included in light fractions. The pool of noncomplexed SOM includes partially decomposed organic debris with a relatively high C:N ratio, which may show a high carbon mineralization rate and nitrogen immobilization potential. It remains to be seen whether density fractionation of samples from long-term arable soils reveal light fraction SOM of significance to SOM turnover in these soils.

Acknowledgment. This work was performed while the author was on sabbatical leave from the Ministry of Agriculture, Danish Institute for Plant and Soil Science. The financial support is gratefully acknowledged.

References

Adams, T. McM. 1980. Macro organic matter content of some Northern Ireland soils. *Record Agric. Res.* 28:1–11.

Adams, T. McM. 1982. The effects of agronomy on C and N distribution in soil organo-mineral fractions. *J. Agric. Sci. Camb.* 98:335–342.

Adams, W.A., and V.I. Stewart. 1969. The effect of ultrasonic dispersion on silurian shale particles. *Soil Sci.* 108:227–228.

Ahl, C. 1984. *Veränderungen der Art und Menge der organischen Substanz in der Ackerkrume von Langzeit-Feldversuchen, gemessen an einigen physikalischen und chemischen Parametern.* Ph.D. thesis, Georg-August-Universität, Göttingen, Germany.

Ahl, C., H.-J. Altemüller, and H. Söchtig. 1983. Einfluss von Bodentyp, Standort und pflanzenbaulichen Massnahmen auf den Anteil organicher Substanz in verschiedenen Mikroaggregatgrössenklassen am Gesamtboden. *Mitteilgn. Dtsch. Bodenkundl. Gesellsch.* 38:177–182.

Ahl, C., H.-J. Altemüller, and H. Söchtig. 1985. Die Tonverteilung auf verschiedene Aggregatgrössenfraktionen des Bodens in Abhängigheit von der organischen Düngung. *Mitteilgn. Dtsch. Bodenkundl. Gesellsch.* 43:325–329.

Ahmed, M., and J.M. Oades. 1984. Distribution of organic matter and adenosine triphosphate after fractionation of soils by physical procedures. *Soil Biol. Biochem.* 16:465–470.

Allison, F.E., M.S. Sherman, and L.A. Pinck. 1949. Maintenance of soil organic matter: I. Inorganic soil colloid as a factor in retention of carbon during formation of humus. *Soil Sci.* 68:463–478.

Amato, M., and J.N. Ladd. 1980. Studies of nitrogen immobilization and mineralization in calcareous soils-V. Formation and distribution of isotope-labelled biomass during decomposition of ^{14}C- and ^{15}N-labelled plant material. *Soil Biol. Biochem.* 12:405–411.

Amato, M., J.N. Ladd, A. Ellington, G. Ford, J.E. Mahoney, A.C. Taylor, and D. Walsgott. 1987. Decomposition of plant material in Australian soils. IV. Decomposition *in situ* of ^{14}C- and ^{15}N-labelled legume and wheat materials in a range of Southern Australian soils. *Aust. J. Soil Res.* 25:95–105.

Anderson, D.W., and E.A. Paul. 1984. Organo-mineral complexes and their study by radiocarbon dating. *Soil Sci. Soc. Amer. J.* 48:298–301.

Anderson, D.W., S. Saggar, J.R. Bettany, and J.W.B. Stewart. 1981. Particle size fractions and their use in studies of soil organic matter: I. The nature and distribution of forms of carbon, nitrogen and sulfur. *Soil Sci. Soc. Amer. J.* 45:767–772.

Anderson, T.H., and K.H. Domsch. 1989. Der Einfluss des Bodengefüges auf mikrobielle Stoffwechselleistungen. *Mitteilgn. Dtsch. Bodenkundl. Gesellsch.* 59:523–528.

Andreux, F., S. Bruckert, A. Correa, and B. Souchier. 1980. Sur une méthode de fractionnement physique et chimique des agrégats des sols: origines possibles de la matière organique des fractions obtenues. *C. R. Acad. Sc. Paris, Série D* 291:381–384.

Angers, D.A., and G.R. Mehuys. 1990. Barley and alfalfa cropping effects on carbohydrate contents of a clay soil and its size fractions. *Soil Biol. Biochem.* 22:285–288.

Armour, J.D., G.S.P. Ritchie, and A.D. Robson. 1990. Extractable zinc in particle size fractions of soils from Western Australia and Queensland. *Aust. J. Soil Res.* 28:387–397.

Arshad, M.A., and L.E. Lowe. 1966. Fractionation and characterization of naturally occurring organo-clay complexes. *Soil Sci. Soc. Amer. Proc.* 30: 731–735.

Atchley, A.A., and L.A. Crum. 1988. Acoustic cavitation and bubble dynamics. pp. 1–64. In K.S. Suslick (ed.). *Ultrasound: Its Chemical, Physical and Biological Effects*. VCH Publishers, New York, USA.

Baldock, J.A., J.M. Oades, A.M. Vassallo, and M.A. Wilson. 1989. Incorporation of uniformly labelled ^{13}C-glucose carbon into the organic fraction of a soil. Carbon balance and CP/MAS ^{13}C NMR measurements. *Aust. J. Soil Res.* 27:725–746.

Baldock, J.A., J.M. Oades, A.M. Vassallo, and M.A. Wilson. 1990. Solid state CP/MAS ^{13}C N.M.R. analysis of particle size and density fractions of a soil incubated with uniformly labelled ^{13}C-glucose. *Aust. J. Soil Res.* 28:193–212.

Balesdent, J., A. Mariotti, and B. Guillet. 1987. Natural ^{13}C abundance as a tracer for studies of soil organic matter dynamics. *Soil Biol. Biochem.* 19: 25–30.

Balesdent, J., G.H. Wagner, and A. Mariotti. 1988. Soil organic matter turnover in long-term field experiments as revealed by carbon-13 natural abundance. *Soil Sci. Soc. Amer. J.* 52:118–124.

Barkoff, E. 1960. Über die Anwendung von Ultraschall zur Steigerung der Reaktionsgeschwindigkeit und zum Dispergieren des Bodens bei Bodenanalytischen Arbeiten. *J. Sci. Agric. Soc. Finland* 32:179–185.
Bernhard-Reversat, F. 1981. Participation of light and organo-mineral fractions of soil organic matter in nitrogen mineralization in a Sahelian savanna soil. *Zbl. Bakt. II. Abt.* 136:281–290.
Bonde, T.A., B.T. Christensen, and C.C. Cerri. 1992. Dynamics of soil organic matter as reflected by natural ^{13}C abundance in particle size fractions of forested and cultivated oxisols organic matter. *Soil Biol. Biochem.* 23:275–277.
Bourget, S.J. 1968. Ultrasonic vibration for particle-size analyses. *Can. J. Soil Sci.* 48:372–373.
Braunack, M.V., and A.R. Dexter. 1989. Soil aggregation in the seedbed: A review. I. Properties of aggregates and beds of aggregates. *Soil Tillage Res.* 14:259–279.
Broersma, K., and L.M. Lavkulich. 1980. Organic matter distribution with particle-size in surface horizons of some sombric soils in Vancouver Island. *Can. J. Soil Sci.* 60:583–586.
Bruckert, S., and G. Kilbertus. 1980. Fractionnement et analyse des complexes organo-mineraux de sols bruns et de chernozems. *Plant Soil* 57:271–295.
Busacca, A.J., J.R. Aniku., and M.J. Singer. 1984. Dispersion of soils by an ultrasonic method that eliminates probe contact. *Soil Sci. Soc. Amer. J.* 48:1125–1129.
Cameron, R.S., and A.M. Posner. 1979. Mineralisable organic nitrogen in soil fractionated according to particle size. *J. Soil Sci.* 30:565–577.
Catroux, G., and M. Schnitzer. 1987. Chemical, spectroscopic, and biological characteristics of the organic matter in particle size fractions separated from an Aquoll. *Soil Sci. Soc. Amer. J.* 51:1200–1207.
Cerri, C., C. Feller, J. Balesdent, R. Victoria, and A. Plenecassagne. 1985. Application du tracage isotopique naturel en ^{13}C, a létude de la dynamique de la matiere organique dans les sols. *C. R. Acad. Sc. Paris, Serie II*, 300:423–428.
Cheshire, M.V., and C.M. Mundie. 1981. The distribution of labelled sugars in soil particle size fractions as a means of distinguishing plant and microbial carbohydrate residues. *J. Soil Sci.* 32:605–618.
Cheshire, M.V., and C.M. Mundie. 1990. Organic matter contributed to soil by plant roots during the growth and decomposition of maize. *Plant Soil* 121: 107–114.
Cheshire, M.V., B.T. Christensen, and L.H. Sørensen. 1990. Labelled and native sugars in particle-size fractions from soils incubated with ^{14}C straw for 6 to 18 years. *J. Soil Sci.* 41:29–39.
Chichester, F.W. 1969. Nitrogen in soil organo-mineral sedimentation fractions. *Soil Sci.* 107:356–363.
Chichester, F.W. 1970. Transformations of fertilizer nitrogen in soil. II. Total and N^{15}-labelled nitrogen of soil organo-mineral sedimentation fractions. *Plant Soil* 33:437–456.
Chiou, C.T., J-F. Lee, and S.A. Boyd. 1990. The surface area of soil organic matter. *Environ. Sci. Technol.* 24:1164–1166.
Christensen, B.T. 1985. Carbon and nitrogen in particle size fractions isolated from Danish arable soils by ultrasonic dispersion and gravity-sedimentation. *Acta Agric. Scand.* 35:175–187.

Christensen, B.T. 1986. Straw incorporation and soil organic matter in macro-aggregates and particle size separates. *J. Soil Sci.* 37:125–135.

Christensen, B.T. 1987a. Decomposability of organic matter in particle size fractions from field soils with straw incorporation. *Soil Biol. Biochem.* 19: 429–435.

Christensen, B.T. 1987b. Use of particle size fractions in soil organic matter studies. *Intecol Bull.* 15:113–123.

Christensen, B.T. 1988. Effects of animal manure and mineral fertilizer on the total carbon and nitrogen contents of soil size fractions. *Biol. Fertil. Soils* 5:304–307.

Christensen, B.T., and S. Bech-Andersen. 1989. Influence of straw disposal on distribution of amino acids in soil particle size fractions. *Soil Biol. Biochem.* 21:35–40.

Christensen, B.T., and L.H. Sørensen. 1985. The distribution of native and labelled carbon between soil particle size fractions isolated from long-term incubation experiments. *J. Soil Sci.* 36:219–229.

Christensen, B.T., and L.H. Sørensen. 1986. Nitrogen in particle size fractions of soils incubated for five years with ^{15}N-ammonium and ^{14}C-hemicellulose. *J. Soil Sci.* 37:241–247.

Christensen, B.T., F. Bertelsen, and G. Gissel-Nielsen. 1989. Selenite fixation by soil particle-size separates. *J. Soil Sci.* 40:641–647.

Christensen, S., and B.T. Christensen. 1991. Organic matter available for denitrification in different soil fractions: effect of freeze/thaw cycles and straw disposal. *J. Soil Sci.* 42:637–647.

Christenson, D.R., and E.C. Doll. 1973. Release of magnesium from soil clay and silt fractions during cropping. *Soil Sci.* 116:59–63.

Churchman, G.J., and K.R. Tate. 1986. Aggregation of clay in six New Zealand soil types as measured by disaggregation procedures. *Geoderma* 37:207–220.

Coleman, D.C., J.M. Oades, and G. Uehara (eds.) 1989. *Dynamics of Soil Organic Matter in Tropical Ecosystems.* NifTAL Project, University of Hawaii at Manoa, USA.

Cooley, J.H. (ed.) 1987. *Soil Organic Matter Dynamics and Soil Productivity.* Intecol Bull. 15. The International Association for Ecology, Athens, Georgia, USA.

Curtin, D., P.M. Huang, and H.P.W. Rostad. 1987. Components and particle size distribution of soil titratable acidity. *Soil Sci. Soc. Amer. J.* 51:332–336.

Dalal, R.C., and R.J. Henry. 1988. Cultivation effects on carbohydrate contents of soil and soil fractions. *Soil Sci. Soc. Amer. J.* 52:1361–1365.

Dalal, R.C., and R.J. Mayer. 1986a. Long-term trends in fertility of soils under continuous cultivation and cereal cropping in Southern Queensland. I. Overall changes in soil properties and trends in winter cereal yields. *Aust. J. Soil Res.* 24:265–279.

Dalal, R.C., and R.J. Mayer. 1986b. Long-term trends in fertility of soils under continuous cultivation and cereal cropping in Southern Queensland. II. Total organic carbon and its rate of loss from the soil profile. *Aust. J. Soil Res.* 24:281–292.

Dalal, R.C., and R.J. Mayer. 1986c. Long-term trends in fertility of soils under continuous cultivation and cereal cropping in Southern Queensland. III. Distribution and kinetics of soil organic carbon in particle-size fractions. *Aust. J. Soil Res.* 24:293–300.

Dalal, R.C., and R.J. Mayer. 1986d. Long-term trends in fertility of soils under continuous cultivation and cereal cropping in Southern Queensland. IV. Loss of organic carbon from different density fractions. *Aust. J. Soil Res.* 24:301–309.

Dalal, R.C., and R.J. Mayer. 1987. Long-term trends in fertility of soils under continuous cultivation and cereal cropping in Southern Queensland. VI. Loss of total nitrogen from different particle-size and density fractions. *Aust. J. Soil Res.* 25:83–93.

Dong, A., G.V. Simsiman, and G. Chesters. 1983. Particle-size distribution and phosphorus levels in soil, sediment, and urban dust and dirt samples from the Menomonee River Watershed, Wisconsin, U.S.A. *Water Res.* 17:569–577.

Dong, A., G.V. Simsiman, and G. Ghesters. 1985. Release of phosphorus and metals from soils and sediments during dispersion. *Soil Sci.* 139:97–99.

Dormaar, J.F. 1983. Chemical properties of soil and water-stable aggregates after sixty-seven years of cropping to spring wheat. *Plant Soil* 75:51–61.

Drake, E.H., and H.L. Motto. 1982. An analysis of the effect of clay and organic matter content on the cation exchange capacity of New Jersey soils. *Soil Sci.* 133:281–288.

Dudas, M.J., and S. Pawluk. 1970. Naturally occurring organo-clay complexes of orthic black chernozems. *Geoderma* 3:5–17.

Dzurec, R.S., T.W. Boutton, M.M. Caldwell, and B.N. Smith. 1985. Carbon isotope ratios of soil organic matter and their use in assessing community composition changes in Curlew Valley, Utah. *Oecologia* 66:17–24.

Edwards, A.P., and J.M. Bremner. 1964. Use of sonic vibration for separation of soil particles. *Can. J. Soil Sci.* 44:366.

Edwards, A.P., and J.M. Bremner. 1965. Dispersion of mineral colloids in soils using cation exchange resins. *Nature* 205:208–209.

Edwards, A.P., and J.M. Bremner. 1967. Dispersion of soil particles by sonic vibration. *J. Soil Sci.* 18:47–63.

Elliott, E.T. 1986. Aggregate structure and carbon, nitrogen, and phosphorous in native and cultivated soils. *Soil Sci. Soc. Amer. J.* 50:627–633.

Elliott, E.T. and C.A. Cambardella. 1991. Physical separation of soil organic matter. *Agric. Ecosyst. Environ.* 34:407–419.

Elonen, P. 1971. Particle-size analysis of soil. *Acta Agralia Fennica* no. 122.

Elustondo, J., D.A. Angers, M.R. Laverdiere, and A. N'Dayegamiye. 1990. Étude comparative de l'ágrégation et de la matiére organique associée aux fractions granulométriques de sept sols sous culture de maïs ou en prairie. *Can. J. Soil Sci.* 70:395–402.

Emerson, W.W. 1959. The structure of soil crumbs. *J. Soil Sci.* 10:235–244.

Emerson, W.W. 1971. Determination of the contents of clay-sized particles in soils. *J. Soil Sci.* 22:50–59.

Emerson, W.W., R.C. Forster, and J.M. Oades. 1986. Organo-mineral complexes in relation to soil aggregation and structure. pp. 521–548. In P.M. Huang and M. Schnitzer (eds.). *Interactions of Soil Minerals with Natural Organics and Microbes*. SSSA, Madison, WI., USA.

Essington, M.E., and S.V. Mattigod. 1990. Element partitioning in size- and density-fractionated sewage sludge and sludge-amended soil. *Soil Sci. Soc. Amer. J.* 54:385–394.

Evans, K.M., R.A. Gill, and P.W.J. Robotham. 1990. The PAH and organic content of sediment particle size fractions. *Water Air Soil Poll.* 51:13–31.

Feller, C. 1979. Une méthode de fractionnement granulométrique de la matière organique des sols. *Cahier ORSTOM, serie Pedologie* 17:339–346.

Filip, Z. 1977. Einfluss von Tonmineralen auf die mikrobielle Ausnutzung der kohlenstoffhaltigen Substanzen und Bildung der Biomasse. *Ecol. Bull. (Stockh.)* 25:173–179.

Fog, K. 1988. The effect of added nitrogen on the rate of decomposition of organic matter. *Biol. Rev.* 63:433–462.

Ford, G.W., and D.J. Greenland. 1968. The dynamics of partly humified organic matter in some arable soils. *Trans. 9th Int. Congr. Soil Sci., Adelaide,* 2: 403–410.

Ford, G.W., D.J. Greenland, and J.M. Oades. 1969. Seperation of the light fraction from soils by ultrasonic dispersion i halogenated hydrocarbons containing a surfactant. *J. Soil Sci.* 20:291–296.

Förstner, U. 1985. Chemical forms and reactivities of metals in sediments. pp. 1–30. In R. Leschber, R.D. Davies and P. L'Hermite (eds.), *Chemical Methods for Assessing Bio-Available Metals in Sludges and Soils.* Elsevier Applied Science Publ., London, UK.

Francis, C.W. 1973. Adsorption of polyvinylpyrrolidone on reference clay minerals. *Soil Sci.* 115:40–54.

Garwood, E.A., C.R. Clement, and T.E. Williams. 1972. Leys and soil organic matter III. The accumulation of macro-organic matter in the soil under different swards. *J. agric. Sci. Camb.* 78:333–341.

Gee, G.W., and J.W. Bauder. 1986. Particle-size analysis. pp. 383–441. In A. Klute (ed.), *Methods of Soil Analysis, Part 1, 2nd Edition.* ASA and SSSA Publ., Madison, WI, USA.

Genrich, D.A., and J.M. Bremner. 1972a. A reevaluation of the ultrasonic-vibration method of dispersing soils. *Soil Sci. Soc. Amer. Proc.* 36:944–947.

Genrich, D.A., and J.M. Bremner. 1972b. Effect of probe condition on ultrasonic dispersion of soils by probe-type ultrasonic vibrators. *Soil Sci. Soc. Amer. Proc.* 36:975–976.

Genrich, D.A., and J.M. Bremner. 1974. Isolation of soil particle-size fractions. *Soil Sci. Soc. Amer. Proc.* 38:222–225.

Gerzabek, M.H., and S.M. Ullah. 1988. Über die Verteilung von ^{137}Cs in den Korngrössenfraktionen zweier kontaminierter Böden. *Die Bodenkultur* 39: 293–297.

Greenland, D.J. 1965a. Interaction between clays and organic compounds in soils. Part I. Mechanisms of interaction between clays and defined organic compounds. *Soils Fert.* 28:415–425.

Greenland, D.J. 1965b. Interaction between clays and organic compounds in soils. Part II. Adsorption of soil organic compounds and its effect on soil properties. *Soils Fert.* 28:521–532.

Greenland, D.J. 1971. Changes in the nitrogen status and physical condition of soils under pastures, with special reference to the maintenance of the fertility of Australian soils used for growing wheat. *Soils Fert.* 34:237–251.

Greenland, D.J., and G.W. Ford. 1964. Seperation of partially humified organic materials from soils by ultrasonic dispersion. *Trans. 8th Int. Congr. Soil Sci., Bucharest*, 3:137–148.

Gregorich, E.G. 1989. *The effects of texture on the stabilization and physical protection of organic matter in soil.* Ph.D. thesis, The Faculty of Graduate Studies, University of Guelph, Ontario, Canada.

Gregorich, E.G., R.G. Kachanoski, and R.P. Voroney. 1988. Ultrasonic dispersion of aggregates: Distribution of organic matter in size fractions. *Can. J. Soil. Sci.* 68:395–403.

Gregorich, E.G., R.G. Kachanoski, and R.P. Voroney. 1989. Carbon mineralization in soil size fractions after various amounts of aggregate disruption. *J. Soil Sci.* 40:649–659.

Gregorich, E.G., R.P. Voroney, and R.G. Kachanoski. 1991. Turnover of carbon through the microbial biomass in soils with different textures. *Soil Biol. Biochem.* 23:799–805.

Hamblin, A.P. 1977. Structural features of aggregates in some East Anglian silt soils. *J. Soil Sci.* 28:23–28.

Hamdy, A.A., and G. Gissel-Nielsen. 1977. Fixation of selenium by clay minerals and iron oxides. *Z. Pflanzenernaehr. Bodenkd.* 140:63–70.

Healy, W.B., and G.G.C. Claridge. 1974. Chemical properties of soil particle size fractions separated by ultrasonic dispersion. *New Zealand J. Sci.* 17: 493–501.

Hinds, A.A., and L.E. Lowe. 1980a. The use of an ultrasonic probe in soil dispersion and in the bulk isolation of organo-mineral complexes. *Can. J. Soil Sci.* 60:389–392.

Hinds, A.A., and L.E. Lowe. 1980b. Distribution of carbon, nitrogen, sulphur and phosphorus in particle-size separates from gleysolic soils. *Can. J. Soil Sci.* 60:783–786.

Hinds, A.A., and L.E. Lowe. 1980c. Dispersion and dissolution effects during ultrasonic dispersion of gleysolic soils in water and in electrolytes. *Can. J. Soil Sci.* 60:329–335.

Huang, P.M., and M. Schnitzer (eds.) 1986. *Interactions of Soil Minerals with Natural Organics and Microbes*. SSSA Publ. Inc., Madison, WI, USA.

Huang, P.M., R. Grover, and R.B. McKercher. 1984. Components and particle size fractions involved in atrazine adsorption by soils. *Soil Sci.* 138:20–24.

IAEA. 1977. *Soil organic Matter Studies*, Vol. I and II. International Atomic Energy Agency, Vienna, Austria.

Janzen, H.H. 1987. Soil organic matter characteristics after long-term cropping to various spring wheat rotations. *Can. J. Soil Sci.* 67:845–856.

Jenkinson, D.S. 1977. Studies on the decomposition of plant material in soil. V. The effects of plant cover and soil type on the loss of carbon from ^{14}C labelled ryegrass decomposing under field conditions. *J. Soil Sci.* 28:424–434.

Jenkinson, D.S., and J.N. Ladd. 1981. Microbial biomass in soil: Measurement and turnover. pp. 415–471. In E.A. Paul and J.N. Ladd (eds.), *Soil Biochemistry, vol. 5*. Marcel Dekker, New York, USA.

Jenkinson, D.S., D.S. Powlson, and R.W.M. Wedderburn. 1976. The effects of biocidal treatments on metabolism in soil—III. The relationship between soil biovolume, measured by optical microscopy, and the flush of decomposition caused by fumigation. *Soil Biol. Biochem.* 8:189–202.

Jenkinson, D.S., P.B.S. Hart, J.H. Rayner, and L.C. Parry. 1987. Modelling the turnover of organic matter in long-term experiments at Rothamsted. *Intecol. Bull.* 15:1–8.

Jensen, E.S., B.T. Christensen, and L.H. Sørensen. 1989. Mineral-fixed ammonium in clay- and silt-size fractions of soils incubated with ^{15}N-ammonium sulphate for five years. *Biol. Fertil. Soils* 8:298–302.

Jensen, V., A. Kjøller, and L.H. Sørensen (eds.) 1986. *Microbial Communities in Soil.* Elsevier Applied Sci. Publ., London, UK.

Kaila, A. 1967. Potassium status in different particle size fractions of some Finnish soils. *J. Sci. Agric. Soc. Finland.* 39:45–56.

Kaila, A., and R. Ryti. 1968. Calcium, magnesium and potassium in clay, silt and fine sand fractions of some Finnish soils. *J. Sci. Agric. Soc. Finland.* 40:1–13.

Kanazawa, S., and Z. Filip. 1986. Distribution of microorganisms, total biomass, and enzyme activities in different particles of Brown soil. *Microb. Ecol.* 12: 205–215.

Khan, A. 1979. Distribution of DTPA-extractable Fe, Zn and Cu in soil particle-size fractions. *Commun. Soil Sci. Plant Anal.* 10:1211–1218.

Koenigs, F.F.R. 1978. Comments on the paper by P.F. North: Towards an absolute measurement of soil structural stability using ultrasound. *J. Soil Sci.* 29:117–120.

Kononova, M.M. 1961. *Soil Organic Matter—Its Nature, Its Role in Soil Formation and in Soil Fertility.* Pergamon Press, Oxford, UK.

Kowalenko, C.G., and G.J. Ross. 1980. Studies on the dynamics of "recently" clay-fixed NH_4^+ using ^{15}N. *Can. J. Soil Sci.* 60:61–70.

Kyuma, K., A. Hussain, and K. Kawaguchi. 1969. The nature of organic matter in soil organo-mineral complexes. *Soil Sci. Plant Nutr.* 15:149–155.

Ladd, J.N., and M. Amato. 1980. Studies of nitrogen immobilization and mineralization in calcareous soils-IV. Changes in the organic nitrogen of light and heavy subfractions of silt- and fine clay-size particles during nitrogen turnover. *Soil. Biol. Biochem* 12:185–189.

Ladd, J.N., J.W. Parsons, and M. Amato. 1977a. Studies of nitrogen immobilization and mineralization in calcareous soils-I. Distribution of immobilized nitrogen amongst soil fractions of different particle size and density. *Soil Biol. Biochem.* 9:309–318.

Ladd, J.N., J.W. Parsons, and M. Amato. 1977b. Studies of nitrogen immobilization and mineralization in calcareous soils-II. Mineralization of immobilized nitrogen from soil fractions of different particle size and density. *Soil Biol. Biochem.* 9:319–325.

Ladd, J.N., M. Amato, and J.W. Parsons. 1977c. Studies of nitrogen immobilization and mineralization in calcareous soils. III. Concentration and distribution of nitrogen derived from the soil biomass. pp. 301–311. In *Soil Organic Matter Studies*, vol. 1. IAEA, Vienna, Austria.

Ladd, J.N., M. Amato, and J.M. Oades. 1985. Decomposition of plant material in Australian soils. III. Residual organic and microbial biomass C and N from isotope-labelled legume material and soil organic matter, decomposing under field conditions. *Aust. J. Soil Res.* 23:603–611.

Lagaly, G. 1984. Clay-organic interactions. *Phil. Trans. R. Soc. Lond. A* 311: 315–332.

Leinweber, P. 1988. *Erfassung und Charakterisierung organisch-mineralischer Komplexe (OMK) und ihrer Differenzierung in Böden von Dauerveldversuchen der DDR.* Ph.D. thesis, Diss. A., Wilhelm-Pieck-Universität, Rostock, Germany.

Leinweber, P., and G. Reuter. 1988. Menge und Qualität organisch-mineralischer Komplexe in Böden unterschiedlicher Standorte. *Tag.-Ber. Akad. Landwirtsch.-Wiss.* DDR. 269:223–235.

Lemieux, G.J. 1964. Efficiency of various shakers in the particle-size analysis of soils. *Can. J. Soil Sci.* 44:228–231.

Leuschner, H.H., R. Aldag, and B. Meyer. 1981. Dichte-Fraktionierung des Humus in A_p-Horizonten von Sandböden mit unterschiedlicher Körnung und Nutzungs-Vorgeschichte. *Mitteilgn. Dtsch. Bodenkundl. Gesellsch.* 32: 583–592.

Lichtfuss, R., and G. Brümmer. 1981. Gehalte an Organischer Substanz, Schwermetallen und Phosphor in Dichtefraktionen von Fluvialen Unterwasserböden. *Geoderma* 25:245–265.

Livens, F.R., and M.S. Baxter. 1988. Particle size and radionuclide levels in some West Cumbrian soils. *Sci. Tot. Environ.* 70:1–17.

Lowe, L.E., and A.A. Hinds. 1983. The mineralization of nitrogen and sulphur from particle size separates of gleysolic soils. *Can. J. Soil Sci.* 63:761–766.

Lynch, D.L., L.M. Wright, and L.J. Cotnoir, jr. 1956. The adsorption of carbohydrates and related compounds on clay minerals. *Soil Sci. Soc. Amer. Proc.* 20:6–9.

Malone, C.R., and M.B. Swartout. 1969. Size, mass, and caloric content of particulate organic matter in old-field and forest soils. *Ecology* 50:395–399.

Martin, A., A. Mariotti, J. Balesdent, P. Lavelle, and R. Vuattoux. 1990. Estimate of organic matter turnover rate in a savanna soil by ^{13}C natural abundance measurements. *Soil Biol. Biochem* 22:517–523.

McGill, W.B., and E.A. Paul. 1976. Fractionation of soil and ^{15}N nitrogen to separate the organic and clay interactions of immobilized N. *Can. J. Soil Sci.* 56:203–212.

McGill, W.B., J.A. Shields, and E.A. Paul. 1975. Relation between carbon and nitrogen turnover in soil organic fractions of microbial origin. *Soil Biol. Biochem.* 7:57–63.

McKeague, J.A. 1971. Organic matter in particle-size and specific gravity fractions of some Ah horizons. *Can. J. Soil Sci.* 51:499–505.

McTainsh, G.H., and N.C. Duhaylungsod. 1989. Aspects of soil particle-size analysis in Australia. *Aust. J. Soil Res.* 27:629–636.

Molloy, L.F., and T.W. Speir. 1977. Studies on a climosequence of soils in tussock grasslands 12. Constituents of the soil light fraction. *New Zealand J. Sci.* 20:167–177.

Molloy, L.F., B.A. Bridger, and A. Cairns. 1977. Studies on a climosequence of soils in tussock grasslands. 13. Structural carbohydrates in tussock leaves, roots, and litter and in the soil light and heavy fractions. *New Zealand J. Sci.* 20: 443–451.

Monnier, G., L. Turc, and C. Jeanson-Luusinang. 1962. Une méthode de fractionnement densimétrique par centrifugation des matières organiques du sol. *Ann. Agron.* 13:55–63.

Morra, M.J., R.R. Blank, L.L. Freeborn, and B. Shafii. 1991. Size fractionation of soil organo-mineral complexes using ultrasonic dispersion. *Soil Sci.* 152:294–303.

Murayama, S. 1981. Persistency and monosaccharide composition of polysaccharides of soil which received no plant materials for certain period under field conditions. *Soil Sci. Plant Nutr.* 27:463–475.

Murayama, S. 1984. Changes in the monosaccharide composition during the decomposition of straws under field conditions. *Soil Sci. Plant Nutr.* 30: 367–381.

Murayama, S., M.V. Cheshire, C.M. Mundie, G.P. Sparling, and H. Shepherd. 1979. Comparison of the contribution to soil organic matter fractions, particularly carbohydrates, made by plant residues and microbial products. *J. Sci. Food Agric.* 30:1025–1034.

Murphy, E.M., J.M. Zachara, and S.T. Smith. 1990. Influence of mineral-bound humic substances on the sorption of hydrophobic organic compounds. *Environ. Sci. Technol.* 24:1507–1516.

Nkedi-Kizza, P., P.S.C. Rao, and J.W. Johnson. 1983. Adsorption of diuron and 2,4,5-T on soil particle-size separates. *J. Environ. Qual.* 12:195–197.

North. P.F. 1976. Towards an absolute measurement of soil structural stability using ultrasound. *J. Soil Sci.* 27:451–459.

Oades, J.M. 1972. Studies on soil polysaccharides: III. Composition of polysaccharides in some Australian soils. *Aust. J. Soil Res.* 10:113–126.

Oades, J.M. 1984. Soil organic matter and structural stability: mechanisms and implications for management. *Plant Soil* 76:319–337.

Oades, J.M. 1989. An introduction to organic matter in mineral soil. pp. 89–159. In: J.B. Dixon and S.B. Weed (eds.), *Minerals in Soil Environments, Second Edition*. SSSA Publ. Inc. Madison, WI, USA.

Oades, J.M. and A.G. Waters. 1991. Aggregate hierarchy in soils. *Aust. J. Soil Res.* 29:815–828.

Oades, J.M., A.M. Vassallo, A.G. Waters, and M.A. Wilson. 1987. Characterization of organic matter in particle size and density fractions from a red-brown earth by solid-state ^{13}C N.M.R. *Aust. J. Soil Res.* 25:71–82.

Oades, J.M., A.G. Waters, A.M. Vasallo, M.A. Wilson, and G.P. Jones. 1988. Influence of management on the composition of organic matter in a red-brown earth as shown by ^{13}C nuclear magnetic resonance. *Aust. J. Soil Res.* 26: 289–299.

Parasher, C.D., and L.E. Lowe. 1970. Isolation of clay-size organo-mineral complexes from soils of the Lower Fraser Valley. *Can. J. Soil Sci.* 50:403–407.

Parton, W.J., D.S. Schimel, C.V. Cole, and D.S. Ojima. 1987. Analysis of factors controlling soil organic matter levels in Great Plains grasslands. *Soil Sci. Soc. Amer. J.* 51:1173–1179.

Paul, E.A., and F.E. Clark. 1989. *Soil Microbiology and Biochemistry*. Academic Press Inc., San Diego, USA.

Paul, E.A., and W.B. McGill. 1977. Turnover of microbial biomass, plant residues and soil humic constituents under field conditions. pp. 149–157. In *Soil Organic Matter Studies*, vol. 1. IAEA, Vienna, Austria.

Plewinsky, B., and R. Kamps. 1984. Sodium metatungstate, a new medium for binary and ternary density gradient centrifugation. *Makromol. Chem.* 185: 1429–1439.

Pritchard, D.T. 1974. A method for soil particle-size analysis using ultrasonic disaggregation. *J. Soil Sci.* 25:34–40.

Ramsay, A.J. 1984. Extraction of bacteria from soil: Efficiency of shaking or ultrasonication as indicated by direct counts and autoradiography. *Soil Biol. Biochem.* 16:475–481.

Reuter, G., and P. Leinweber. 1988. Konzeption und Methodik der Untersuchung organischmineralischer Komplexe (OMK) in Böden. *Tag.-Ber., Akad. Landwirtsch.-Wiss. DDR.* 269:213–222.

Richter, M., I. Mizuno, S. Aranguez, and S. Uriarte. 1975. Densimetric fractionation of soil organo-mineral complexes. *J. Soil Sci.* 26:112–123.

Saly, R. 1967. Use of ultrasonic vibration for dispersing soil samples. *Soviet Soil Sci.* 2:1547–1559.

Scharpenseel, H.W., K. Tsutsuki, P. Becker-Heidmann, and J. Freytag. 1986. Untersuchungen zur Kohlenstoffdynamik und Bioturbation von Mollisolen. *Z. Pflanzenernaehr. Bodenk.* 149:582–597.

Scheffer, B. 1977. Stabilization of organic matter in sand mixed cultures. pp. 359–363. In *Soil Organic Matter Studies*, Vol. II, IAEA, Vienna, Austria.

Schimel, D.S. 1986. Carbon and nitrogen turnover in adjacent grassland and cropland ecosystems. *Biogeochemistry* 2:345–357.

Schlesinger, W.H. 1986. Changes in soil carbon storage and associated properties with disturbance and recovery. pp. 194–220. In J.R. Trabalka and D.E. Reichle (eds.) *The Changing Carbon Cycle—A Global Analysis*. Springer-Verlag, New York, USA.

Schnitzer, M., and K.C. Ivarson. 1982. Different forms of nitrogen in particle size fractions separated from two soils. *Plant Soil* 69:383–389.

Schnitzer, M., and P. Schuppli. 1989a. The extraction of organic matter from selected soils and particle size fractions with 0.5 M NaOH and 0.1 M $Na_4P_2O_7$ solutions. *Can. J. Soil Sci.* 69:253–262.

Schnitzer, M., and P. Schuppli. 1989b. Method for the sequential extraction of organic matter from soils and soil fractions. *Soil Sci. Soc. Amer. J.* 53: 1418–1424.

Schnitzer, M., J.A. Ripmeester, and H. Kodama. 1988. Characterization of the organic matter associated with a soil clay. *Soil Sci.* 145:448–454.

Schulten, H.R., and M. Schnitzer. 1990. Aliphatics in soil organic matter in fine-clay fractions. *Soil Sci. Soc. Amer. J.* 54:98–105.

Seech, A.G., and E.G. Beauchamp. 1988. Denitrification in soil aggregates of different sizes. *Soil Sci. Soc. Amer. J.* 52:1616–1621.

Shaymukhametov, M.S., N.A. Titova, L.S. Travnikova, and Y.M. Labenets. 1985. Use of physical fractionation methods to characterize soil organic matter. *Soviet Soil Sci.* 16:117–128.

Shiel, R.S. 1986. Variation in amounts of carbon and nitrogen associated with particle size fractions of soils from the Palace Leas meadow hay plots. *J. Soil Sci.* 37:249–257.

Skjemstad, J.O., and R.C. Dalal. 1987. Spectroscopic and chemical differences in organic matter of two Vertisols subjected to long periods of cultivation. *Aust. J. Soil Res.* 25:323–335.

Skjemstad, J.O., R.C. Dalal, and P.F. Barron. 1986. Spectroscopic investigations of cultivation effects on organic matter of Vertisols. *Soil Sci. Soc. Amer. J.* 50:354–359.

Skjemstad, J.O., R.P. Le Feuvre, and R.E. Prebble. 1990. Turnover of soil organic matter under pasture as determined by ^{13}C natural abundance. *Aust. J. Soil Res.* 28:267–276.

Sollins, P., G. Spycher, and C.A. Glassman. 1984. Net nitrogen mineralization from light- and heavy-fraction forest soil organic matter. *Soil Biol. Biochem.* 16:31–37.

Somasiri, S., S.Y. Lee, and P.M. Huang. 1971. Influence of certain pedogenic factors on potassium reserves of selected Canadian Prairie soils. *Soil Sci. Soc. Amer. Proc.* 35:500–505.

Spycher, G., and J.L. Young. 1977. Density fractionation of water-dispersible soil organic-mineral particles. *Commun. Soil Sci. Plant Anal.* 8:37–48.

Spycher, G., and J.L. Young. 1979. Water-dispersible soil organic-mineral particles: II. Inorganic amorphous and crystalline phases in density fractions of clay-size particles. *Soil Sci. Soc. Amer. J.* 43:328–332.

Spycher, G., P. Sollins, and S. Rose. 1983. Carbon and nitrogen in the light fraction of a forest soil: Vertical distribution and seasonal patterns. *Soil Sci.* 135:79–87.

Stevenson, F.J., E.T. Elliott, C.V. Cole, J. Ingram, J.M. Oades, C. Preston, and P.J. Sollins. 1989. Methodologies for assessing the quantity and quality of soil organic matter. pp. 173–199. In D.C. Coleman, J.M. Oades and G. Uehara (eds.) *Dynamics of Soil Organic Matter in Tropical Ecosystems.* NifTAL Project, University of Hawaii at Manoa, USA.

Stevenson, I.L. 1958. The effect of sonic vibration on the bacterial plate count of soil. *Plant Soil* 10:1–8.

Stotzky, G. 1986. Influence of soil mineral colloids on metabolic processes, growth, adhesion, and ecology of microbes and viruses. pp. 305–428. In P.M. Huang and M. Schnitzer (eds.) *Interactions of Soil Minerals with Natural Organics and Microbes.* SSSA Publ., Madison, WI, USA.

Stotzky, G., and R.G. Burns. 1982. The soil environment: Clay-humus-microbe interactions. pp. 105–133. In R.G. Burns and J.H. Slater (eds.) *Experimental Microbial Ecology.* Blackwell Sci. Publ., Oxford, UK.

Suslick, K.S. (ed.) 1988a. *Ultrasound: Its Chemical, Physical and Biological Effects.* VCH Publishers, Inc., New York, USA.

Suslick, K.S. 1988b. Homogeneous sonochemistry. pp. 123–163. In K.S. Suslick (ed.) *Ultrasound: Its Chemical, Physical and Biological Effects.* VCH Publishers, New York, USA.

Suslick, K., and S. Doktycz. 1990. Sounding out new chemistry. *New Scientist* 3:50–53.

Syers, J.K., R. Shah, and T.W. Walker. 1969. Fractionation of phosphorus in two alluvial soils and particle-size separates. *Soil Sci.* 108:283–289.

Swift, M.J., O.W. Heal, and J.M. Anderson. 1979. *Decomposition in Terrestial Ecosystems.* Blackwell Sci. Publ., Oxford, UK.

Sørensen, L.H. 1967. Duration of amino acid metabolites formed in soils during decomposition of carbohydrates. *Soil Sci.* 104:234–241.

Sørensen, L.H. 1972. Stabilization of newly formed amino acid metabolites in soil by clay minerals. *Soil Sci.* 114:5–11.

Sørensen, L.H. 1975. The influence of clay on the rate of decay of amino acid metabolites synthesized in soils during decomposition of cellulose. *Soil Biol. Biochem.* 7:171–177.

Sørensen, L.H. 1981. Carbon-nitrogen relationships during the humification of cellulose in soils containing different amounts of clay. *Soil Biol. Biochem.* 13:313–321.

Sørensen, L.H. 1983. Size and persistence of the microbial biomass formed during the humification of glucose, hemicellulose, cellulose, and straw in soils containing different amounts of clay. *Plant Soil* 75:121–130.

Sørensen, L.H. 1987. Organic matter and microbial biomass in a soil incubated in the field for 20 years with ^{14}C-labelled barley straw. *Soil Biol. Biochem.* 19: 39–42.

Tan, K.H., and P.S. Troth. 1981. Increasing sensitivity of organic matter and nitrogen analysis using soil separates. *Soil Sci. Soc. Am. J.* 45:574–577.

Tanner, C.B., and M.L. Jackson. 1947. Nomographs of sedimentation times for soil particles under gravity or centrifugal acceleration. *Soil Sci. Soc. Amer. Proc.* 12:60–65.

Tate, K.R., and G.J. Churchman. 1978. Organo-mineral fractions of a climosequence of soils in New Zealand tussock grasslands. *J. Soil Sci.* 29: 331–339.

Theng, B.K.G., G.J. Churchman, and R.H. Newman. 1986. The occurrence of interlayer clay-organic complexes in two New Zealand soils. *Soil Sci.* 142: 262–266.

Theodorou, C. 1990. Nitrogen transformations in particle size fractions from a second rotation pine forest soil. *Commun. Soil Sci. Plant Anal.* 21:407–413.

Thorburn, P.J., and R.J. Shaw. 1987. Effects of different dispersion and fine-fraction determination methods on the results of routine particle-size analysis. *Aust. J. Soil Res.* 25:347–360.

Tiessen, H., and J.W.B. Stewart. 1983. Particle-size fractions and their use in studies of soil organic matter: II. Cultivation effects on organic matter composition in size fractions. *Soil Sci. Soc. Amer. J.* 47:509–514.

Tiessen, H., and J.W.B. Stewart. 1988. Light and electron microscopy of stained microaggregates: the role of organic matter and microbes in soil aggregation. *Biogeochemistry* 5:312–322.

Tiessen, H., J.W.B. Stewart, and J.O. Moir. 1983. Changes in organic and inorganic phosphorus composition of two grassland soils and their particle size fractions during 60–90 years of cultivation. *J. Soil Sci.* 34:815–823.

Tiessen, H., R.E. Karamanos, J.W.B. Stewart, and F. Selles. 1984a. Natural nitrogen-15 abundance as an indicator of soil organic matter transformation in native and cultivated soils. *Soil Sci. Soc. Amer. J.* 48:312–315.

Tiessen, H., J.W.B. Stewart and H.W. Hunt. 1984b. Concepts of soil organic matter transformations in relation to organo-mineral particle size fractions. *Plant Soil* 76:287–295.

Tinsley, J., and J.F. Darbyshire (eds.) 1984. Biological Processes and Soil Fertility. Martinus Nijhoff/Dr. W. Junk Publ. The Hague, The Netherlands.

Tisdall, J.M., and J.M. Oades. 1982. Organic matter and water-stable aggregates in soils. *J. Soil Sci.* 33:141–163.

Tsutsuki, K., and S. Kuwatsuka. 1989. Degration and stabilization of the humus in buried humic Ando soils. *Sci. Tot. Environ.* 81/82, 437–446.

Turchenek, L.W., and J.M. Oades. 1974. Size and density fractionation of naturally occurring organo-mineral complexes. *Trans. 10th Int. Congr. Soil Sci., Moscow* 2:65–72.

Turchenek, L.W., and J.M. Oades. 1978. Organo-mineral particles in soils. pp. 138–144. In W.W. Emerson et al. (eds.) *Modification of Soil Structure.* Wiley, Chichester.

Turchenek, L.W., and J.M. Oades. 1979. Fractionation of organo-mineral complexes by sedimentation and density techniques. *Geoderma* 21:311–343.

Van der Linden, A.M.A., J.A. Van Veen, and M.J. Frissel. 1987. Modelling soil organic matter levels after long-term applications of crop residues, and farmyard and green manures. *Plant Soil* 101:21–28.

Van Veen, J.A., and P.J. Kuikman. 1990. Soil structural aspects of decomposition of organic matter by micro-organisms. *Biogeochemistry* 11:213–234.

Van Veen, J.A., and E.A. Paul. 1981. Organic carbon dynamics in grassland soils. 1. Background information and computer simulation. *Can. J. Soil Sci.* 61:185–201.

Van Veen, J.A., J.N. Ladd, and M. Amato. 1985. Turnover of carbon and nitrogen through the microbial biomass in a sandy loam and a clay soil incubated with [^{14}C (U)] glucose and [^{15}N] $(NH_4)_2SO_4$ under different moisture regimes. *Soil Biol. Biochem.* 17:747–756.

Vaughan, D., and R.E. Malcolm (eds.) 1985. *Soil Organic Matter and Biological Activity*. Martinus Nijhoff/Dr. W. Junk Publ., Dordrecht, The Netherlands.

Verberne, E.L.J., J. Hassink, P. de Willigen, J.J.R. Groot, and J.A. van Veen. 1990. Modelling organic matter dynamics in different soils. *Neth. J. Agric. Sci.* 38:221–238.

Vitorello, V.A., C.C. Cerri, F. Andreux, C. Feller, and R.L. Victoria. 1989. Organic matter and natural carbon-13 distribution in forested and cultivated oxisols. *Soil Sci. Soc. Amer. J.* 53:773–778.

Voroney, R.P., E.A. Paul, and D.W. Anderson. 1989. Decomposition of wheat straw and stabilization of microbial products. *Can. J. Soil Sci.* 69:63–77.

Walker, P.H., and J. Hutka. 1973. Grain fragmentation in preparing samples for particle-size analysis. *Soil Sci. Soc. Amer. Proc.* 37:278–280.

Wang, T.S.C., S.W. Li, and Y.L. Ferng. 1978. Catalytic polymerization of phenolic compounds by clay minerals. *Soil Sci.* 126:15–21.

Wang, T.S.C., M.C. Wang, and Y.L. Ferng. 1983. Catalytic synthesis of humic substances by natural clays, silts and soils. *Soil Sci.* 135:350–360.

Wang, T.S.C., P.M. Huang, C.-H. Chou, and J.-H. Chen. 1986. The role of soil minerals in the abiotic polymerization of phenolic compounds and formation of humic substances. pp. 251–281. In P.M. Huang and M. Schnitzer (eds.), *Interactions of Soil Minerals with Natural Organics and Microbes*. SSSA, Madison, Wisconsin, USA.

Watson, J.R. 1970. *Studies on clay-organic nitrogen complexes in soils*. Ph.D. thesis, Department of Soil Science, University of Aberdeen, UK.

Watson, J.R. 1971. Ultrasonic vibration as a method of soil dispersion. *Soils Fert.* 34:127–134.

Watson, J.R., and J.W. Parsons. 1974a. Studies of soil organo-mineral fractions. I. Isolation by ultrasonic dispersion. *J. Soil Sci.* 25:1–8.

Watson, J.R., and J.W. Parsons. 1974b. Studies of soil organo-mineral fractions. II. Extraction and characterization of organic nitrogen compounds. *J. Soil Sci.* 25:9–15.

Weissler, A., and E.J. Hine. 1962. Variations of cavitation intensity in an ultrasonic generator. *J. Acoust. Soc. Amer.* 34:130–131.

Whitehead, D.C., H. Buchan, and R.D. Hartley. 1975. Components of soil organic matter under grass and arable cropping. *Soil Biol. Biochem.* 7:65–71.

Williams, B.L. 1983. The nitrogen content of particle size fractions separated from peat and its rate of mineralization during incubation. *J. Soil Sci.* 34: 113–125.

Williams, B.L., M.V. Cheshire, and G.P. Sparling. 1987. Distribution of ^{14}C between particle size fractions and carbohydrates separated from a peat incubated with ^{14}C-glycine. *J. Soil Sci.* 38:659–666.
Young, J.L., and G. Spycher. 1979. Water-dispersible soil organic-mineral particles: I. Carbon and nitrogen distribution. *Soil Sci. Soc. Amer. J.* 43: 324–328.
Zhang, H., M.L. Thompson, and J.A. Sandor. 1988. Compositional differences in organic matter among cultivated and uncultivated Argiudolls and Hapludalfs derived from loess. *Soil Sci. Soc. Amer. J.* 52:216–222.
Zhu, Y., G. Pardini, G. Poggio, and P. Sequi. 1983. Distribution of phosphorus in particle-size fractions of soils treated with organic wastes. *Agrochimica* 27: 105–111.

Soil Water Repellency

M.G. Wallis and D.J. Horne

Advances in Soil Science, Volume 20

I. Introduction

Water repellent soils exhibit hydrophobic properties when dry, resisting or retarding water infiltration into the soil matrix (Brandt, 1969a). Infiltration rates may be reduced by an order of magnitude, even in soils which visually appear to wet "normally" (Wallis et al., 1991). Repellent soils have been reported in many countries (DeBano and Letey, 1969) and may occupy large areas, such as the sandy soils of South and Western Australia (Bond, 1969a). However, severe repellency is relatively uncommon, and hence the condition is generally regarded as interesting but inconsequential. Indeed, the assumption of nonwater-repellent behavior is the norm in soil physics (Philip, 1969).

Over the years a number of soil scientists have suggested that repellency may be more widespread than is commonly acknowledged (Bond, 1964; Gilmour, 1968; DeBano, 1969b; Marshall and Holmes, 1979; Nakaya, 1982). Wallis et al. (1991) evaluated a sensitive technique for measurement of repellency (Tillman et al., 1989) on 14 New Zealand soils which ranged in texture from sand to clay. All the soils were found to be repellent at field moisture conditions, which led Wallis et al. (1991) to suggest that repellency is the norm rather than the exception, with the degree of repellency variable. Furthermore, it was demonstrated that repellency would affect the hydrologic balance of all the soils under heavy summer rainfall and under many irrigation systems.

A large body of research on repellency exists. As early as 1910, Schreiner and Shorey referred to repellent soils in California. Jamison (1945) cited "large bodies of difficultly wettable soil which remained unwetted even during the rainy season" under Florida citrus trees. DeBano (1969a) described the 1960s as a decade of increasingly widespread interest in the problem of soil wettability. Much research was conducted at the University of California, Riverside, where an international conference was held on the subject (DeBano and Letey, 1969). Research in California remained active throughout the 1970s and interest spread to other states and countries. Repellency has now been studied in natural soils, where it may be induced by fire (DeBano et al., 1976), in coal-mine spoils (Miyamoto et al., 1977; Richardson and Woollenhaupt, 1983), and in sports turf where it is a problem in numerous countries (Charters, 1980; Shiels, 1982; Danneberger and White, 1988; Wallis et al., 1989a,b). Although research into repellency has waned somewhat over the past two decades, there are now over 150 papers on the topic which have not been reviewed.

The purpose of this review is to:

1. Draw attention to the hydrologic and agronomic problem of repellency
2. Identify techniques for repellency measurement and discuss their relative merits

3. Summarize the processes by which repellency is thought to develop
4. Discuss measures for ameliorating repellency
5. Identify future research priorities

Most importantly, it is hoped that this review will stimulate debate and research in an underrecognized field.

II. The Effects of Repellency

A. Infiltration, Plant Growth, Runoff, and Erosion

The primary effect of repellency is a reduction in the rate of water infiltration. In contrast to "wettable soils," in which infiltration rates decline during water penetration, the infiltration rate of severely repellent soils is slow during the initial phases of infiltration and increases with time (DeBano, 1969c; Bond, 1964; Wallis et al., 1991). When the repellent layer is at depth, which may occur after fire, infiltration markedly decreases when the wetting front approaches the repellent layer (DeBano, 1969a).

DeBano (1971) studied infiltration into repellent soil and ignited (nonrepellent) soil. Horizontal infiltration was 25 times slower in the repellent soil, which also displayed a more diffuse wetting front, i.e., the repellent soil did not wet completely when the wetting front passed. He found that the diffusivity of repellent soil was smaller than that of nonrepellent soil at all water contents. The largest differences in diffusivity occurred at both the lower and higher water contents. DeBano (1971) suggested that the substances in the soil responsible for repellency had the greatest effect on water movement at low water contents, but could not explain the large diffusivity difference at high water contents.

Repellency is most strongly expressed as soil dries (Bond, 1964; King, 1981; Wallis et al., 1990a). However, King (1981) and Wallis et al. (1990a) found that repellency severity declined somewhat when the soils became very dry (Figure 1). Wallis et al. (1990a) suggested that the decreased repellency severity was due to molecular conformational changes in the organic matter responsible for repellency (Section II.A); however, the mechanism is unknown.

Wallis et al. (1990a) found that dry, elevated areas of repellent sand had an infiltration rate of approximately $37\,mm\,h^{-1}$ compared with lower-lying, moist areas of soil with an infiltation rate of approximately $204\,mm\,h^{-1}$ (Figure 2). Measurement of the repellent soil using the "repellency index" (Section III.B) has shown that the rate of water infiltration may decline to less than 1% of the "potential" rate if the soil was nonrepellent (Wallis et al. 1991).

Reduced water infiltration due to repellency may affect the available soil water for seed germination and plant growth (Bond, 1972; McGhie,

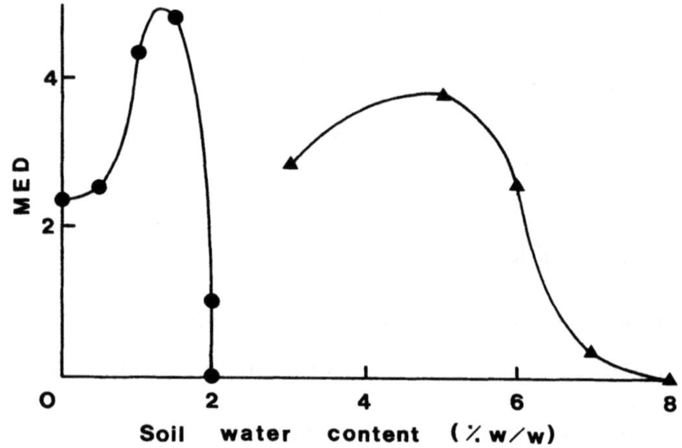

Figure 1. Variation of soil water repellency (molarity of ethanol droplet, MED) with soil water content. (▲, Himatangi sand, after Wallis et al. 1990a; ●, sand 'B' following 168 h moistening at 9°C, after King 1981). Curves fitted by hand.

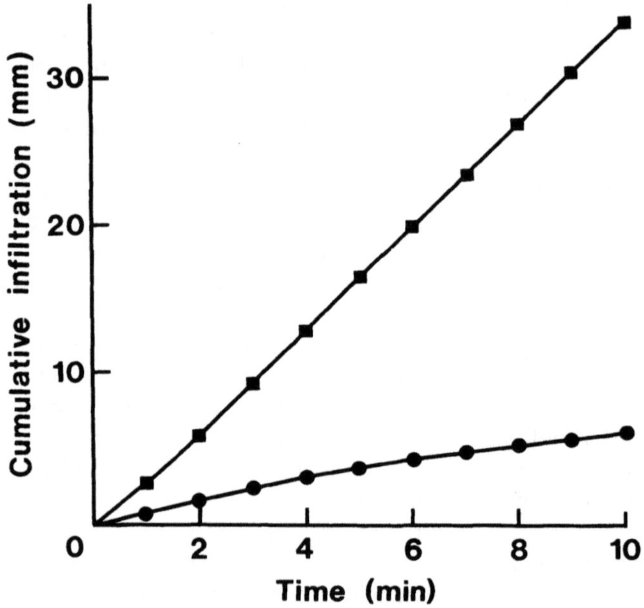

Figure 2. Mean water infiltration into repellent (●) and less repellent (■) Himatangi sand from a +10 mm head disc permeameter (three replicates per site; L.S.D. (0.01) = 10.6). After Wallis et al. (1990a).

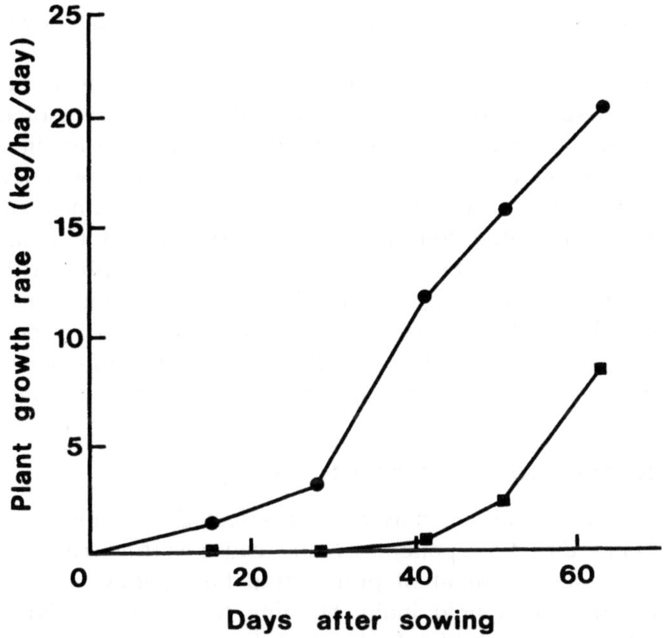

Figure 3. Growth rate of ryegrass (*Lolium perenne*) sown in glasshouse trays of water repellent Himatangi sand (control, ■) compared with trays treated with a nonionic wetting agent (●). L.S.D. (0.01) = 2.03, 1.73, 6.17, 7.13, and 6.52 kg ha^{-1} day^{-1} at 15, 28, 41, 51, and 63 days respectively. After Wallis et al. (1990b).

1983; Wallis et al., 1990b) (Figure 3). Bond (1972) measured delayed, reduced germination of barley on repellent sands in Australia which decreased the average grain yield to 50% of that on less repellent sites. Shepherd and Wallis (unpublished data) measured a yield effect of similar magnitude in a repellent humic clay (Fluvaquentic Haplaquoll) under maize in New Zealand.

The reduction in water infiltration has important hydrologic implications (John, 1978). Water repellency may be responsible for increased surface water redistribution and erosion on many hill and steepland soils, particularly following heavy summer and autumn rainfall events (Wallis et al., 1991). Osborn et al. (1964) found that accelerated rill erosion occurred on repellent plots compared with plots treated with a wetting agent to increase infiltration (Section V.C). They also noted significantly lower vegetative establishment on the repellent soil. John (1978) found that accelerated erosion caused by heat-induced repellency (Section IV.F) could be a serious problem on normally nonrepellent Andisols in New Zealand. The influence of repellency upon runoff and erosion deserves more attention, considering both the on-site and downstream consequences.

The creation of synthetic repellent soil to artificially increase water runoff has been suggested for areas which suffer from a seasonal water deficit (Fink and Myers, 1969). They found that a spray application of sodium methyl silanolate in a small field trial produced repellency which caused 94% of rainfall to run off from 25 storms over 5 months compared with 41% from untreated soil. Fink (1970) pointed out that an effective material should also stabilize the soil or be compatible with a soil stabilizer, must be durable against weathering, and be inexpensive and easy to apply. While Fink (1976) found that a mixture of a petroleum resin and paraffin wax was stable and durable in laboratory tests, the economics and practicality of field application have yet to be evaluated.

B. Soil Water Movement and Aggregate Stability

Water movement within soil may also be affected by water repellency. Raats (1973) noted that a repellent soil may cause wetting front instability and lead to the development of preferential flow pathways. Hendrickx et al. (1988a,b) used a field iodine-staining technique to show that a repellent soil in the Netherlands developed preferential flow paths for water and solute movement. As a consequence, the repellent soil had a greater potential for groundwater contamination. Wallis and Horne (unpublished data) have also found evidence of preferential flow pathways in repellent sand (Figure 4). Water repellency may be a significant but neglected factor in the development of models used to predict water and solute flow through soil.

Giovannini et al. (1983) found that the natural hydrophobic substances in a repellent clay loam appeared to behave as cementing agents of soil aggregates. The hydrophobic substances increased aggregate stability but prevented water from entering the aggregates, restricted infiltration, and caused intense surface runoff. Synthetic repellent chemicals have been developed for engineering uses which exploit these properties (Brandt, 1969b; Bozer et al., 1969). Certain hydrophobic chemicals stabilize soil by limiting water absorption. One such chemical is 4-tert-butylcatechol, which can make soils more easily compacted than in their natural state (Brandt, 1969b). Presumably, the cost of such chemicals has prevented their adoption by engineers.

The potential of increased aggregate stability of repellent soil to decrease problems with surface soil crusts was investigated in laboratory experiments with a clay loam (Rawitz and Hazan, 1978). Mulches of coarse synthetic repellent aggregates reduced surface soil crusting, increased water infiltration, and improved seedling emergence by 40 to 100% compared with the control of bare soil. They pointed out that

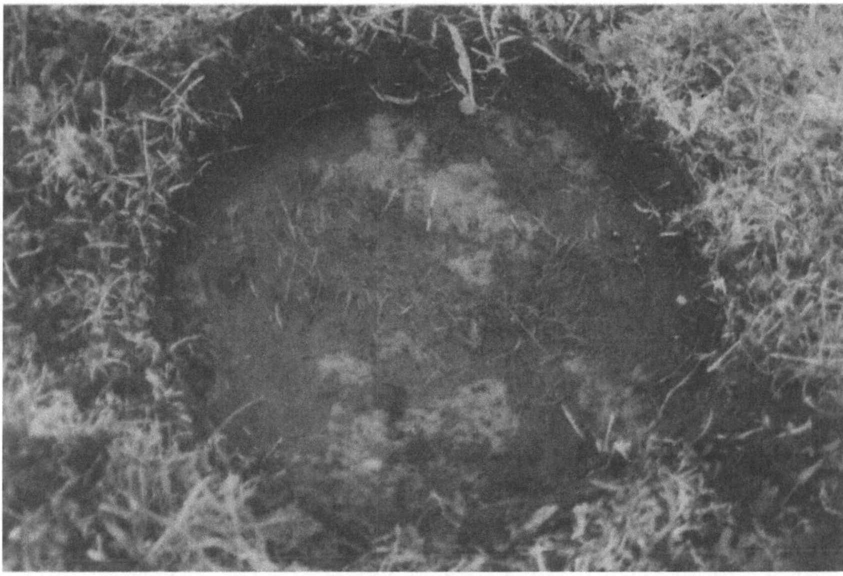

Figure 4. Apparent preferential flow pathways at 50 mm depth in water repellent Himatangi sand after water was ponded in a ring infiltrometer (diameter 400 mm). Note the high concentration of plant roots within the wetted soil (photo M.G. Wallis).

field-testing was necessary and that the procedure was expensive at the application rates evaluated.

Synthetic hydrophobic substances have also been used to reduce evaporation from soil. Lemon (1956) cited Russian research in the 1940s in which a "naptha soap compound" rendered soil repellent and reduced evaporation by "6 or 7 times." DeBano (1969a) cited a 15% reduction in evaporation from a sand which was made repellent with ammonium hydroxide extracts of chaparral litter.

In regions with severe water limitations, the use of repellent surface mulches for moisture conservation may prove viable, provided that the mulches may be manufactured simply and economically. Hillel and Berliner (1974), Fairbourn and Gardner (1975), and Rawitz and Hazan (1978) found that surface soil mulches of coarse repellent aggregates allowed adequate water infiltration but reduced evaporation to 5, 73, and 80% of that of the control soils, respectively. The coarse repellent aggregates (2 to 5 mm) actually increased water infiltration by allowing more rapid water movement through the interaggregate voids (Hillel and Berliner, 1974). The effect of natural repellent soil layers upon evaporation is uncertain and warrants further research. Studies incorporating wetting agents have yielded conflicting results (Section V.C).

Water movement and root exploration may be restricted by repellent layers in the reconstituted soil profile of coal-mine spoils (Richardson and Woollenhaupt, 1983). Synthetic water repellent layers have also been evaluated for the reduction of salt contamination in the reclamation of saline soils. De Jong (1983) found that layers of crude oil and soil prevented salt contamination of "cover soil" in soil columns. The oil-soil mixture was found to be more effective than oil alone at both preventing capillary rise from the saline soil and preventing downward movement of applied water into the saline soil. The technique requires field evaluation and is likely to be inappropriate in areas with a seasonal surplus of rainfall over evaporation due to waterlogging and associated problems in the cover soil.

III. Definition and Measurement of Repellency

In order to study repellency development or amelioration one must use appropriate measurement techniques. Ideally the technique should be simple and inexpensive and should provide a rapid quantitative measure of practical significance. Over the years many techniques have been developed to measure repellency, during which time the understanding and definition of repellency has evolved. This section discusses the inherent concepts and assumptions of each measurement technique and outlines their relative merits and limitations.

A. Contact Angle

Zisman (1964) provided a comprehensive review of the ways in which the nature of a liquid and a plane solid surface affect the resultant contact angle, H between them. The affinity or repellency of a solid surface for water originates from the attractive forces between them (adhesion). If adhesion exceeds the attraction between the individual water molecules (cohesion) then the water will spread out on the solid surface. Conversely, if adhesive forces are less than cohesive forces, the solid surface will repel water. When this happens the water "balls up" on the surface, forming a contact angle, H.

The mechanical equilibrium of a water droplet on a solid surface (Figure 5) was defined by Young (1805) as Eq. 1.

$$\gamma_{SV} - \gamma_{SL} = \gamma_{LV} \cos H \tag{1}$$

In which γ_{sv} is the tension at the solid-vapor interface ($\mathrm{N\,m^{-1}}$), γ_{sl} the tension of the solid-liquid interface ($\mathrm{N\,m^{-1}}$), and γ_{lv} the tension of the liquid-vapor interface; the liquid "surface tension" ($\mathrm{N\,m^{-1}}$).

The liquid-solid contact angle can be geometrically measured for a plane solid surface. In porous media such as soils, the contact angle is

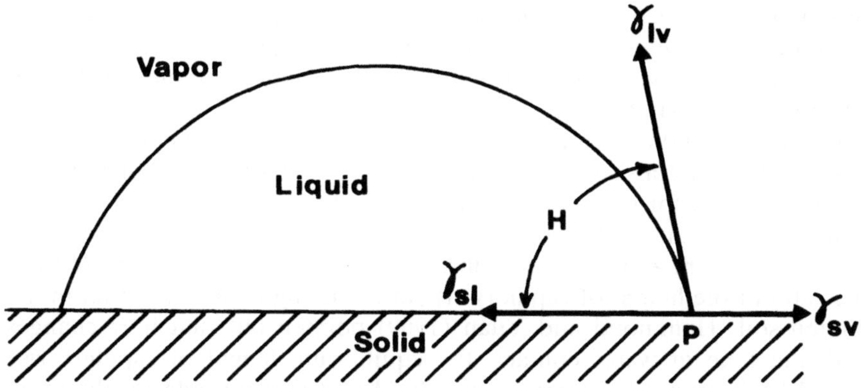

Figure 5. A liquid droplet at mechanical equilibrium on a solid surface. The forces (interfacial tensions, γ) about point P are balanced, as described in Eq. 1. After Zisman (1964). Reprinted with permission from W.A. Zisman, Advances in Chemistry Series 43:1–51. Copyright 1964 American Chemical Society.

conceptually the effective angle formed by a water meniscus in contact with the soil pore walls (DeBano, 1969a). Bond and Hammond (1970), Fink and Myers (1969), and Mallik and Rahman (1985) directly measured the contact angle of water drops lying on soil surfaces. The technique is only applicable to extremely repellent soils because in less repellent soils the water drop will penetrate and make geometric measurement impossible. However, even in extremely repellent soils, surface roughness and pore size distribution will affect the measured value (Letey, 1969). At best, only an "apparent contact angle" may be measured (Philip, 1971). A full discussion of this aspect is given in Section III.A.3.

1. Capillary Rise

Letey et al. (1962a) evaluated an infiltration and a capillary rise technique to indirectly determine the apparent contact angle, H. They assumed that the soil was characterized by cylindrical tubes and used Poiseuille's approximation in which the rate of solution capillary rise into a soil column is related to the physical properties of the solution, the effective pore radius, r (m) and H. The height of capillary rise at equilibrium, h (m) is represented by Eq. 2.

$$h = \frac{2\gamma \cos H}{\rho g r} \tag{2}$$

where γ is the surface tension of the solution ($\mathrm{N\,m^{-1}}$), ρ the density of the solution ($\mathrm{kg\,m^{-3}}$), and g the acceleration due to gravity ($\mathrm{m\,s^{-2}}$).

Neither r nor H are known. Ethanol was found to have the same H for all soil materials, which Letey et al. (1962a) assumed to be zero. Capillary rise of ethanol was used to calculate *r*, which was then used to calculate *H* when other solutions were used. Emerson and Bond (1963) used ignited soil samples to solve for *r* by assuming that the ignited sample had $H = 0°$ when wet with water.

Letey et al. (1962a) found that the capillary rise technique produced reasonable and consistent measurements of *H*. They chose an arbitrary equibrium time for maximum capillary rise of 24 h. Nakaya et al. (1977b) immersed capillary tubes of ignited and nonignited soil in water to a depth of 300 mm and, after equilibration was reached, calculated *H* from the ratio of the height of capillary rise in the nonignited soil to that in the ignited soil. Hammond and Yuan (1969) pointed out that if *H* changes with the time of water contact, the equilibrium height of capillary rise would be influenced. Emerson and Bond (1963) were able to obtain an estimate of the initial *H* by plotting the rate of capillary rise over the first approximately 15 min against the reciprocal of the height of rise and extrapolating to a zero rate of rise.

The general perception that soils with $H < 90°$ should be considered "nonrepellent," which was based upon water penetration into porous media (Watson and Letey, 1970; Section III.D), was challenged by Nakaya (1982). He contended that all soils with $H > 0°$ are repellent to some degree. DeBano (1969a) pointed out that many researchers have reported *H* to be approximately 50° to 60° in soils which appeared to absorb water normally, and he stated that "wetting angles other than zero would be expected to affect the moisture movement processes in the soil governed by unsaturated flow." Wallis et al. (1991) have demonstrated that the repellency of such soils may be hydrologically significant (Section III.B).

The major disadvantages of the capillary rise techniques are as follows:

1. They are very time consuming. King (1981) reported that only 12 samples per person-day could be tested using Emerson and Bond's (1963) technique.
2. The methods of Letey et al. (1962a) and Nakaya et al. (1977b) do not indicate the initial *H*, since water is in contact with the soil for an arbitrary period such as 24 h (the "reaction time"; Zisman, 1964). However, Emerson and Bond's (1963) method does allow an estimate of the initial *H* to be made.
3. The need to consider hysteresis effects; there is a difference between an advancing *H* and a receding *H* (Hammond and Yuan, 1969).
4. Observations on porous media such as soils yield, at best, *apparent* contact angles which cannot be related directly to the actual contact angle at interfaces within the medium (Philip, 1971). This limitation arises because the pore geometry affects the measured "contact angle."
5. They are not suitable for intact soil.

2. Infiltration

Letey et al. (1962a) evaluated an infiltration technique for H calculation which involved measurement of water and ethanol movement into soil columns. They found that the method worked well with soils but not with pure sands, which they suggested may have been due to sand pores trapping air while transmitting water downward. A constant but unspecified supply potential was used for the infiltration technique. A further limitation of the technique which was recognized by Letey et al. (1962a) was the measurement of wetting front positions with the implicit assumption that the soil or sand was saturated behind the front. DeBano (1971) found that repellent soils have a more diffuse wetting front than nonrepellent soils.

3. Intrinsic Diffusivity and Sorptivity

Philip (1957) developed the theory of intrinsic diffusivity and sorptivity (Eqs. 3 and 4).

$$D^*(\theta) = D(\theta)\left[\frac{\mu}{\gamma \cos H}\right] \tag{3}$$

$$S^*(\theta) = S(\theta)\left[\frac{\mu}{\gamma \cos H}\right]^{1/2} \tag{4}$$

where $D^*(\theta)$ is the intrinsic diffusivity (m), $D(\theta)$ the liquid diffusivity ($m^2 s^{-1}$), μ the liquid dynamic viscosity ($N s m^{-2}$), γ the liquid surface tension ($N m^{-1}$), $S^*(\theta)$ the intrinsic sorptivity ($m^{1/2}$), $S(\theta)$ the liquid sorptivity ($m s^{-1/2}$). D^*, D, S^*, and S are functions of the soil volumetric water content, θ.

Philip (1957) described D^* and S^* as properties of the porous medium, independent of the properties of the absorbing liquid. DeBano (1969a) and Hammond and Yuan (1969) applied Eqs. 3 and 4 respectively to water and ethanol. They assumed that ethanol wet the soil with $H = 0°$ ($\cos H = 1.0$) (Letey et al., 1962a), which enabled calculation of H for water infiltration.

DeBano (1969a) found that H values measured using the intrinsic diffusivity technique were similar to those measured by capillary rise (Letey et al., 1962a) for a repellent soil. He stated that the diffusivity technique had advantages over other methods for H measurement because the diffusivity and therefore H values are linked to specific values of soil water content. Bahrani et al. (1973) also used Eq. 4 to calculate H and asserted that the technique had advantages over the capillary rise and direct observation methods because soil treatment to create wettability (e.g., by ignition; Emerson and Bond, 1963) was unnecessary and the "reaction time" (Zisman, 1964) for solid-liquid contact was 30 to 60 min compared with 24 h using the technique of Letey et al. (1962a).

Despite potential advantages, researchers have found that H calculation from Eqs. 3 and 4 may be erroneous. DeBano (1969a) found a discrepancy between the H values of a nonrepellent soil measured by intrinsic diffusivity and capillary rise which he could only explain as measurement errors. In contrast Hammond and Yuan (1969) found that the values of H calculated from intrinsic sorptivity were in good agreement with those measured by the capillary rise technique of Emerson and Bond (1963). Kijne (1967) used the capillary rise method of Letey et al. (1962a) to calculate H. He then applied Eqs. 3 and 4 and found that neither D^* nor S^* were constant for a range of infiltrating solutions. Mustafa et al. (1970) found that Eqs. 3 and 4 adequately described infiltration of water and organic liquids into glass beads and nonrepellent soil. The equations also held for infiltration of organic liquids into repellent soil, but not for water into a repellent soil.

Philip (1969) corrected his earlier work. He pointed out that D^* and S^* depend upon H in a very complex way so that the inclusion of H in Eqs. 3 and 4 was unjustified. He therefore redefined D^* and S^* as Eqs. 5 and 6 respectively:

$$D^*(\theta) = D(\theta)\left[\frac{\mu}{\gamma}\right] \tag{5}$$

$$S^*(\theta) = S(\theta)\left[\frac{\mu}{\gamma}\right]^{1/2} \tag{6}$$

Philip (1971) quoted Farrell's (1963) criticism that "the contact angle can only be dissociated from the geometry of the soil if the porous medium conforms to (an) idealized capillary tube model. For any porous medium in which the walls of the pore channels diverge and converge, a unique relationship between the relative sorptivity and contact angle does not exist." Philip (1971) went on to point out that even for circular-cylindrical pores, scaling is exact only in the limit where pore diameter is negligibly small compared with pore length, therefore observations on porous media such as soils yield, at best, *apparent* contact angles which cannot be related directly to the contact angle at interfaces within the medium.

Bond and Hammond (1970) used scanning electron micrographs of water drops lying upon sand grains to explore the relationship between the *real* contact angle of individual grains and the *apparent* contact angle measured when water drops were placed between sand particles. They found that the *apparent* contact angle was equal to the *real* contact angle plus the angle of pore divergence. Surface roughness of the sand grains did not explain the differences between measured values of *real* and *apparent* contact angle.

For the reasons outlined by Philip (1971), measurement of an apparent contact angle to quantify repellency is inappropriate and misleading.

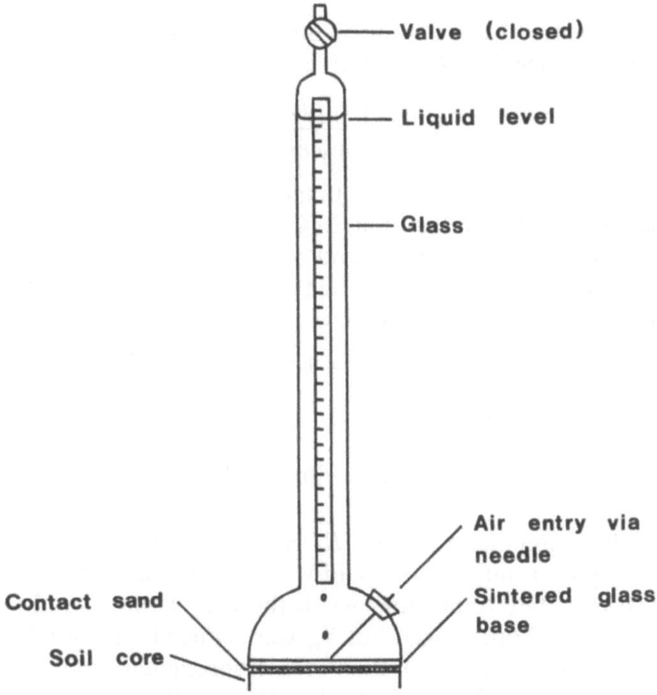

Figure 6. The disc permeameter used to measure water and ethanol sorptivity in undisturbed soil cores for calculation of the intrinsic sorptivity repellency index (RI).

Tillman et al. (1989) have developed an index based upon the intrinsic sorptivity of the soil medium as defined in Eq. 6 which involves calculation of a dimensionless "repellency index" (RI).

B. Repellency Index (RI)

The RI is based upon the theory of intrinsic sorptivity described above and was seen to offer two major advantages over other techniques for repellency measurement (Tillman et al., 1989). First, sorptivity is a well-defined soil parameter with physical significance. Second, sorptivity may be rapidly measured in the laboratory or in situ in the field. Furthermore, compared with contact angle measurements, sorptivity measurements using permeameters are relatively straightforward (Clothier and White, 1981; Perroux and White, 1988). Tillman et al. (1989) measured infiltration of water and 95% ethanol at a −0.4 kPa supply potential which removed the effect of soil pores with diameter >0.74 mm. Air entry into the apparatus via a hypodermic needle maintained a constant supply potential so that a constant pore size range conducted liquid (Figure 6).

Sorptivity to water, S_W and to 95% ethanol, S_E was calculated and Eq. 6 applied, from which Eq. 7 was derived:

$$S_W = \left[\frac{(\mu_E/\gamma_E)^{1/2}}{(\mu_W/\gamma_W)^{1/2}}\right] S_E \tag{7}$$

where μ_E is the viscosity of 95% ethanol at 20°C ($0.0012\,N\,s\,m^{-2}$), γ_E the surface tension of 95% ethanol at 20°C ($0.023\,N\,m^{-1}$), μ_W the viscosity of water at 20°C ($0.0010\,N\,s\,m^{-2}$), and γ_W the surface tension of water at 20°C ($0.073\,N\,m^{-1}$). Therefore, when values for the relevant parameters are inserted, in nonrepellent media:

$$S_W = 1.95 S_E \tag{8}$$

So the RI was defined as

$$RI = 1.95\left(\frac{S_E}{S_W}\right) \tag{9}$$

such that in nonrepellent media, $RI = 1.0$. Tillman et al. (1989) found that Eq. 9 held for initially dry, acid-washed sand, as expected (i.e., $RI = 1.0$). However, they found that when the acid-washed sand was initially moist S_E remained unchanged but S_W was reduced, giving RI values >1.0 but <1.95. They postulated that in a nonrepellent soil, regardless of initial water content, $S_E < S_W$, so that $RI < 1.95$.

Tillman et al. (1989) measured the *RI* of a moist sandy loam which appeared to absorb water "normally." They found that *RI* was 140 during the first 30 s of water infiltration but declined to approximately 7 as time increased due to increased water sorption. Increased water infiltration rate with time in repellent soil has been reported by DeBano (1969c) and may be due to adsorption of vapor water molecules (Malik et al., 1981; DeBano, 1969a).

Wallis et al. (1991) evaluated the RI index on a wide range of soils in New Zealand and compared the index with the WDPT and MED techniques. They found that the RI measured all soils repellent ($RI > 1.95$) at field moisture conditions, and was more sensitive than the WDPT or MED tests and therefore particularly useful for evaluating soils with low degrees of repellency. A further advantage of the RI is that it is a physically significant parameter which can be used to calculate actual and "potential" short-time water infiltration which may then be compared with rainfall and irrigation intensities. Wallis et al. (1991) undertook such comparisons and found that repellency was hydrologically significant, even in soils which appeared to wet normally. The RI may be measured in situ or on undisturbed cores at either field-moist or air-dry conditions. Elimination of macropore flow using a negative supply potential reduces the effect of disturbance when cores are used.

C. Infiltration

Measurements of water infiltration rates have been used to indicate the severity of repellency since the earliest studies of the problem. Letey et al. (1962c) excavated basins to measure in situ infiltration under ponded conditions. Ring infiltrometers have also been used widely to measure in situ infiltration (Bond, 1964; Wilkinson and Miller, 1978; Taylor and Blake, 1982; King, 1981; McGhie and Tipping, 1984; Wallis et al., 1990a). Other authors have used a range of rainfall simulation devices to measure in situ infiltration, runoff, and erosion rates (Krammes and Hellmers, 1963; Gilmour, 1968; Hussain et al., 1969; Meeuwig, 1971; McGhie and Tipping, 1984). Laboratory infiltration studies have been used with either undisturbed soil cores (Wilkinson and Miller, 1978) or repacked soil columns (Pelishek et al., 1962; Morgan et al., 1966; Mustafa and Letey, 1970; Miyamoto, 1985).

King (1981) compared the results of a small ring infiltrometer (SRI) test with measurement of the apparent contact angle, H by capillary rise (Emerson and Bond 1963) on 101 sandy soils from South Australia. He found that the relationship between $\log_{10}$SRI and H was nearly linear for $H > 81°$, but was curvilinear when the whole range $60° < H < 101°$ was examined. The variability in the relationship between $\log_{10}$SRI and H increased when $H > 92°$, which King proposed was due to water being forced between the soil and the walls of the infiltrometer ring in the SRI test and the capillary tube in the contact angle test. Despite the efforts of King (1981) to define a relationship between infiltration rate and more precise measurements of repellency, infiltration measurements generally only provide information regarding the effect of repellency when compared with measures of the same, nonrepellent soil. They may be used to compare untreated and wetting-agent treated soil (e.g., McGhie and Tipping, 1984) or to compare soil at different proximate locations, such as within a golf green (Wilkinson and Miller, 1978) or within a soil type (Wallis et al., 1990a).

Indirect "measurements" of infiltration may be used to determine the effect of repellency. Osborn et al. (1967) inferred the effect of repellency upon infiltration by measuring germination and establishment of seeds placed on untreated and wetting-agent treated repellent and nonrepellent soil in pots at 0° or 30° slope. They found a large effect of repellency upon the germination and establishment of seeds sown on the slope, but not upon those sown on level soil. Wallis et al. (1990b) improved the technique of Osborn et al. (1967) by sowing seed below the soil surface, and they were able to measure the effect of repellency upon seedling emergence in level trays of soil.

Fink and Myers (1969) and Fink (1970) assessed the "infiltration resistance" of repellent soil by measuring the applied pressure necessary to initiate water flow into the pore structure. The breakthrough pressure data for sands treated with increasing quantities of synthetic repellent chemicals provided information identical to the contact angle data which was directly measured from a large water drop on the soil surface. Fink and Myers (1969) noted that the breakthrough pressure method was more difficult and time-consuming than the contact angle technique. Consequently, the breakthrough pressure technique has not been used in other studies.

Infiltration into repellent soil produces a high spatial variability of soil water content over small distances (Section IV.G). Sawada et al. (1989) have developed the "SOWADIN" index to measure the distribution of water in sections of soil following infiltration with and without applications of wetting agents. The SOWADIN index uses computer-assisted tomography applied to gamma-ray attenuation measurements and is therefore not widely applicable outside research institutions which have the appropriate equipment.

D. Water Drop Penetration Time (WDPT)

The WDPT method simply consists of placing a water drop on the soil surface and recording the time taken for the water to penetrate the sample (Letey, 1969). Normally, soil samples are first sieved, placed upon a dish, and the soil surface smoothed by hand to provide standard conditions, since surface roughness and pore geometry affect WDPT (Wessel, 1988). Richardson and Hole (1978) modified the WDPT test by using 2 ml of water instead of a single drop. They asserted that this improved repeatability and increased the ease of observation. Fink (1970) compared the height of approximately 30 mm diameter water drops resting upon soil with that of a similar drop resting upon a smooth, nonporous solid surface with a 90° apparent contact angle to produce a "relative repellency" measure. Neither of the modified techniques have been widely used and many authors have simply increased replication to improve estimates of WDPT.

There have been a number of developments in the interpretation of the WDPT test. Letey (1969) considered that the WDPT divides soils into two broad categories; those with apparent contact angles, H, above and those with H below, 90°. If the water drop forms a ball on the soil surface, $H > 90°$, and the soil is considered repellent. Watson and Letey (1970) pointed out that water will only penetrate a porous medium once $H < 90°$. Normally the water drop will eventually penetrate the soil, indicating that H must decrease over time (Letey, 1969). Therefore, the method may be a better index of repellency persistence than an actual estimate of the initial contact angle (Letey, 1969; Watson and Letey, 1970).

The apparent decline of H over time may be explained by the occurrence of a solid-liquid interaction (John, 1978) which may involve adsorption of vapor water molecules (Malik et al., 1981). DeBano (1969a) monitored temperature fluctuations at the wetting front of repellent and nonrepellent soil which indicated that vapor mechanisms of water movement were more important in the repellent soil.

A number of authors have attempted to correlate WDPT with measurements of H. Watson and Letey (1970) found that the WDPT method ranked a range of natural and synthetic repellent soils in the same order of repellency severity as the capillary rise method of Letey et al. (1962a) for H measurement. Scholl (1975), King (1981) and Wessel (1988) have reported significant linear relationships between the logarithm of WDPT and H. In contrast, Richardson and Hole (1978) and McGhie and Posner (1981) have recorded a poor relationship between WDPT and H (measured by the techniques of Letey et al. (1962a) and Emerson and Bond (1963) respectively). The study of McGhie and Posner (1981) was an "unnatural" situation, using nonrepellent soil amended with ground plant material, compared with the other studies which utilized a range of repellent soils.

The reported relationships between WDPT and H have been markedly dissimilar. Scholl (1975) used Letey et al.'s (1962a) technique to measure H and correlated $H = 90°$ with WDPT of approximately 10,000 s. King (1981) used Emerson and Bond's (1963) technique for H measurement and correlated $H = 90°$ with WDPT = 260 s. Wessel (1988) directly measured H and correlated $H = 90°$ with WDPT < 1 s. The differences in the correlations may be due to the different techniques for H measurement employed which have widely different "reaction times" for solid-liquid contact. However, the correlation differences tend to substantiate the perceptions of Letey (1969) and Watson and Letey (1970) that the WDPT is a better measure of repellency persistence than of initial resistance to wetting.

As a consequence, the distinction between "repellent" and "norepellent" soils based upon WDPT is arbitrary. Watson and Letey (1970), John (1978) and Miller and Letey (1975) chose WDPT = 5 s as the division; Richardson and Hole (1978) used 8 s; Savage et al. (1972) used 1 s. In some cases further classification of repellency severity has also been made. For example, Roberts and Carbon (1972) assigned WDPT > 300 s to soils with $H > 90°$, WDPT 1 to 300 s with H approximately 70 to 90° and WDPT < 1 s with $H < 70°$.

The WDPT method is only capable of measuring a narrow range of repellency within a few degrees span of $H = 90°$ (King, 1981). He found that WDPT < 1 s for $H < 75°$, which indicates the limitations of the technique for quantifying low degrees of repellency. WDPT > 1 h have been measured (Savage et al., 1972; Roberts and Carbon, 1972; Singer and Ugolini, 1976; Wallis et al., 1990a), which indicates the limitation of the technique for quantifying severe repellency.

The WDPT method may also have limited application for very coarse media. Richardson and Hole (1978) measured a low WDPT on a coarse sand with a high H value. Hillel and Berliner (1974) found that infiltration may be enhanced through a system of large repellent aggregates which allow rapid macropore flow and reduce aggregate sorption. However, in some studies WDPT has proven more useful than measurements of *H*. Richardson and Hole (1978) measured a very high WDPT of a litter layer which had a low *H*, which may indicate that the capillary rise technique is not suitable for repellency measurements of such material.

Although the WDPT method has the several limitations described, it is simple and rapid, and has therefore been used in many studies. King (1981) reported that 200 samples may be tested per person-day. Furthermore, the technique is suited to measurements in situ or on undisturbed samples. Giovannini and Lucchesi (1984) measured in situ WDPT = 800 to 900 s and undisturbed soil WDPT = 700 to 760 s in an Italian forest soil. In situ WDPT measurements have also been used in turf evaluation (Wilkinson and Miller, 1978; Charters, 1980; Templeton and Rodriguez, 1988) and a large-scale survey (DeByle, 1973). The simplicity of the WDPT method enables its use by farmers and turf managers to identify and broadly characterize a repellent soil condition.

E. Ninety-Degree Surface Tension (NDST)

The liquid-solid contact angle, *H*, depends upon both the nature of the solid and the surface tension of the liquid, as defined in Eq. 1. Zisman (1964) reported a linear relationship between cos *H* of plane surfaces and the surface tension of a homologous series. He found that the highest surface tension at which there was complete wetting (i.e., $H = 0°$) was more or less characteristic of the solid surface itself. He called this value the "critical surface tension."

Letey (1969) suggested that the critical surface tension may be developed as an index of repellency which could be used to determine optimum surfactant concentration for soil application. Watson and Letey (1970) found that the linear relationship between cos *H* and liquid surface tension of Zisman (1964) held adequately for porous media. They developed the "ninety-degree surface tension" (NDST) method in which the surface tension at $\cos H = 0$ ($H = 90°$) of soil samples was determined. The method consisted of preparing a series of aqueous ethanol solutions with varying surface tension, and finding the surface tension at which a drop will remain on the soil for an arbitrary time period of 5 s. Watson and Letey noted that this method had application for determining the optimum dilution rate for surfactant addition to soil.

Miyamoto and Letey (1971) used a similar procedure to that of Watson and Letey (1970) to measure the NDST except that a range of pure liquids were used. This appeared to be a backward step, since a similar range of liquids were evaluated by Watson and Letey (1970) and produced a lower correlation coefficient between cos *H* and the liquid surface tension than the aqueous ethanol series. Furthermore, the aqueous ethanol series is less expensive, readily obtained, and nontoxic (Watson and Letey, 1970). Miyamoto and Letey (1971) discussed the use of liquids with a higher surface tension than water for NDST evaluation of "wettable" soils. They pointed out that such liquids were rare and expensive, so that the NDST technique was best for repellent materials but not for extremely wettable materials.

Letey et al. (1975) divided the WDPT by NDST to derive the "water repellency index" (WRI). WRI < 0.1 was categorized as a wettable soil; WRI from 0.1 to 1.0 slightly repellent; WRI from 1.0 to 10 moderately repellent; and WRI > 10 very repellent. John (1978) used the NDST and WDPT methods and calculated WRI for a range of New Zealand soils. He found that the NDST method was rapid, simple, and sensitive for indicating the initial repellency of a soil, and in conjunction with WDPT (a measure of repellency persistence) could provide a good indication of the repellent qualities of a soil. However, he found that the relationship between NDST and WDPT was complex. For WDPT < 90 s the relationship was approximately exponential. For WDPT > 90 s there was no apparent relationship with NDST. As a consequence, the WRI index has not been widely used, although the WDPT and a modified NDST method (MED) have been applied in many studies.

F. Molarity of an Ethanol Droplet (MED)

The MED test was developed by King (1981) from the NDST test of Watson and Letey (1970). The MED test measures the molarity of an aqueous ethanol droplet required for soil infiltration within 10 s. The ethanol lowers the liquid-solid contact angle, increasing the rate of infiltration into repellent soil.

King (1981) compared the MED test with measurement of the apparent contact angle, H (Emerson and Bond, 1963), WDPT and infiltration measured with a small ring infiltrometer (SRI). MED was highly correlated with H ($R^2 = 0.75$), however variability in the relationship increased in soil with measured $H > 92°$ (MED > 2.0). King proposed that this was due to water being forced between the soil and the walls of the capillary tube in the contact angle test. King found that the MED test was not useful for media with low degrees of repellency, since MED $= 0$ when $H \leqslant 81°$. For such soils King noted that the WDPT test can be treated as an extension of the MED test below MED $= 0$, however, he found that WDPT declined to < 1 s when $H < 75°$. The $\log_{10}$ WDPT data was correlated with H by a quadratic relationship ($R^2 = 0.85$).

The most important advantages of the MED test are its simplicity and rapidity, especially for severely repellent soil which may have WDPT in excess of 1 h (Wallis et al., 1991). The major disadvantage of the test is that it is unsuitable for soils with a low degree of repellency.

G. Thermal Analysis

In a discussion of measurement techniques at the University of California, Riverside, Conference on repellent soils, the possibility of using the heat of wetting to indicate the level of free energy associated with the solid matrix was discussed (Hammond and Yuan, 1969). This possibility was not explored until the Russians Lishtvan and Zuyev (1983) measured the heat of wetting to characterize the affinity of peats for water. Mallik and Rahman (1985) found that repellent soil thermally released water more easily than nonrepellent soil. Measures of thermal release were generally well correlated with WDPT and direct measures of *H*. Differences between thermal release, WDPT, and *H* were ascribed to the fact that WDPT and *H* measure the interaction of water with the solid surface, whereas thermal release is a measure of both surface and bulk properties of the soil. Application of thermal analysis for repellency measurement is likely to be limited because it demands specialized equipment, and it only provides semiquantitative data. Further evaluation of thermal analysis is necessary to compare its relative merits with other techniques on a range of soils.

IV. Development of Repellency

A. An Organic Coating on Soil Particles

Although numerous researchers have attempted to determine the causes of repellency, research findings have not been entirely conclusive, and would suggest that there are a number of causes which are site-specific. However, a number of mechanisms appear common to most soils which are important to understand. Fundamentally, most workers agree that a coating of organic matter on the surface of soil particles imparts repellency (Bond, 1969a). Studies employing high magnification of repellent sand grains have shown the presence of an organic coating which was not present on sand grains from nonrepellent soil (Wilkinson and Miller, 1978; Rankin and Ross, 1982) (Figure 7a,b).

Bozer et al. (1969) found that synthetic materials which decreased soil wettability included many organic compounds with amphiphilic characteristics (with both hydrophilic and hydrophobic groups). They postulated that the hydrophilic end adsorbs to the hydrophilic soil surface, probably by complex formation, by hydrogen bonding and/or adsorption by Van der Waal's dipole-dipole interaction. The hydrophobic portion of the compounds would then effectively create the "new surface" of the soil particle.

Ma'shum and Farmer (1985) produced evidence which indicated that the molecular orientation of organic matter determines repellency of the soil. They found that prolonged shaking of repellent soil detached organic

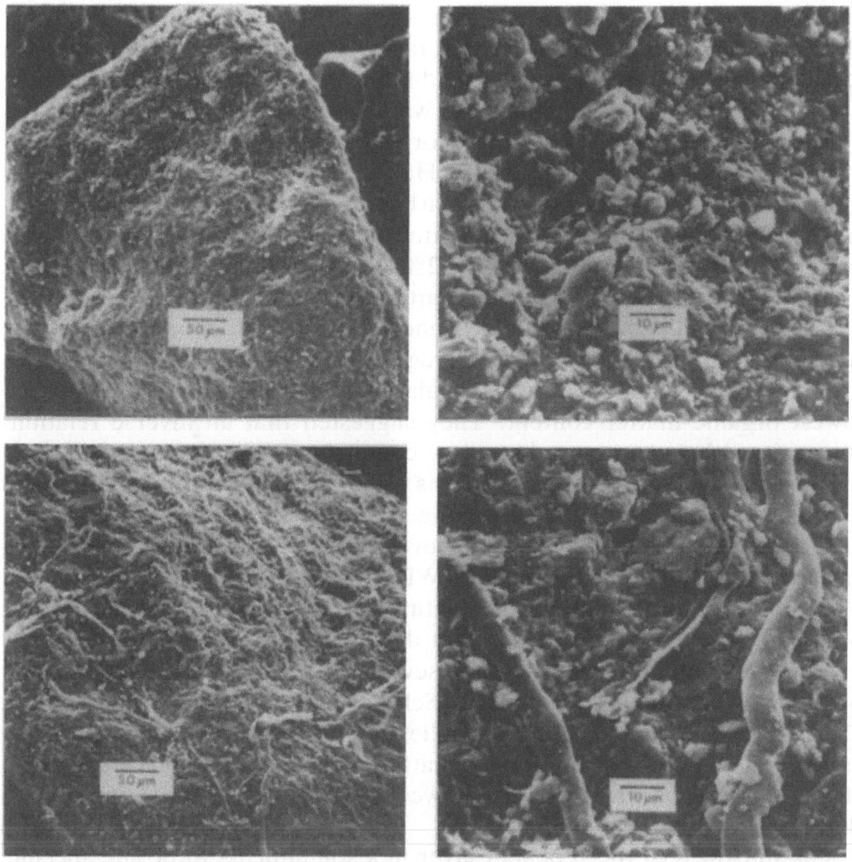

Figure 7. A. Two scanning electron micrographs (uppermost) of "clean" sand grains from a non-repellent golf green site, with a general absence of an organic coating. B. Two scanning electron micrographs (lowermost) of sand particles from a repellent golf green site, with an organic coating and tubular strands of fungal hyphae. After Rankin and Ross (1982).

matter coatings from sand particles and reduced repellency. Wallis et al. (1990a) made similar observations (Section V.A).

There is conflicting evidence regarding the thickness of the repellent coating necessary to induce maximum repellency. Fink (1970) found that repellency increased rapidly with increasing addition of synthetic repellent chemicals to a nonrepellent sand, until a point was reached when repellency of the soil did not increase. He assumed that at this point complete, "monolayer" coverage of the sand particle surfaces was achieved. Nakaya et al. (1977b) made similar observations when humic acid was added at increasing concentration to a nonrepellent sand. In contrast, Ma'shum et al. (1988) found that 16 times the amount of cetyl alcohol necessary for monolayer coverage was necessary before maximum repellency developed. They suggested that the "excess" cetyl alcohol may have been necessary to coat applied water as well as the sand grains.

A surprisingly small amount of organic matter is necessary to produce severe repellency. Bond and Harris (1964) found that Australian sands with <0.1% total organic carbon had apparent contact angles, H (Emerson and Bond, 1963) greater than 90°. DeBano et al. (1970) reported $H = 85°$ in sands with 0.02% organic matter. DeBano (1969a) pointed out that the total organic matter content bears no relation to the degree of repellency since some sands with >5% C were more easily wetted than others with 0.1% C. Savage et al. (1969b) found that the two most severely repellent sands inoculated with fungal cultures had the lowest organic matter content. They suggested that an inverse relationship existed between organic matter content and repellency, which was difficult to substantiate upon the basis of so few observations. Singer and Ugolini (1976) attempted to correlate the repellency of a number of soil and litter horizons with %C, however, there was a large amount of variation about the regression line (WDPT ranged from 0.2 s to >100 s at 2% C). It would appear that the nature rather than the quantity of soil organic matter is the most important determinant of repellency severity.

However, within a soil profile the severity of repellency is related to the quantity of organic matter present. Scholl (1975) and Singer and Ugolini (1976) reported a decrease in repellency with depth which was closely associated with decreasing organic matter. Wallis et al. (1990a) reported a strong correlation ($R^2 = 0.79$) between %C and the MED index with depth in a sandy soil under pasture.

The thickness of the repellent layer in a soil appears to be site-specific, however some generalizations may be made. DeBano and Rice (1973) stated that repellency usually occurs within 100 to 150 mm from the soil surface, which was supported by observations of DeByle (1973), Reeder and Jurgensen (1979), and Wallis et al. (1990a). However, in burned soils the repellent layer may be only 10 to 20 mm thick and at any depth within the topsoil (Section IV.F). In sports turf the repellent layer may be restricted to 20 to 50 mm (Wilkinson and Miller, 1978; Tucker et al., 1990), however, Wallis et al. (1989b) found that soil from New Zealand golf courses was repellent to an average depth of 100 mm. Singer and Ugolini (1976) found that repellency extended to >300 mm in some forested soils.

Generally the more extreme cases of repellency are found in sandy soils. Sand grains are more readily coated by organic material due to their relatively low specific surface area (McGhie, 1987). In an analysis of heat-induced repellency, DeBano et al. (1970) found that the thickness of the repellent layer increased as the percentage of silt and clay decreased. The thickest and most intensely repellent layer was produced in a uniform sand of particle diameter <425 μm. The effect of texture upon repellency was attributed to the large differences in specific surfaces of the soils, which ranged from 7.7×10^{-3} to $55\,m^2\,g^{-1}$.

In sports turf, widespread repellency may be related to the use of sand as a medium. Rieke (1981) pointed out that with the increased emphasis upon high-sand-content soil mixes in the construction of new greens and the widespread acceptance of sand topdressing on old greens, the potential for repellency development will increase.

Studies of repellency were mainly restricted to sandy soils for many years, but repellency is not restricted to sands. Bond (1969a,b) noted that small repellent pockets were found in Western Australian loams which contained up to 20% clay. Giovannini et al. (1983) reported severe repellency in heavy textured Italian soils (40% clay) and McGhie and Posner (1980) reported severe repellency in Australian soils with >20% clay. McGhie and Posner proposed that clay aggregates had become coated with hydrophobic material, a mechanism which Wallis et al. (1991) supported in research with a New Zealand humic clay.

1. Peat

Despite general recognition of the problem of repellency in peats there have been surprisingly few studies, and consequently the mechanism of repellency development in peats is poorly understood. Van't Woudt (1959) first reported repellency of peat bogs in New Zealand and later explained (van't Woudt, 1969) that when peats are drained the surface layers may undergo irreversible drying. Lishtvan and Zuyev (1983) reported that peat water absorption was reduced when dried beyond a certain limit (typically below 0.30 g/g). They argued that changes in the peat pore system rather than the development of repellency were responsible for this change; however, they reported that dry peats could be readily wetted with organic liquids of low surface tension which indicated that repellency had developed. Nakaya et al. (1977a) measured severe repellency (WDPT > 300 s) of a dry Japanese peat soil.

The nature of the repellency mechanism may differ between peat and other soils due to the ways in which water is held in peats (Lishtvan and Zuyev, 1983). Flaig (1986) pointed out that the physical nature of peat differs fundamentally from that of soil, primarily as a result of the anaerobic conditions under which humification of peat-forming plants occurs. In peat, the decomposition products of plant constituents such as lignins and the mechanical structure of mosses and graminaceous plants yield macroporous, spongelike products which absorb large quantities of water very strongly (Kwak et al., 1986). However, it has been suggested (Kwak et al., 1986) that peat surfaces are predominantly hydrophobic, resulting in restricted infiltration when the peat dries.

Peat development and composition may affect the affinity of peat for water. Kay and Goit (1977) found that the most decomposed sphagnum moss peats had the highest water adsorption, which they assigned to increased specific surface area and changes in organic matter composition. Flaig (1986) pointed out that chemical alteration during decomposition, especially of the lignins, makes the organic surfaces less repellent, which may at least partly account for Kay and Goit's (1977) observation. Lishtvan and Zuyev (1983) found that the water-affinity of a number of Russian peats developed from sedges changed insignificantly during mineralization. The contrasting results with those of Kay and Goit (1977) may be due to the different plant origins of the peat bogs (Section IV.B.1).

Van't Woudt (1969) stated that repellent surface peat layers necessitated removal of the layer until a moist layer is exposed. However, he pointed out that such management was unsustainable since the peat surface is gradually lowered and productivity decreased. Water-table management or irrigation could be used to prevent excessive drying of the surface peat layers. Lishtvan and Zuyev (1983) reported that they had used wetting agents to improve peat wettability (Section V.C).

2. Thatch

In sports turf, thatch may be a contributing factor to the development of repellent soil (Waddington, 1969). Thatch is the layer of partially decomposed organic material which comprises the uppermost layer of most turf profiles, and is widely believed by turf managers to be responsible for the development of repellency in turf (commonly called "dry patch" because of its "patchy" occurrence). Thatch is known to reduce infiltration into turf when dry (Taylor and Blake, 1982). However, thatch is not a prerequisite for dry patch development (Danneberger and White, 1988). Wilkinson and Miller (1978) found that thatch removal from dry patch cores did not significantly increase water infiltration. Irrigation water penetrated the turf and thatch but would not penetrate the thatch/soil interface. They concluded that repellency extended into the soil and was not merely a function of the thatch layer. Wallis et al. (1989b) found that although samples of thatch were severely repellent (MED 2.6 to 5.0), the soil alone from 31 New Zealand golf greens was also severely repellent (MED 3.3 to 3.5).

Thatch control by a variety of mechanical means is an integral part of turf management and would be expected to reduce the severity of repellency. However, thatch control alone will not prevent the development and expression of repellency in turf. Techniques to ameliorate "dry patch" in sports turf are discussed in Section V.

B. Origin of the Organic Coating

1. Vegetation

The relationship between repellency and the principal origin of the organic coating, the vegetation cover, has occupied scientists since the studies of Jamison (1942, 1945) on citrus groves in Florida. Bornemisza (1964) added citrus leaves (*Citrus* sp.) to a nonrepellent sand and found that repellency was produced. The technique of Bornemisza (1964) has also been used by Roberts and Carbon (1972), McGhie and Posner (1980, 1981), and Reeder and Jurgensen (1979). Conclusions which may be drawn from these studies are limited because the plant material had not undergone natural decomposition and incorporation into the soil, and the contribution of plant roots to soil organic matter was neglected. Studies other than those listed above have been based upon field surveys. One limitation of field studies is that they may not account for the contribution of past vegetation types to repellency development. Table 1 summarizes studies which have examined the relationship between plants and repellency.

Table 1. Plants and soil repellency

Vegetation type	Associated (or not associated) with water repellency
Citrus spp.	Jamison 1942, 1945; Wander 1949
Deciduous trees	
Acer saccharum (maple)	(Richardson and Hole 1978)
Quercus sp. (oak)	(Richardson and Hole 1978) (Reeder and Jurgensen 1979)
Carya sp. (hickory)	(Richardson and Hole 1978) (Reeder and Jurgensen 1979)
Populus tremuloides (quaking aspen)	Reeder and Jurgensen 1979
Heath plants and conifers	
Unspecified heath	van't Woudt 1959
Primarily *Calluna vulgaris*	Mallik and Rahman 1985
Vaccinium sp. and other *Erica* species	Richardson and Hole 1978
Tsuga canadensis (hemlock)	Richardson and Hole 1978
Pinus resinosa (red pine)	Richardson and Hole 1978; Reeder and Jurgensen 1979
Pinus sp. (unidentified)	Richardson 1984
Pinus pinaster	Giovannini and Lucchesi 1984
Pinus monophylla	Holzhey 1969a
Pinus jeffreyii (Jeffrey pine)	Hussain et al. 1969
Pinus radiata (Radiata pine)	(Bond 1964)
Pinus strobus (white pine)	(Reeder and Jurgensen 1979)
Pinus banksiana (Jack pine)	Reeder and Jurgensen 1979

Table 1. *Continued*

Vegetation type	Associated (or not associated) with water repellency
Juniperus spp.	Holzhey 1969a
Juniperus osteosperma	Scholl 1971
Larix occidentalis (Western larch)[a]	DeByle 1973
Pseudotsuga menziesii (Douglas fir)	DeByle 1973
Picea engelmanii (Engelman spruce)	DeByle 1973
Eucalypts	
Eucalyptus spp.	Gilmour 1968; Bond 1964; McGhie and Posner 1981
E. marginata (Jarrah)	Roberts and Carbon 1972
E. astringens (brown mallet)	McGhie and Posner 1980, 1981
Cereal crops	
Triticum aestivum (Wheat)	(McGhie and Posner 1981)
A. sativa (Oats)	(McGhie and Posner 1981)
H. vulgare (Barley)	McGhie and Posner 1981
Legumes	
Medicago sativa (Lucerne)	Bond 1964; Roberts and Carbon 1972; (McGhie and Posner 1981)
Lupinus cosentinii (W.A. blue lupin)	(McGhie and Posner 1981)
Trifolium subterraneum (subterranean clover)	Roberts and Carbon 1972; McGhie and Posner 1981
Grasses	
Phalaris sp.	Bond 1964
E. calcycina (perennial veldt grass)	McGhie and Posner 1981
Agrostis spp. (Bentgrasses)	Wilkinson and Miller 1978
S. secundatum (Buffalo grass)	McGhie and Posner 1981; (McGhie and Posner 1981)
Cynodon sp. (Couch grass)	(McGhie and Posner 1981)
Desert scrub	
Larrea divaricata	Adams et al. 1969
Prosopis juliflora	Adams et al. 1969
Cercidium floridum	Adams et al. 1969

[a] Deciduous conifer.

In general, deciduous tree species have not been associated with repellency (the larch, a deciduous conifer, is an exception). Fire was a compounding factor in the study by Reeder and Jurgensen (1979) (Section IV.F). Most coniferous plants, in particular *Pinus* species, have been associated with repellency development. However, not all *Pinus* species impart repellency; therefore, the likely effect of a species cannot be readily estimated from existing studies, even from plants within the same genus. Comparison of the existing studies is difficult because most have used relative measures of repellency such as in situ infiltration.

A large range of *Eucalyptus* species have been found to be associated with repellency in Australian field and laboratory studies. McGhie and Posner (1980) found that the leaf drip from *E. astringens* did not induce repellency in a nonrepellent sand, whereas comminuted litter induced severe repellency. Jamison (1942) had suggested that repellency under citrus trees was related to leaf drip, but later suggested that colloidal organic matter might be more important (Jamison 1945).

McGhie and Posner (1981) evaluated the effect of three cereal crops upon repellency. Study of plant rotation systems showed that repellency increased under subterranean clover pasture but was reduced under wheat. They suggested that in suitable farming systems, crops such as wheat could be grown to reduce repellency, however, they demonstrated that the change to wheat would need to be made before severe repellency developed.

Leguminous species have been found to differ in their effect upon repellency. The contrasting findings of Bond (1964), Roberts and Carbon (1972), and McGhie and Posner (1981) with lucerne may be due to differing experimental procedures. Bond (1964) measured field infiltration rates, whereas McGhie and Posner (1981) added freshly ground plant-top material to nonrepellent sand, and Roberts and Carbon (1972) used a range of extractants and adjusted the pH of plant sap before addition to nonrepellent sand. For the reasons outlined previously, the field study of Bond (1964) would be the most reliable of the three studies.

The relationship between repellency and native or introduced grass species has been explored in natural, agricultural, and turf ecosystems. Holzhey (1969a,b) found that grasses in the Southern Californian foothills developed a thin root zone with high organic-matter content in which seasonal repellency was most persistent. Forbs usually did not develop such a zone, and were most often associated with shorter-lived seasonal repellency. In a study of water infiltration into a South Australian sand under a range of plant species Bond (1964) found that the soil under *Phalaris* sp. pasture was most severely repellent. Sites bare of vegetation were more dry and repellent than adjacent sites under annual grasses. The bare sites must have previously supported plants which contributed to the organic coating. Bond (1969a) reported that bare coastal and desert dune sands were nonrepellent, but became repellent after the establishment of semipermanent or permanent vegetation. Even dunes which had a sparse cover of sedges were found to be repellent.

In turf, "dry patch" has been found to be prevalent on bentgrass greens (Wilkinson and Miller, 1978). However, this may be due to the relatively shallow-rooted nature of the species and subsequent poor drought tolerance, rather than to the composition of the plant (Bolton, 1990). McGhie and Posner (1981) found that ground tops of buffalo grass did not induce repellency in nonrepellent sand. However, they measured severe repellency on 10-year-old lawns of buffalo grass and couch. This was a further example of the superiority of field evaluation of the influence of vegetation upon repellency.

Adams et al. (1969) found a relationship between repellency and the establishment of annual plants in California desert scrub vegetation. They found that soil beneath burned *Larrea divaricata*, *Prosopis juliflora*, and *Cercidium floridum* hummocks was repellent and supported no or few annual plants up to 4 years after fire.

2. Soil Microorganisms

In 1917, Schantz and Piemeisel showed that fairy rings in pastures and crops were due to basidiomycete fungi, and that the presence of basidiomycete mycelia was associated with poor soil water absorption. Bond and Harris (1964) found that repellent Australian sands always had a zone of copious fungal proliferation, with a regular occurrence of basidiomycete fungi. Some basidiomycete mycelia themselves had a waxlike, repellent surface. However, they noted that whether the basidiomycetes themselves were a major source of the repellent organic material remained uncertain, because many other hyphomycete fungi and bacteria were also found in the repellent soils.

Bornemisza (1964) found that mycelia of the fungus *Aspergillus niger* added to a nonrepellent fine sand induced repellency. In a similar experiment, Savage et al. (1969b) innoculated sterile sand with cultures of a range of fungal spores. However, in contrast to Bornemisza he found that only *Penicillium nigricans* and *Aspergillus sydowi* caused mild repellency while *A. niger* and five other fungi did not develop measurable repellency in a 53-day incubation. Savage et al. (1972) found that heating a fungal sand culture prepared by the method of Savage et al. (1969b) produced repellent organic substances which were very similar to those responsible for repellency in a soil susceptible to heat-induced repellency (Section IV.F).

McGhie and Posner (1981) found that the effect of fungi upon repellency varied as widely as the effect on repellency of the different plant species. They found that cultures of soil fungi added to a nonrepellent sand and mixtures of plant material could either increase or decrease repellency, depending upon the relative repellency of the fungi and the plant. Bond (1964) also suggested that different degrees of repellency in soils under a range of plant species may be due to differing suites of microorganisms growing in association with the plants. Similarly, Richardson and Hole (1978) implicated fungal mycelia in the development of repellency under mor forest humus which were not prevalent under less repellent mull humus from deciduous forests. DeByle (1973) found that mor forest litter was extremely repellent 1 year after burning only if it contained fungal mycelia. However, Roberts and Carbon (1972) found that repellency could be induced in a nonrepellent sand with the addition of microorganism cultures, with or without the presence of plant material.

The specific microorganism(s) involved with the development of repellency may vary between soils. Jex et al. (1985) found that the repellency of an incubated soil was most strongly correlated with the population of actinomycetes rather than with the number of fungi or bacteria. Scanning electron micrographs of the repellent sand grains also revealed high populations of actinomycetes.

The contribution of fungi in the development of repellency in sports turf has received some attention. Wilkinson and Miller (1978) undertook an extensive study of the causes of "dry patch" in Ohio, USA, sand golf greens. They found that the organic coatings on repellent sand grains appeared to incorporate fungal mycelia, but no difference was found between the fungal species composition of non-dry patch and dry patch soil. "Fairy rings," which are a common problem in turf and are produced by basidiomycete fungi, consist of a circle of lush turf growth, stimulated by nitrogen mineralization produced by the radially extending fungal mycelia. Within the lush ring of growth is a dense mycelial "zone of inhibition," where turf growth suffers (McIver, 1988). Toxins, primarily hydrogen cyanide, are released from the fungi (Lebeau and Hawn, 1963), and the turf develops repellency (Wilkinson and Miller, 1978) (Figure 8). Whitely and Baldwin (1990) found that, compared with unaffected soil, the zone of inhibition had lower soil water content and, consequently, higher soil strength which was likely to restrict root penetration.

It appears that although a direct contribution of soil microorganisms to the development of repellency may be shown in incubation studies, in the field they normally play a more indirect role through organic matter decomposition and hence the accumulation of the organic coating on repellent soil particles. Soil microorganisms growing in association with certain plants may produce repellent decomposition products. When dense populations of a specific microorganism develop, such as in the case of basidiomycete fungi in "fairy rings" (Schantz and Piemiesel, 1917) or fungi in forest litter (DeByle, 1973), the microorganism may directly affect the development of repellency.

C. Chemistry of the Organic Coating

Many extractants have been used in attempts to isolate and characterize the repellent organic coating. Methanol, ether, ethanol, and benzene have had variable effectiveness, which reflects different extraction conditions but may also indicate differences in the nature of the repellent coating of the soils studied (Jamison, 1945; Wander, 1949; Savage et al.,

Figure 8. Water repellent turf ("dry patch") in the "fairy ring" pattern characteristic of basidiomycete fungi. Note that soil cores have been removed from the dry patch ring. Also note that dry patch is most often found in irregular patterns (photo P. Rieke).

1969b; Ma'shum and Farmer, 1985; Ma'shum et al., 1988; Roberts and Carbon, 1972; McGhie and Posner, 1980; Giovannini et al., 1983; Jackson and Gillingham, 1985). Other extractants such as acetone (Savage et al., 1969b; Roberts and Carbon, 1972; McGhie and Posner, 1980) and chloroform (McGhie and Posner, 1980; Ma'shum and Farmer, 1985) have been found consistently ineffective. Chloroform treatment actually increased repellency, which was ascribed to either molecular reorientation of repellent organic matter, e.g., alkyl chains, or redistribution of repellent material on exposed nonrepellent surfaces. Cold water has been found ineffective, however, boiling hot water or stirring with heated water has been found to remove repellency (Roberts and Carbon, 1972; McGhie and Posner, 1980). The physical disturbance of boiling and stirring may have reduced repellency (Ma'shum and Farmer, 1985; Wallis et al., 1990a).

Removal and identification of the organic matter fraction responsible for repellency is difficult. Although the "effective" extractants described may render repellent soil nonrepellent, the extracted molecules may bear little resemblance to the molecules responsible for the hydrophobic layer surrounding soil particles due to possible cleavage of covalent linkages and changes in molecular conformation. Ma'shum et al. (1988) found that an amphiphilic mixture of isopropanol and 15.7 *M* ammonia extracted organic materials responsible for repellency in a range of repellent Australian soils after 8 h under soxhlet reflux conditions. After extraction the repellent soils measured MED = 0, and the extracted materials restored repellency of acid-washed or ignited sands to levels comparable to the original, repellent soils. Although Ma'shum et al. (1988) claimed that "complete" extraction had been achieved, soils with MED = 0 may have hydrologically significant repellency, as discussed in Section III.B and shown by Wallis et al. (1991).

The nature of the organic coating on repellent soil particles has been the subject of much investigation and debate. Bozer et al. (1969) found that synthetic substituted phenols could induce repellency when added to nonrepellent sand. The position and nature of the substitution groups on the benzene ring determined the repellency characteristics of the compound. Chlorosilanes, amines, and nitrogen-containing heterocyclics also exhibited repellency. Savage et al. (1969b) also found that amines induced repellency when added to a nonrepellent sand. The substituted phenols and other organic compounds identified by Bozer et al. (1969) are naturally occurring compounds in soil, with phenolic compounds very common in natural resins. From infrared spectra of the extract from a repellent soil, Giovannini and Lucchesi (1984) hypothesized that an ester between phenolic acids and polysaccharide-like substances was responsible. However, Roberts and Carbon (1972) found that an Australian sand remained repellent after a soxhlet extraction of resins with hot ethanol.

Savage et al. (1972) captured the volatilized products of a Californian soil susceptible to heat-induced repellency (Section IV.F). Adsorption chromatography with a benzene-acetone solvent isolated three fractions which produced repellency when added to a nonrepellent sand. The elution pattern was as follows: fraction I, very pale yellow, 4 to 7.5% of extract; fraction II, dark brown, 14 to 39%; fraction III, dark brown, 7%. Giovannini and Lucchesi (1984) also obtained an elution pattern with three fractions following adsorption chromatography of a benzene extract: fraction I, pale yellow, 25%; fraction II, yellowish green, 25%; fraction III, pale brown, 40%. It would appear that there were some similarities in the organic matter fractions responsible for repellency in the two soils.

Ma'shum et al. (1988) used thin layer chromatography to identify three elution fractions of a benzene/ethanol (2:1 vol/vol) extract. However, they did not describe the color or relative proportions of the fractions. An isopropanol/ammonia extract produced only two fractions which occupied different positions on the chromatogram than the fractions of the benzene/ethanol extract. Savage et al. (1972) found that thin layer chromatographic analysis indicated that the three fractions did not comprise pure materials. Elemental, infrared, and NMR analyses indicated that fraction I was a mixture of aliphatic hydrocarbons, and fractions II and III were also aliphatic but contained up to 15% oxygen as carbonyl groups.

Savage et al. (1972) found that the captured volatilized products of heated cellulose appeared and behaved in a similar manner to the captured volatilized material from a repellent soil. They suggested that cellulose may be one source of the repellent material. However, Roberts and Carbon (1972) found that an Australian sand remained repellent after extraction of the cellulosic fractions with acid solutions.

Giovannini and Lucchesi (1984) found that differential thermal analysis (DTA) revealed an exothermic peak at 490°C in both a repellent clay loam and materials extracted with acetylacetone/benzene, but not in the extracted soil. They suggested that the 490°C peak was diagnostic for repellency, since the height of the peak correlated with the severity of repellency induced by each of the extracted fractions upon a nonrepellent sand. However, the 490°C peak was absent in a DTA study of a burned repellent soil in Scotland (Mallik and Rahman, 1985). DTA studies may prove useful for elucidating properties of the repellent coating, but need to be conducted upon a wide range of soils.

Giovannini and Lucchesi (1984) used the infrared spectrum of the repellent extract to infer the nature of atomic bonding within the material. they found evidence of aromatic compounds, esters, polysaccharides, transolefinic hydrocarbons, and methoxyl groups. They cited the presence of methoxyl groups as an indication that the repellent fraction had a low degree of humification.

Some contention has arisen over whether the repellent coating is within the humic or fulvic acid fraction. Miller and Wilkinson (1977) found that the organic coating surrounding dry patch sand grains produced an infrared spectrum closely resembling those of fulvic acids (Stevenson and Goh, 1971). However, Roberts and Carbon (1972) found that the humic acid (HA) fraction and not the fulvic acid (FA) fraction of the organic coating induced repellency when added to a nonrepellent sand. Adhikari and Chakrabarti (1976) found that addition of one soil HA and two microbial HAs increased the repellency of an Indian soil (from WDPT 20 s to WDPT 73 to 160 s). The microbial HAs produced the greatest effect, which was increased slightly by the addition of iron. Nakaya et al. (1977b) added a range of soil HAs to a nonrepellent sand ($H = 0°$; Emerson and Bond, 1963). All of the five HAs tested produced repellency in the sand (H 71° to 97°). Increasing HA concentrations on the sand successively reduced the capillary rise of water. In a contrasting study, Savage et al. (1969a) evaluated the effect of three microbial, one soil, and one peat HA upon a nonrepellent sand and sandy loam. Only one of the microbial HAs produced slight repellency (WDPT < 15 s) at soil pH 6, and they concluded that HA substances probably contribute little to repellency.

The soil humic fraction (HA and FA) represents a complex colloidal system consisting of molecules with a wide range of molecular weights (Stevenson, 1982). The humic colloids have hydrophilic properties, however, close association with soil particles may reduce their affinity for water due to changes in molecular orientation and the relative positions of charged sites. Debate regarding the relative contribution of HA and FA to repellency is somewhat unproductive, because HA and FA are classified on the basis of extraction conditions and comprise many organic compounds.

A number of studies have implicated fatty acids with repellency. Wander (1949) suggested that alcohol extracts from repellent citrus soils indicated the presence of fatty acids. He found that stearic acid created extreme repellency in a nonrepellent soil after treatment with calcium and magnesium hydroxide and concluded that repellency was due to Ca and Mg soaps of fatty acids. Savage et al. (1969b) found that palmitic acid induced repellency in sands after heating. Ma'shum et al. (1988) found that palmitic acid and another highly polar lipid, cetyl alcohol, induced severe repellency (MED approximately 6) in an acid-washed sand without heating. Calculations with cetyl alcohol showed that measurable repellency only occurred after sufficient material was added to produce a close-packed monolayer on the sand grains. A 16-fold excess of material was necessary to produce maximum repellency, which led Ma'shum et al. (1988) to suggest that the lipids may need to coat the surfaces of water droplets as well as soil particles.

Wilkinson and Miller (1978) speculated that Ca and Mg soaps of fatty acids were responsible for repellency due to the solubility properties of organic material extracted from repellent sites on golf greens. Tucker et al. (1990) rejected the speculation of Wilkinson and Miller (1978), because they measured no difference in the extractable Ca and Mg content between "dry patch" and "non-dry patch" cores. However, Wallis et al. (1989b) have found that "non-dry patch" areas of a green (i.e., those areas not displaying the symptoms of repellency) have a repellency severity equal to that of the dry patch areas once dry. Therefore the criticism of Tucker et al. (1990) is invalid.

Ma'shum et al. (1988) used spectroscopic and chromatographic examination of an *iso*-propanol/ammonia extract to identify the 16 to 32 carbon atom and free and esterified fatty acids within the extracted material. The fatty acids had a bimodal distribution with maxima at C_{16} and C_{22} and comprised 1 to 18% of the extracted carbon. The ^{13}C-NMR and infrared spectra of the most hydrophobic extract suggested that molecules with extensive polymethylene chains were responsible. The infrared spectra were similar to previously published spectra for organic coatings extracted from sand grains of repellent golf greens (Miller and Wilkinson, 1977), repellent soils (Giovannini and Lucchesi, 1984), and repellent plant litter (McGhie, 1980).

The role of fatty acids in other repellent soils deserves further research. The use of a standard extraction technique would facilitate comparisons between studies, and the *iso*-propanol/ammonia solvent used by Ma'shum et al. (1988) appears to have advantages over other extractants used to date. Fatty acids comprised 1 to 18% of the extract in the study of Ma'shum et al. (1988). The characterization of other components in the repellent coating would improve our understanding of repellency development and expression.

D. The Influence of Soil pH

Whether soil pH has an influence upon the development of repellency is uncertain. Van't Woudt (1959) found that the application of lime to repellent New Zealand soils successively reduced repellency. When $2\,t\,ha^{-1}$ was worked into the 0 to 150 mm soil layer, repellency was removed; the resulting pH values were not, however, reported. Bond (1969a) reported that calcareous and acid sands of Australia were equally affected by repellency. Savage et al. (1969a) found that repellency induced by addition of a microbial humic acid to a nonrepellent sand and sandy loam was influenced by pH, however, the results were not consistent between the two mediums.

Two New Zealand scientists, Jackson and Gillingham (1985), reported that lime addition in several field trials produced rapidly increased soil moisture levels. They cited researchers in a range of districts which had reported similar observations. The moisture response was most evident in late summer and early autumn, and particularly after dry periods. They found that liming had reduced repellency as measured by capillary rise and water drop penetration time, but proposed no mechanisms for the effect.

A possible mechanism for the reduction of repellency by liming is as follows. Fulvic acid (FA) is soluble in water at any pH that may occur in soils, whereas humic acid (HA) is water soluble only at $pH > 6.5$ (Chen and Schnitzer, 1978). Furthermore, Chen and Schnitzer (1978) demonstrated that both HA and FA solutions significantly reduced the surface tension of water, which would decrease the solid-liquid contact angle and increase infiltration into repellent soil. The surface tension of HA and FA solutions also decreased with increased pH. Liming soil to increase pH would therefore increase the ability of resident FA (and HA at $pH > 6.5$) to increase infiltration into repellent soil. Chen and Schnitzer (1978) noted that if there is a deficiency of FA in the soil solution, repellency would be more severe.

The balance between quantities of FA and HA in the soil may therefore be implicated in the development of repellency, along with soil acidification. However, as reported earlier there is conflicting evidence regarding the contribution of HA and FA to repellency. FA is known to leach with soil development, because of its solubility across the pH range (Schnitzer and Desjardins, 1969). As soils develop, topsoil FA would tend to decrease (Goh et al., 1976) with a consequent increase in repellency.

Although increasing soil pH is relatively cheap and straightforward, any beneficial effect of repellency reduction must be balanced against the influence of soil pH upon soil biological activity and the availability of plant nutrients.

E. The Influence of Soil Relative Humidity

The reported effects of soil relative humidity (RH) upon repellency are inconsistent. At RH $<$ 100% the repellency of dry sands has been found to increase, decrease, or remain constant (Yuan and Hammond, 1968; Jex et al., 1985; Roberts and Carbon, 1971). At 100% RH the repellency of dry sands has been found to either increase (Jex et al., 1985) or decrease (Roberts and Carbon, 1971). The repellency of sands wetted prior to incubation at 100% RH has been found to decrease (Roberts and Carbon, 1971; Jex et al., 1985). The effects of humidity upon repellency should be regarded within the context of normal soil RH values. Soil RH exceeds 99.56% at matric potentials of less than -6 bars (Hillel, 1980). Unless the soil becomes very dry RH is therefore always very close to 100%, so that the studies that imposed very low RH values are irrelevant to field soil conditions.

F. The Effect of Fire

High temperatures have been used to render repellent soils non-repellent (Emerson and Bond, 1963). However, DeBano (1969c) found that the relationship between heat and repellency was more complex. The repellency of a soil from a burned watershed greatly increased at >200°C but was destroyed at >430°C in a muffle furnace. Cory and Morris (1969) made similar observations, however other studies have found that repellency is lost at temperatures of 200° to 315°C (DeByle, 1973; Savage, 1974; Scholl, 1975; Nakaya et al., 1977a,b; John, 1978).

Water repellency in burned soils was first recognized by workers at the University of California, Riverside (Osborn et al., 1964). Krammes and Osborn (1969) reported that 60% of the burned watersheds in Southern California that supported predominantly chamise chaparral bush *Adenostoma fasculatum* were repellent. Osborn et al. (1964) used a field wetting-agent treatment to identify repellency as an important factor that influenced runoff, erosion, and plant establishment in the burned watersheds. Since this time, repellency in burned soils has been reported in other states of America (Hussain et al., 1969; DeByle, 1973; Scholl, 1975; Reeder and Jurgensen, 1979) and in other countries including

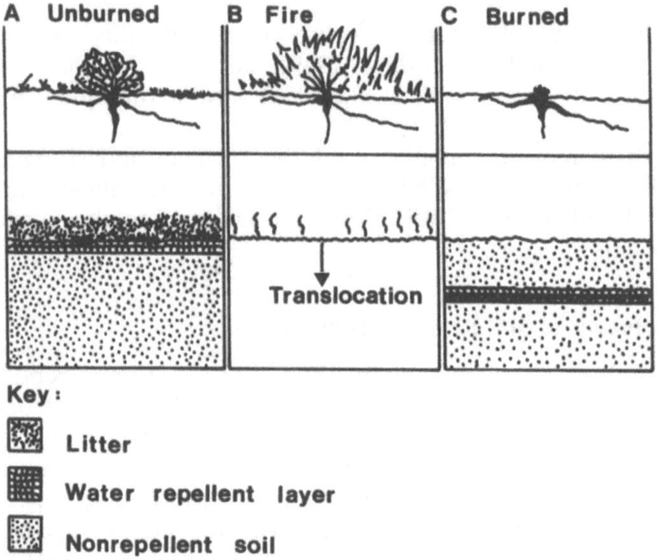

Figure 9. Water repellency before, during, and after fire. A. Before fire the water repellent substances accumulate in the litter layer and mineral soil immediately beneath it. B. Fire burns vegetation litter layer, causing the water repellent substances to move downward along temperature gradients. C. After a fire a water repellent layer is located below and parallel to the soil surface. After DeBano (1969c).

Japan, New Zealand, Italy, and Scotland (Nakaya et al., 1977a,b; John, 1978; Giovannini and Lucchesi, 1983; Mallik and Rahman, 1985).

Water repellency in burned soils differs from "natural" repellency. In a naturally repellent soil hydrophobic material naturally accumulates and may take many years to reach significant levels (DeBano, 1969d). "Heat-induced" repellency can develop rapidly under wild fires or controlled burn-offs (John, 1978). Fire may increase or decrease the existing natural repellency of a soil (DeBano, 1969c; Giovannini and Lucchesi, 1983) or fire may induce repellency in soil which was previously nonrepellent. DeBano (1969c) found that burning plant litter over a nonrepellent sand induced repellency at depth in the sand. On unburned watershed areas repellency was found only in the upper 50 mm layer (Figure 9). In lightly burned areas the upper 25 mm of soil was essentially nonrepellent, and repellency was concentrated in the 25 to 75 mm layer. On the severely burned areas repellency was not detectable in the first 65 mm of the profile, and repellency appeared to have been translocated into the 75 to 150 mm layer. Reeder and Jurgensen (1979) found that the heat-induced repellency of Michigan forest soils occurred mainly within 0 to 50 mm depth but was occasionally found up to 150 mm depth.

DeBano (1969c) proposed that the mechanism of heat-induced repellency comprised volatilization, downward distillation, and condensation of repellent organic material. Giovannini and Lucchesi (1983) found that the organic matter in a burned repellent soil was decreased by 36% in the topsoil, which they suggested was due to volatile loss. Organic matter in the subsoil increased by 50% but did not induce repellency, which they suggested was due to the fine texture of the soil (clay loam topsoil, clay subsoil). DeByle (1973) found that upwards movement and deposition of organic material could also occur. Heating repellent coniferous litter at 315°C removed the repellency of open soil samples but induced repellency at the surface of enclosed soil. Lower temperatures did not affect repellency and higher temperatures completely removed repellency from the open and enclosed soil. Scholl (1975) found that fires of temperature up to 270°C at 5 mm depth induced repellency at the soil surface, but a 331°C fire made the soil surface nonrepellent and increased repellency at greater depth. Heating the soil in closed vs open containers showed that volatile material was lost from the soil at temperatures >270°C. Scholl (1975) contended that below 270°C sufficient material was trapped and condensed within the soil to increase surface repellency.

Savage et al. (1969b) found that the substance responsible for repellency of fungal sand cultures was not extractable after heating but was extractable under the same conditions before heating, which indicated that a heat-induced chemical change in the substance had occurred. Savage et al. (1972) captured volatile substances released from a burned soil which was susceptible to heat-induced repellency. The captured substances induced some degree of repellency when added to a nonrepellent sand, however, they induced severe repellency when heated on the sand at 300°C for 10 min. Savage (1974) found that the primary movement of organic substances from burning litter into underlying soil occurred during fire and induced natural fractionation of the substance. Heating at over 300°C decreased the ease of extracting the fractions, which lends support to Savage et al.'s (1969b) theory of in situ heat "fixing." The remainder of material which was not "fixed" was volatilized at 300°C.

The effect of fire temperature appears to be site-specific and may depend upon the nature of the burned material (Section IV.F). Savage et al. (1969b) measured the heat-induced repellency of ten naturally repellent soils found under a range of vegetation. Variable effects of heat and solvent extractants led Savage et al. (1969b) to suggest that a range of substances were responsible for repellency in the soils. DeBano et al. (1976) found that the severity of heat-induced repellency was not simply related to the %C of the organic matter deposited on the soil, which supported the theory of DeBano et al. (1970) that only a fraction of the translocated organic matter was responsible for repellency.

Soil texture, water content, and the time of burning have been found to affect heat-induced repellency. DeBano (1969c) reported field observations that the thickest and most severely repellent layers were usually found in coarse-textured soils. DeBano et al. (1970) found that the thickness of the repellent layer induced by burning plant litter over a range of soils was increased as the proportion of silt and clay decreased. DeBano et al. (1976) found that a short burn over dry sand produced the most severe repellency. A longer burn translocated the repellent materials deeper in the sand but removed surface repellency. Burning over wet sand concentrated repellency in the surface layer.

Like natural repellency, heat-induced repellency is spatially variable. DeByle (1973) found that 5% of 0 to 50 mm soil samples from a coniferous Montana forest were repellent (WDPT > 5 s) before fire, but 30% were repellent after fire. Reeder and Jurgensen (1979) reported that 23% of 0 to 150 mm soil samples from unburned forest plots were repellent (WDPT > 10 s), but 40% of samples removed from 53 fire sites with a range of vegetation and soils were repellent. Krammes and Osborn (1969) reported that burned chapparal soil had a high spatial variability of moisture content.

It would appear that heat-induced repellency may be relatively short-lived. DeByle (1973) found that 1 year after fire the number of repellent samples tested had declined from 30 to 7%. Reeder and Jurgensen (1979) found that most (65%) of the soils which were repellent after fire were nonrepellent after 1 year. DeBano and Rice (1973) reported that repellent layers had been found in soils >10 years after fire, but suggested that repellency was greatly reduced after 1 year and became negligible after 5 to 10 years.

A number of techniques have been suggested for the management of soils with heat-induced repellency. DeBano and Rice (1973) suggested that a cool-burning fire would minimize repellency severity or that slash could be piled to concentrate the problem. DeBano et al. (1976) suggested that the least repellency-forming conditions of a short burn over wet soil could be attained by dessicating standing brush and burning when the soil is wet. DeBano and Rice (1973) suggested that the repellent layer could be broken up with a seed drill or "sheepsfoot roller" if terrain permitted. DeBano and Rice (1973) evaluated wetting agents for alleviation of heat-induced repellency and found that the chemicals at the rates used were unsuccessful. In contrast Osborn et al. (1964) found that a wetting agent was effective and resulted in "a very favorable benefit to cost ratio" (Krammes and Osborn, 1969). A comprehensive discussion of wetting agents is given in Section V.C.

G. Spatial and Temporal Variability of Expression

An important aspect of repellency development is high spatial variability of expression. Bond (1964) found that the moisture content within a repellent soil after rainfall varied from 1 to 8% over 10 mm distance.

Wallis and Horne (unpublished data) have measured moisture content differences of even greater magnitude over 50 mm distances in repellent sands. Mallik and Rahman (1985) reported high variability in the WDPT of an undisturbed podzolic soil. Highly variable WDPT and MED values on undisturbed soil samples have been measured by Wallis and Horne (unpublished data).

A number of studies have found that rain or irrigation water penetrates repellent soils through channels and leaves the remainder of the soil dry (Jamison, 1945; Bond, 1964; Gilmour, 1968; Hendrickx et al., 1988a,b). Hendrickx et al. (1988a,b) found that the channels were initiated at soil surface locations where repellency was least. The channels effectively created preferential flow pathways for water and solute movement which extended into the subsoil. Raats (1973) described the theory whereby repellent soil may cause wetting front instability which leads to the development of preferential flow pathways.

The cause(s) of high spatial variability of repellency are unknown, but may include the nature and distribution of past and/or present vegetation species and microbial populations (Section IV.B). Microscale variability (over a few mm or cm) may be due to microtopographic, physical, and chemical variability (Bond, 1972; Mallik and Rahman, 1985). Regardless of the possible causes, high spatial variability of repellency causes patchy plant establishment and growth (Bond, 1972) and makes field research difficult (Wallis et al., 1990b).

The expression of water repellency may vary over time because repellency is expressed most strongly when the soil is dry (DeBano, 1971). However, a number of studies have found that zones of soil may remain dry and repellent throughout extended periods of rainfall (Jamison, 1945; Bond, 1964; Hendrickx et al., 1988a,b). Hendrickx et al. (1988a,b) took soil moisture measurements over winter which demonstrated that the dry, repellent zones of soil were wetted by capillary rise as the water table rose rather than by direct rainfall infiltration. Topography may affect the seasonal expression of repellency through both the soil position in relationship to the water table and the influence of slope angle upon surface water redistribution.

V. Amelioration of Repellency

A. Cultivation

Cultivation to ameliorate repellency has been recommended to dilute the repellent soil fraction with nonrepellent soil (Holzhey, 1969a). Cultivation may also cause abrasion of repellent soil particles. Ma'shum and Farmer (1985) and Wallis et al. (1990a) found that abrasion of repellent sand in end-over-end shakers significantly reduced repellency (Figure 10).

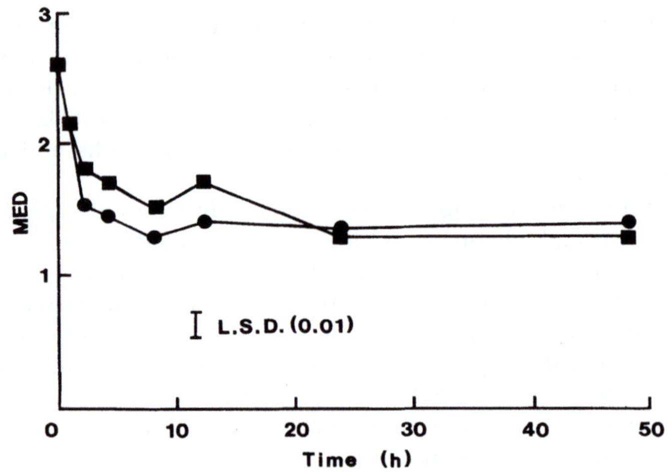

Figure 10. Water repellency (molarity of ethanol droplet, MED) of air-dry Himatangi sand immediately after periods of agitation in an end-over-end shaker (●) and at 72 h after agitation commenced (■). After Wallis et al. (1990a).

However, Wallis et al. (1990a) found that the repellency of the shaken sand increased somewhat after standing, which they proposed was due to changes in the molecular conformation of the organic matter responsible for repellency. They pointed out that this may limit the opportunities for using cultivation as a field treatment for repellency and that in soils such as the one studied a deep layer of severely repellent soil would limit the practicality of dilution recommended by Holzhey (1969a). Furthermore, repellent sands are often prone to wind erosion when cultivated, and dilution of the topsoil would reduce the water- and nutrient-holding capacity of the surface soil.

B. Amendment of Soil Texture

Clay amendments were used by Dutch farmers in the 1950s to improve water- and nutrient-holding capacity, prevent wind erosion, and reduce repellency. (Hendrickx et al., 1988b) Jamison (1945) found that mixing repellent soil with "wettable" soil greatly reduced or eliminated repellency. Roberts (1966) found that only 2.5% of three fine-particle amendments (fly ash, clay, and finely ground silica) were necessary to overcome the effects of repellency upon seedling emergence in an Australian sand. McGhie and Posner (1981) reported that the incorporation of 3% silt/clay effluent from an aluminum refinery was sufficient to lower the WDPT of a sand from minutes to seconds. Addition of the fine particles may enable relatively long-term control of repellency because the high specific surface area particles would be less readily coated by a hydrophobic organic coating (McGhie, 1987).

Recently Ma'shum et al. (1989) have found that the addition of sodic kaolinite and illite clays reduced the repellency of a natural and a synthetic repellent soil. The equivalent of 2.5 to 5 t ha^{-1} to 100 mm depth was required to make a severely repellent sand nonrepellent. Many repellent soils in South Australia have sodic clay subsoils, which may be displaced to the soil surface using a specially designed plough implement. Field evaluation of this technique is currently being conducted (Oades, personal communication). For intensive horticulture such as sports turf, natural sodic clays or clays fortified with sodium could be imported and incorporated in the growth medium or used as a texture amendment. The viability of such techniques would be subject to the availability of sodic clays and the cost and practicality of its incorporation in the root zone mix.

Waddington (1969) noted that topdressing sports turf with soil and/or the removal of small cores had been shown to decrease the incidence of soil repellency. Bond (1978) investigated the addition of cores of non-repellent sand and sandy loam to overcome repellency in turf on sandy soils. He found that coring (core holes 15 mm diameter, 75 mm deep, and 50 mm apart) without backfilling increased water infiltration rates and reduced the visible occurrence of dry patch for <18 months. Addition of sand to the core holes was also beneficial for <18 months, while addition of sandy loam produced beneficial effects for >30 months. He pointed out that the use of fine textured soil on golf greens would not be readily accepted because of potential drainage problems, but noted that these problems should not occur if the soil is added in narrow vertical columns (cores) rather than in a continuous layer. Furthermore, addition of silt and/or clay would increase the water-holding capacity of the soil. Wilkinson and Miller (1978) and Danneberger and White (1988) have found that application of a wetting agent in conjunction with coring was more effective than coring alone.

C. Soil Wetting Agents

Wetting agents are surfactant chemicals used on repellent soils to increase water infiltration (Watson, 1969). Due to their cost, wetting agent use has been principally confined to turf culture. The addition of a wetting agent to a solution will decrease the surface tension of the liquid in the same manner as addition of ethanol in the NDST and MED repellency tests (Section III.E,F). Decreased liquid surface tension will decrease the infiltration rate into a nonrepellent soil. Pelishek et al. (1962) found that a wetting agent reduced infiltration into a clean quartz sand with an apparent contact angle, $H = 43^{\circ}$ (Letey et al., 1962a). However, in repellent soils surfactant application will also lower the liquid-solid contact angle. This beneficial effect will outweigh the detrimental effect of decreased surface tension, and the infiltration rate of the solution containing the surfactant will be increased (Mustafa, 1969). Miyamoto (1985) found that wetting agents improved infiltration into natural soils with *H* 56 to 81° which indicated that wetting agents may be effective for soils that appear nonrepellent (low *H* and WDPT).

In addition to the initial improvement in the infiltration rate, an effective wetting agent must also have a residual effect (Pelishek et al., 1962; McGhie, 1987) which will maintain increased water infiltration rates. Letey et al. (1962b) found that initial infiltration into an extremely repellent soil was more rapid for wetting agent solutions than for water. Subsequent water applications to plots treated with wetting agent also infiltrated more rapidly. Wetting agents vary in their effect upon infiltration. McGhie and Tipping (1984) found that only 4 of 42 products tested were very effective at increasing the initial infiltration rate and producing a residual effect. The effectiveness of wetting agents was found to be dependent on the application rate used and may also be affected by the dilution rate used for application (Wallis et al., 1990b).

By increasing water infiltration into repellent soil, effective wetting agents have been shown to increase plant establishment and growth (Osborn et al., 1964; Osborn et al., 1967; Osborn, 1969; Wallis et al., 1990b). Wetting agents may be used to reduce runoff and subsequent erosion. Osborn et al. (1964, 1976) found that a wetting agent treatment reduced debris movement by 95% and surface runoff by 32% from five small storms of moderate intensity on a hill soil in Southern California. The establishment of grass was improved fourfold by treatment with a wetting agent, and the stabilizing effect of the vegetation was found to reduce erosion in the longer term.

The chemistry of wetting agents determines their effects upon infiltration and other soil properties. Wetting agent molecules exhibit amphipatic character, that is, they contain a hydrophobic hydrocarbon group and a polar group which is hydrophilic. This gives rise to amphipatic adsorption, in which the hydrophobic groups are oriented away from the water and the polar groups toward it (Black, 1969). When the linear, hydrophobic portion of the molecule forms the anion in aqueous solution, it is said to be anion-active or anionic. If the molecule ionizes so that the hydrophobic portion forms the cation, it is cationic. The nonionic class is characterized by nonionized hydrophilic end-groups that are usually polar-active in nature (Law and Kunze, 1966). Nonionics have been most widely adopted for soil application because they are least phytotoxic and the degree of solubilization, which is important in controlling adsorption behavior, can be varied more precisely than with ionic compounds (Black, 1969). Nonionics may not be the ideal class of surfactant in mediums other than soil. Miyamoto (1978) found that an anionic wetting agent out-performed two nonionic wetting agents for application to repellent coal-mine spoils.

Figure 11. Discoloration of turf grass due to the phytotoxicity of a wetting agent applied during hot, dry conditions without subsequent irrigation (photo M.G. Wallis).

The properties of a wetting agent depend upon the size and polarity of the hydrophilic and hydrophobic groups in the molecule. The HLB value is an index of the balance between the hydrophilic and lipophilic (lipid-loving or hydrophobic) groups in a molecule. Black (1969) reported that wetting agents require an HLB of 7 to 9. Weil et al. (1979) tested a range of polyethoxylate chemicals and confirmed that the best wetting agents had HLB values of 7 to 9. In contrast McGhie and Tipping (1984) found that several chemicals with HLB 7 to 9 were totally ineffective. However, most of the effective wetting agents had an HLB of 8.3 to 8.5.

Wetting agents may have phytotoxic effects due to foliar contact or root uptake. Kaufman and Williamson (1981) found that soil applications of two nonionic wetting agents reduced the stomatal number and/or conductance of *Poa pratensis* turf grass. Petrovic et al. (1985) found that a nonionic wetting agent applied at normal application rates to *Poa annua* sports turf reduced seed formation by 73% on average. The wetting agent was not washed off the leaves and despite the apparent plant growth regulatory effect did not decrease the visual quality rating nor the clipping yield of the turf. Effron et al. (1990) cited a range of turf grass research which has shown that wetting agent phytotoxicity, such as turf discoloration and leaf-tip burn, have resulted from the use of high application rates, especially under hot humid conditions, when low dilution rates and/or no watering in have been used (Figure 11). Wetting agents may be applied at rates exceeding recommended label rates with little risk of phytotoxicity provided that the chemical is washed off the leaves within a few minutes of application (Karnok and Tucker, 1989).

Phytotoxicity due to root uptake may be difficult to avoid compared with problems arising from foliar contact. Parr and Norman (1964, 1965) showed that nonionic surfactants can inhibit root elongation when applied to soil at low concentration. Endo et al. (1969) found that two nonionic surfactants were more toxic in solution culture than in soil. The effects included shoot growth and rooting depth inhibition, and reduction of seed germination. Surfactant adsorption was assumed to account for the reduction in toxicity in soil, since the most strongly adsorbed wetting agent was also the least phytotoxic. They concluded that the critical factor determining the degree of phytotoxicity in a soil system is the extent to which the surfactant is adsorbed by the soil.

Law and Kunze (1966) and Law et al. (1966) found that nonionic wetting agents were adsorbed by kaolinite clay in amounts somewhat greater than anionics but less than cationics. Montmorillonite clays were found to adsorb the most wetting agent, which was attributed to the oxygen ions of the mineral surface hydrogen-bonding with the polar-active groups of the nonionic compounds. Nonionics were adsorbed with greater energy than that of water and thus displaced water from the surfaces. Valoras et al. (1969) found that adsorption of nonionic wetting agents increased to a maximum with increasing wetting agent concentration and then levelled off. Greater adsorption occurred on soils with a higher specific surface. The limited adsorption capacity of sandy soils may cause wetting agent phytotoxicity in sports turf or naturally repellent soils, however, such effects have not been investigated. Fortunately, a number of wetting agents used in turf are known to be strongly adsorbed, even to sandy loam soil (Miller et al., 1975).

Adsorption of a wetting agent has also been found to affect the degree of leaching and hence longevity in the soil. Valoras et al. (1969) found that one nonionic wetting agent appeared to be irreversibly adsorbed because none of the adsorbed chemical could be leached. Another nonionic was leached until a minimum adsorbed concentration was reached, after which very little was removed by further leaching. Miller and Letey (1975) studied the same chemicals and made similar observations. On the basis of a leaching study Krammes and DeBano (1967) suggested that one wetting agent may remain effective for at least 1 year in California. Osborn et al. (1969) evaluated the longevity of a range of nonionic and ionic wetting agents applied to repellent peat pots which were periodically leached over a period of 3 years. The effectiveness of all of the wetting agents was found to decline over the 3 years, however, all products retained some effect throughout the experiment.

Field testing is necessary to ascertain the longevity of wetting agents because the prevailing chemical, biological, and physical conditions may differ with those in laboratory experiments (Wallis et al., 1990b). Wetting agents are biodegradable, and therefore application must be repeated over time to ensure continued effectiveness. The effective life of wetting agents applied to turf or natural soils in field experiments has received little attention.

Mustafa and Letey (1969) found that the application of two wetting agents either before or with irrigation water decreased the aggregate stability of a repellent sandy loam. The wetting agents also decreased the aggregate stability of nonrepellent soil when applied with the irrigation water, but the residual effect increased the aggregate stability of a nonrepellent sandy loam and clay loam. Mustafa and Letey (1969) pointed out that the favorable effect of wetting agents upon the liquid-solid contact angle of repellent soil would normally overshadow the aggregate destabilizing effect. Furthermore, the limited structural development of sandy soils would reduce the importance of any wetting agent effect upon structure. In structured soils, these effects may be more significant.

Wetting agents have also been found to affect soil hydraulic conductivity (K). Watson et al. (1969) found that high concentrations (500 and 5000 ppm) of wetting agent decreased K of salt-affected soils significantly, which they suggested was due to clay swelling and dispersion. However, low wetting agent concentrations did not affect K. Miller et al. (1975) found that wetting agents did not affect K of a wettable soil, but decreased K of a repellent soil. Higher surfactant concentrations reduced K to greater degrees. The authors assigned this effect to the formation of wetting agent micelles which could block soil pores, aggregate destabilization, and particle migration.

The effect of wetting agents upon evapotranspiration is unclear. Morgan et al. (1966) found that a wetting agent did not significantly affect evapotranspiration from turf. In contrast, Law (1964) found that a wetting agent reduced evaporation from a bare soil surface by 50%. The contrasting findings may be due to the absence of a plant cover in Law's (1964) study. He suggested that the wetting agent probably reduced capillary flow of water to the surface layer of the soil and caused the formation of a dry diffusion barrier.

McGhie (1983) developed a method in Australia for wetting agent use in "broad-acre" agriculture. He applied wetting agent in bands behind a direct drill to reduce the volume of surfactant applied. Nonionic surfactant was applied at $20 \, l \, ha^{-1}$ (within-band application), which gave an actual usage of 3 to $4 \, l \, ha^{-1}$. Several wetting agents were found to improve plant establishment on repellent sands at a number of sites using this technique. McGhie considered that the band-spraying technique had excellent potential, but commented that farmers should experiment with their own soil to find the optimum rate required. Wallis et al. (1990b) found that two band-sprayed wetting agents increased pasture growth in a repellent New Zealand sand.

The advantages of using wetting agents for amelioration of repellency in sports turf or natural soils should outweigh any disadvantages if correct application techniques are used. However, questions remain regarding root-uptake phytotoxicity and the effects of wetting agents upon evaporation. Although wetting agents may be effective, they are expensive and provide only temporary control of soil repellency.

D. Irrigation

In certain agronomic and turf systems, irrigation may be used to maintain soil in a moist state and thereby avoid the expression of repellency which develops as the soil dries (Bond, 1964; DeBano, 1971; Wallis et al., 1990a). In sports turf, irrigation and syringing (light watering used primarily to reduce turf heat-stress) are employed to maintain adequate soil moisture levels during hot dry periods and reduce the severity of dry patch (Danneberger and White, 1988; Moore, 1981). Wallis et al. (1989b) and Way (1990) found that "dry patches" on New Zealand golf courses were related to areas receiving poor irrigation coverage (Figure 12). Wallis et al. (1989b) suggested that the uniformity of irrigation water application is an important determinant of "dry patch" development, and that no amount of applied wetting agent would overcome the problem if a poorly designed and/or managed irrigation system was in place. It may be for this reason that Tucker et al. (1990) found no relationship between dry patch severity and the use of wetting agents in a survey of golf courses in Georgia, USA. Wallis et al. (1989b) pointed out that irrigation needed to be scheduled to match plant requirements, the soil should not be allowed to dry out early in the growth season, and frequent, light irrigations which only wet the thatch and surface soil layer should be avoided.

Wallis et al. (1990b) investigated the interaction of irigation with the use of wetting agents upon a severely repellent sand. They found that delays of up to 14 days until initial irrigation following wetting agent application did not reduce the effectiveness of two wetting agents in a glasshouse. However, they pointed out that delays between wetting agent application and initial irrigation in the field could reduced wetting agent performance due to ultraviolet light degradation and soil displacement by wind. Shorter irrigation return intervals consistently improved plant growth in untreated and wetting-agent treated repellent sand. Wallis et al. (1990b) noted that, so long as each watering is adequate, short return intervals would have facilitated infiltration into the repellent sand by maintaining higher surface soil water contents. They pointed out that, in contrast to turf irrigation and other intensive horticultural enterprises, short irrigation return intervals are often impractical and uneconomic in extensive agriculture. Wallis et al. (1991) also pointed out that the effects of repellency are likely to be most pronounced under irrigation systems with high application rates.

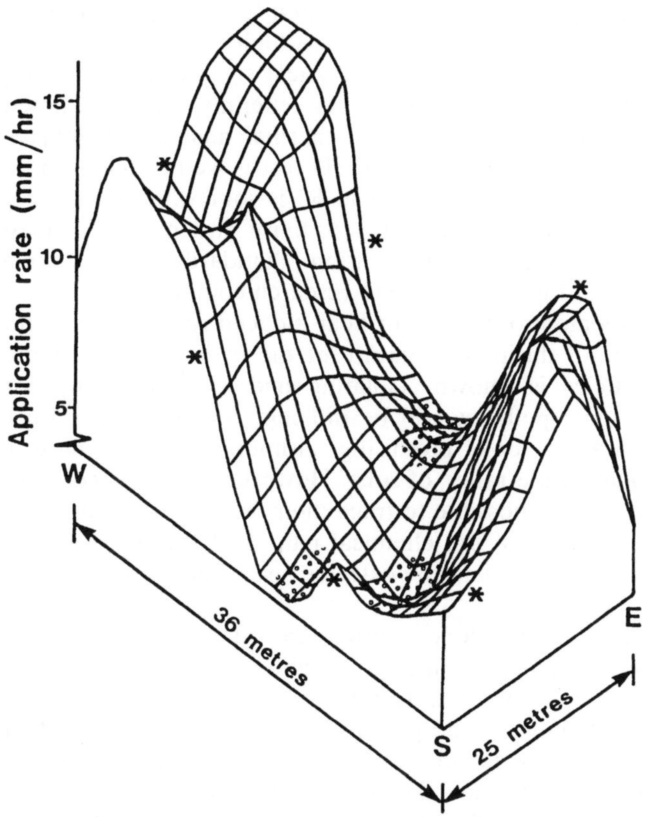

Figure 12. Contour map of the irrigation application rate on green 11, Manawatu golf course, Palmerston North, New Zealand. Shaded zones represent dry patch areas, which occurred in the areas of low application rate. The positions of sprinkler heads are indicated (*). After Wallis et al. (1989b).

VI. Conclusions and Future Research Priorities

Soil water repellency has been found to affect water movement into and within many soils. Recent research suggests that all soils may be repellent to some degree, challenging the common perception that repellent soils are only an interesting aberration. The dramatic effects of severe repellency upon infiltration, subsequent surface water redistribution, and plant growth are readily observed and have been well documented. However, less severe repellency may produce subtle effects which are nonetheless hydrologically significant. Further research is warranted to quantify such effects, in particular the generation of preferential flow pathways, given worldwide concern over groundwater contamination. The effect of repellency upon evaporation also requires clarification.

Many techniques have been developed for measuring repellency. The WDPT test is most simple but is unsuitable for severely repellent soil. Contact angle measures have no physical significance and are generally time-consuming. Infiltration measures generally must be related to comparative, "nonrepellent" soil, and thermal analysis techniques provide only semiquantitative data. The MED test is well suited for rapid assessment of severely repellent soil. The RI is somewhat more involved than the MED test, but has greater sensitivity and provides measurements of practical significance to infiltration. Standardization of measurement techniques would greatly facilitate research comparisons.

The development of natural or heat-induced repellency will not be fully understood until the precise nature of the hydrophobic organic coatings on soil particles is known. Isolating and identifying the components responsible for repellency will be difficult. Soxhlet extraction using *iso*-propanol/ammonia looks promising, although this procedure has yet to be tried upon a wide range of soils. However, once the compound(s) are identified it may be possible to trace their origin to plant and/or microbial species. Knowledge of the chemical nature of the hydrophobic coatings should also help to explain the extreme spatial and temporal variability of repellency.

The options for repellency control are often constrained by economic considerations. Wetting agents provide temporary relief, however, further research of their longevity and phytotoxicity is required. Soil texture amendment has the potential for longer-term benefits and is the subject of ongoing research. Irrigation may be used to avoid repellency expression by maintaining the soil water content above a critical level. Irrigation is also normally used in conjunction with wetting agent treatments. The irrigation water must be applied evenly, and due consideration should be given to the application rate and return interval. The complete control of repellency may not be possible until its development is fully understood.

References

Adams, S., B.R. Strain, and M.S. Adams. 1969. Water repellent soils and annual plant cover in a desert scrub community of Southeastern California. Proc. Symp. Water Rep. Soils, Univ. Calif., Riverside.

Adhikari, M., and G. Chakrabarti. 1976. Contribution of natural and microbial humic acids to water repellency in soil. J. Indian Soc. Soil Sci. 24:217–219.

Bahrani, B., R.S. Mansell, and L.C. Hammond. 1973. Using infiltrations of heptane and water into soil columns to determine soil-water contact angles. Soil Sci. Soc. Amer. Proc. 37:532–534.

Black, W. 1969. Basic chemistry of surface-active agents. Proc. Symp. Water Rep. Soils, Univ. Calif., Riverside.

Bolton, S. 1990. The management of dry patch on golf greens. Turf Craft Australia May:29, 54.

Bond, R.D. 1964. The influence of the microflora on the physical properties of soils. II Field studies on water repellent sands. Aust. J. Soil Res. 2:123–131.

Bond, R.D. 1969a. The occurrence of water-repellent soils in Australia. Proc. Symp. Water Rep. Soils, Univ. Calif., Riverside.

Bond, R.D. 1969b. Factors responsible for water repellence of soils. Proc. Symp. Water Rep. Soils, Univ. Calif., Riverside.

Bond, R.D. 1972. Germination and yield of barley when grown in a water-repellent sand. Agron. J. 64:402–403.

Bond, R.D. 1978. Addition of cores of loam. In W.W. Emerson, R.D. Bond, and A.R. Dexter (eds.). Modification of Soil Structure. Wiley.

Bond, R.D., and L.C. Hammond. 1970. Effect of surface roughness and pore shape on water repellence of sandy soils. Soil and Crop Sci. Soc. Flor. Proc. 30:308–315.

Bond, R.D., and J.R. Harris. 1964. The influence of the microflora on physical properties of soils. I. Effects associated with filamentous algae and fungi. Aust. J. Soil Res. 2:111–122.

Bornemisza, E. 1964. Wettability of soils in relation to their physico-chemical properties. Ph.D. dissertation. Univ. Florida. Diss. Abstr. 25:3771.

Bozer, K.R., G.H. Brandt, and J.B. Hemwall. 1969. Chemistry of materials that make soils hydrophobic. Proc. Symp. Water Rep. Soils, Univ. Calif., Riverside.

Brandt, G.H. 1969a. Water movement in hydrophobic soils. Proc. Symp. Water Rep. Soils, Univ. Calif., Riverside.

Brandt, G.H. 1969b. Soil Physical properties altered by adsorbed hydrophobic materials. Proc. Symp. Water Rep. Soils, Univ. Calif., Riverside.

Charters, R. 1980. Dry patch—a soil condition. Australasian Turf Care Feb: 31–32.

Chen, Y., and M. Schnitzen. 1978. The surface tension of aqueous solutions of soil humic substances. Soil Sci. 125:7–15.

Clothier, B.E., and I. White. 1981. Measurement of sorptivity and soil water diffusivity in the field. Soil Sci. Soc. Amer. J. 45:241–245.

Cory, J.T., and R.J. Morris. 1969. Factors restricting infiltration rates on decomposed granitic soils. Proc. Symp. Water Rep. Soils, Univ. Calif., Riverside.

Danneberger, K., and S. White. 1988. Treating localised dry spots. Golf Course Mgt. Feb:6–8.

DeBano, L.F. 1969a. Water movement in water-repellent soils. Proc. Symp. Water Rep. Soils, Univ. Calif., Riverside.

DeBano, L.F. 1969b. Observations on water-repellent soils in Western United States. Proc. Symp. Water Rep. Soils, Univ. Calif., Riverside.

DeBano, L.F. 1969c. The relationship between heat treatment and water repellency in soils. Proc. Symp. Water Rep. Soils, Univ. Calif., Riverside.

DeBano, L.F. 1969d. Water-repellent soils: a worldwide concern in management of soils and vegetation. Agr. Sci. Rev. 17(2):11–18.

DeBano, L.F. 1971. The effect of hydrophobic substances on water movement in soil during infiltration. Soil Sci. Soc. Amer. Proc. 35:340–343.

DeBano, L.F., and J. Letey. 1969. Proc. Symp. Water Rep. Soils, Univ. Calif., Riverside.

DeBano, L.F., and R.M. Rice. 1973. Water repellent soils: their implications in forestry. J. Forestry 71:220–223.

DeBano, L.F., L.D. Mann, and D.A. Hamilton. 1970. Translocation of hydrophobic substances into soil by burning organic litter. Soil Sci. Soc. Amer. Proc. 34:130–133.

DeBano, L.F., S.M. Savage, and D.A. Hamilton. 1976. The transfer of heat and hydrophobic substance during burning. Soil Sci. Soc. Amer. J. 40:779–782.

DeByle, N.V. 1973. Broadcast burning of logging residues and the water repellency of soils. Northwest Sci. 47:77–87.

DeJong, E. 1983. The effect of a subsurface hydrophobic layer on water and salt movement. Can. J. Soil Sci. 63:57–65.

Effron, D., A. Moore, and D. Powell. 1990. Perspectives on soil wetting agents. Aquatrols Corporation, Pennsavken, NJ, U.S.A.

Emerson, W.W., and R.D. Bond. 1963. The rate of water entry into dry sand and calculation of the advancing contact angle. Aust. J. Soil Res. 1(1):9–16.

Endo, R.M., J. Letey, N. Valoras, and J.F. Osborn. 1969. Effects of nonionic surfactants on monocots. Agron. J. 61:850–854.

Fairbourn, M.L., and H.R. Gardner. 1975. Water-repellent soil clods and pellets as mulch. Agron. J. 67:377–380.

Farrell, D.A. 1963. The hydraulics of surface irrigation. Ph.D. Thesis, Univ. Melbourne.

Fink, D.H. 1970. Water repellency and infiltration resistance of organic-film-coated soils. Soil Sci. Soc. Amer. Proc. 34:189–194.

Fink, D.H. 1976. Laboratory testing of water-repellent soil treatments for water harvesting. Soil Sci. Soc. Amer. J. 40:562–566.

Fink, D.H., and L.E. Myers. 1969. Synthetic hydrophobic soils for harvesting precipitation. Proc. Symp. Water Rep. Soils, Univ. Calif., Riverside.

Flaig, W. 1986. Some basic principles of water management in peat. In C.H. Fuchsman (ed.). Peat and Water. Elsevier, London and New York.

Gilmour, D.A. 1968. Water repellence of soils related to surface dryness. Aust. Forestry 32:143–148.

Giovannini, G., and S. Lucchesi. 1983. Effect of fire on hydrophobic and cementing substances of soil aggregates. Soil Sci. 136(4):231–236.

Giovannini, G., and S. Lucchesi. 1984. Differential thermal analysis and infra-red investigations on soil hydrophobic substances. Soil Sci. 137(6):457–463.

Giovannini, G., S. Lucchesi, and S. Cervelli. 1983. Water repellent substances and aggregate stability in hydrophobic soil. Soil Sci. 135(2):110–113.

Goh, K.M., T.A. Rafter, J.D. Stout, and T.W. Walker. 1976. The accumulation of soil organic matter and its carbon isotope content in a chronosequence of soils developed on aeolian sand in New Zealand. J. Soil Sci. 27:89–100.

Hammond, L.C., and T.L. Yuan. 1969. Methods of measuring water repellency of soils. Proc. Symp. Water Rep. Soils, Univ. Calif., Riverside.

Hendrickx, J.M.H., L.W. Dekker, M.H. Bannink, and H.C. van Ommen. 1988a. Significance of soil survey for agrohydrological studies. Agr. Water Mgt. 14: 195–208.

Hendrickx, J.M.H., L.W. Dekker, E.J. van Zuilen, and O.H. Boersma. 1988b. Water and solute movement through a water repellent sand soil with grass-cover. Int. Conf. and Workshop on Validation of Flow and Transport Models for the Unsaturated Zone. May 23–26, Ruidoso, NM, U.S.A.

Hillel, D. 1980. Fundamentals of Soil Physics. Academic Press, New York.

Hillel, D., and P. Berliner. 1974. Water proofing surface-zone soil aggregates for water conservation. Soil Sci. 118:131–135.

Holzhey, C.S. 1969a. Water-repellent soils in Southern California. Proc. Symp. Water Rep. Soils, Univ. Calif., Riverside.
Holzhey, C.S. 1969b. Soil morphological relationships and water repellence. Proc. Symp. Water Rep. Soils, Univ. Calif., Riverside.
Hussain, S.B., C.M. Skau, S.M. Bashir, and R.O. Meeuwig. 1969. Infiltrometer studies of water repellent soils on the east slope of the Sierra Nevada. Proc. Symp. Water Rep. Soils, Univ. Calif., Riverside.
Jackson, B.L.J., and A.G. Gillingham. 1985. Lime Soil Moisture Relationships. In B.L.J. Jackson and D.C. Edmeades. (eds.). Proc. Workshop on Lime in New Zealand Agriculture.
Jamison, V.C. 1942. The slow reversible drying of surface sandy soils beneath citrus trees in central Florida. Soil Sci. Soc. Amer. Proc. 7:36–41.
Jamison, V.C. 1945. The penetration of irrigation and rain water into sandy soils of Central Florida. Soil Sci. Soc. Amer. Proc. 10:25–29.
Jex, G.W., B.H. Bleakley, D.H. Hubbell, and L.L. Munro. 1985. High humidity-induced increase in water repellency in some sandy soils. Soil Sci. Soc. Amer. J. 49:1177–1182.
John, P.H. 1978. Heat-induced water repellency in some New Zealand pumice soils. N.Z. J. Sci. 21:401–407.
Karnok, K., and K. Tucker. 1989. The cause and control of localised dry spots on bentgrass greens. Golf Course Mgt. August:28, 32, 34.
Kaufman, J.E., and D.R. Williamson. 1981. Chemical manipulation of stomatal number and behavior in Merion Kentucky bluegrass (*Poa pratensis* L.). In R.W. Sheard (ed.). Proc. Int. Turfgrass Res. Conf. IV. Ontario Agr. College, Univ. Guelph, Ontario.
Kay, B.D., and J.B. Goit. 1977. Thermodynamic characterisation of water adsorbed on peat. Can. J. Soil Sci. 57:497–501.
Kijne, J.W. 1967. Influence of soil conditioners on infiltration and water movement in soils. Soil Sci. Soc. Amer. Proc. 31:8–13.
King, P.M. 1981. Comparison of methods for measuring severity of water repellence of sandy soils and assessment of some factors that affect its measurement. Aust. J. Soil Res. 19:275–285.
Krammes, J.S., and L.F. DeBano. 1967. Leachability of a wetting agent treatment for water-resistant soils. Soil Sci. Soc. Amer. Proc. 37:709–711.
Krammes, J.S., and H. Hellmers. 1963. Tests of chemical treatments to reduce erosion from burned watersheds. J. Geophys. Res. 68(12):3667–3672.
Krammes, J.S., and J. Osborn. 1969. Water-repellent soils and wetting agents as factors influencing erosion. Proc. Symp. Water Rep. Soils, Univ. Calif., Riverside.
Kwak, J.C.T., A.L. Ayub, and J.D. Sheppard. 1986. The role of colloid science in peat dewatering: principles and dewatering studies. In C.H. Fuchsman (ed.). Peat and Water. Elsevier, London and New York.
Law, J.P. Jr. 1964. The effect of fatty alcohol and a nonionic surfactant on soil moisture evaporation in a controlled environment. Soil Sci. Soc. Amer. Proc. 28:695–699.
Law, J.P. Jr., and G.W. Kunze. 1966. Reactions of surfactants with montmorillonite: Adsorption mechanisms. Soil Sci. Soc. Amer. Proc. 30: 321–327.
Law, J.P. Jr., M.E. Bloodworth, and J.R. Runkles. 1966. Reactions of surfactants with montmorillonitic soils. Soil Sci. Soc. Amer. Proc. 30:327–332.

Lebeau, J.B., and E.J. Hawn. 1963. A simple method for the control of fairy ring caused by *Marasmius oreades*. J. Sports Turf Res. Inst. 11:23–25.

Lemon, E.R. 1956. The potentialities for decreasing soil moisture evaporation loss. Soil Sci. Soc. Amer. Proc. 20:120–125.

Letey, J. 1969. Measurement of contact angle, water drop penetration time, and critical surface tensions. Proc. Symp. Water Rep. Soils, Univ. Calif., Riverside.

Letey, J., J. Osborn, and R.E. Pelishek. 1962a. Measurement of liquid-solid contact angles in soil and sand. Soil Science 93(3):149–153.

Letey, J., J. Osborn, and R.E. Pelishek. 1962b. The influence of the water-solid contact angle on water movement in soil. Bull. Int. Ass. Sci. Hydrology 3: 75–81.

Letey, J., N. Welch, R.E. Pelishek, and J. Osborn. 1962c. Effect of wetting agents on irrigation of water repellent soils. Calif. Agric. 16(112):12–13.

Letey, J., J.F. Osborn, and N. Valoras. 1975. Soil water repellency and the use of nonionic surfactants. Univ. Calif. Contribution No. 154. Univ. Calif., Riverside.

Lishtvan, I.I., and T.T. Zuyev. 1983. Effect of mineralisation on the hydrophilic properties of peat-bog soils. Soil Physics 38:92–96.

Malik, R.S., C.H. Laroussi, and L.W. DeBacker. 1981. Penetration coefficient in porous media. Soil Sci. 132(6):394–401.

Mallik, A.U., and A.A. Rahman. 1985. Soil water repellency in regularly burned Calluna heathlands: comparison of three measuring techniques. J. Env. Mgt. 20:207–218.

Marshall, T.J., and J.W. Holmes. 1979. Soil Physics. Cambridge University Press, Cambridge.

Ma'shum, M., and V.C. Farmer. 1985. Origin and assessment of water repellency of a sandy South Australian soil. Aust. J. Soil Res. 23:623–626.

Ma'shum, M., M.E. Tate, G.P. Jones, and J.M. Oades. 1988. Extraction and characterisation of water-repellent materials from Australian soils. J. Soil Science 39:99–110.

Ma'shum, M., J.M. Oades, and M.E. Tate. 1989. The use of dispersible clays to reduce water-repellency of sandy soils. Aust. J. Soil Res. 27:797–806.

McGhie, D.A. 1980. The contribution of the Mallet Hill surface to run-off and erosion in the Narrogin region of Western Australia. Aust. J. Soil Res. 18: 299–307.

McGhie, D.A. 1983. Water repellent soils on farms. West Australia Dept. Agr. Farmnote 117/83.

McGhie, D.A. 1987. Non-wetting soils in Western Australia. N.Z. Turf Mgt. J. 2:13–16.

McGhie, D.A., and A.M. Posner. 1980. Water repellence of a heavy textured Western Australian surface soil. Aust. J. Soil Res. 18:309–323.

McGhie, D.A., and A.M. Posner. 1981. The effect of plant top material on the water repellence of fired sands and water repellent soils. Aust. J. Agr. Res. 32:609–620.

McGhie, D.A., and P.I. Tipping. 1984. The development and testing of sufactants for water repellent soils. Aust. Soil Sci. Conf., Brisbane.

McIver, I. 1988. Controlling fairy rings. Proc. 15th Turfgrass Mgt. Seminar, Sanctuary Cove, Queensland, Sept. 6–7. Aust. Turfgrass Res. Inst., Concord West, N.S.W.

Meeuwig, R.O. 1971. Infiltration and water repellency in granitic soils. USDA Forest Service Res. Paper INT-111. USDA Ogden, Utah.
Miller, R.H., and J.F. Wilkinson. 1977. Nature of the organic coating on sand grains of non-wettable golf greens. Soil Sci. Soc. Amer. J. 41:1203–1204.
Miller, W.W., and J. Letey. 1975. Distribution of nonionic surfactant in soil columns following application and leaching. Soil Sci. Soc. Amer. Proc. 39: 17–22.
Miller, W.W., N. Valoras, and J. Letey. 1975. Movement of two nonionic surfactants in wettable and water-repellent soils. Soil Sci. Soc. Amer. Proc. 39:11–16.
Miyamoto, S. 1978. Effects of wetting agents on water infiltration into water-repellent coal mine spoils. Soil Sci. 125:184–187.
Miyamoto, S. 1985. Effects of wetting agents on water infiltration into poorly wettable sand, dry sod and wettable soils. Irr. Sci. 6:271–279.
Miyamoto, S., and J. Letey. 1971. Determination of solid-air surface tension of porous media. Soil Sci. Soc. Amer. Proc. 35:856–859.
Miyamoto, S., A. Bristol, and W.L. Gould. 1977. Wettability of coal-mine spoils in north-western New Mexico. Soil Sci. 123:258–263.
Moore, R. 1981. Wetting agents: a tool for influencing water behaviour in soil. Golf Course Mgt. July:26, 28.
Morgan, W.C., J. Letey, S.J. Richards, and N. Valoras. 1966. Physical soil amendments, soil compaction, irrigation and wetting agents in turfgrass management 1: Effects on compactability, water infiltration rates, evapotranspiration and number of irrigations. Agron. J. 58:525–527.
Mustafa, M.A. 1969. Nonionic surfactant effect upon soil aggregate stability. Proc. Symp. Water Rep. Soils, Univ. Calif., Riverside.
Mustafa, M.A., and J. Letey. 1969. The effect of two nonionic surfactants on aggregate a stability of soils. Soil Sci. 107(5):343–347.
Mustafa, M.A., and J. Letey. 1970. Factors influencing effectiveness of two surfactants on water-repellent soils. Calif. Agr. June:12–13.
Mustafa, M.A., J. Letey, and C.L. Watson. 1970. Evaluation of the intrinsic-penetrability and -diffusivity concepts to predict horizontal infiltration in porous media. Soil Sci. Soc. Amer. Proc. 34:369–372.
Nakaya, N. 1982. Water repellency of soils. Min. Agr., For. and Fish., Japan. JARQ 16(1):24–28.
Nakaya, N., S. Motomura, and H. Yokoi. 1977a. Some aspects on water repellency of soils. Soil Sci. Plant Nutr. 23(4):409–415.
Nakaya, N., H. Yokoi, and S. Motomura. 1977b. The method for measuring of water repellency of soil. Soil Sci. Plant Nutr. 23(4):417–426.
Osborn, J.F. 1969. The effect of wetting agents and water repellency on the germination and establishment of grass. Proc. Symp. Water Rep. Soils, Univ. Calif., Riverside.
Osborn, J.F., R.E. Pelishek, J.S. Krammes, and J. Letey. 1964. Soil wettability as a factor in erodibility. Soil Sci. Soc. Amer. Proc. 28:294–295.
Osborn, J.F., J. Letey. and L.F. De Bano. 1967. Seed germination and establishment as affected by non-wettable soils and wetting agents. Ecology 48:494–497.
Osborn, J.F., J. Letey, and N. Valoras. 1969. Surfactant longevity and wetting characteristics. Calif. Agr. March:6–8.

Parr, J.F., and A.G. Norman. 1964. Effects of nonionic surfactants on root growth and cation uptake. Plant Phys. 39:502–507.

Parr, J.F., and A.G. Norman. 1965. Considerations in the use of surfactants in plant systems: a review. Bot. Gazette 126(2):86–96.

Pelishek, R.E., J. Osborn, and J. Letey. 1962. The effect of wetting agents on infiltration. Soil Sci. Soc. Amer. Proc. 26:595–598.

Perroux, K.M., and I. White. 1988. Designs for Disc Permeameters. Soil Sci. Soc. Amer. J. 52:1205–1215.

Petrovic, A.M., R.A. White, and M. Kligerman. 1985. Annual bluegrass growth and quality as influenced by treatments of growth retardants and wetting agents. Agron. J. 77:670–674.

Philip, J.R. 1957. The theory of infiltration: 4. Sorptivity and algebraic infiltration equations. Soil Sci. 84:257–264.

Philip, J.R. 1969. Theory of Infiltration. IV. Adv. in Hydroscience 5:216–291.

Philip, J.R. 1971. Limitations on scaling by contact angle. Soil Sci. Soc. Amer. Proc. 35:507–509.

Raats, P.A.C. 1973. Unstable wetting fronts in uniform and nonuniform soils. Soil Sci. Soc. Amer. Proc. 37:681–685.

Rankin, P., and C. Ross. 1982. Dry patch—so what's new? Sports Turf Rev. 139:61–66. N.Z. Turf Culture Inst.

Rawitz, E., and A. Hazan. 1978. The effect of stabilised, hydrophobic aggregate layer properties on soil water regime and seedling emergence. Soil Sci. Soc. Amer. J. 42:787–793.

Reeder, C.J., and M.F. Jurgensen. 1979. Fire-induced water repellency in forest soils of upper Michigan. Can. J. Forest Res. 9:369–373.

Richardson, J.L. 1984. Field observation and measurement of water repellency for soil surveyors. Soil Survey Horizons 25(2):32–36.

Richardson, J.L., and F.D. Hole. 1978. Influence of vegetation on water repellency in selected Western Wisconsin soils. Soil Sci. Soc. Amer. J. 42: 465–467.

Richardson, J.L., and N.C. Wollenhaupt. 1983. Water repellency of degraded lignite in a reclaimed soil. Can. J. Soil Sci. 63:405–407.

Rieke, P.E. 1981. Wetting agents: applications vary for different soils. Golf Course Mgt. July:27, 29, 30.

Roberts, F.J. 1966. The effects of sand type and fine particle amendments on the emergence and growth of subterannean clover (*Trifolium subterraneum* L.) with particular reference to water relations. Aust. J. Agr. Res. 17:657–672.

Roberts, F.J., and B.A. Carbon. 1971. Water repellence in sandy soils of South-Western Australia. I. Some studies related to field occurrence. Queensl. Stn. Rec. Div. Pl. Ind. CSIRO (Austr.) 10:13–20.

Roberts, F.J., and B.A. Carbon. 1972. Water repellence in sandy soils of South-Western Australia: II. Some chemical characteristics of the hydrophobic skins. Aust. J. Soil Res. 10:35–42.

Savage, S.M. 1974. Mechanism of fire-induced water repellency in soil. Soil Sci. Soc. Amer. Proc. 38:652–657.

Savage, S.M., J.P. Martin, and J. Letey. 1969a. Contribution of humic acid and a polysaccharide to water repellency in sand and soil. Soil Sci. Soc. Amer. Proc. 33:149–151.

Savage, S.M., J.P. Martin, and J. Letey. 1969b. Contribution of some soil fungi to natural and heat-induced water repellency in sand. Soil Sci. Soc. Amer. Proc. 33:405–409.

Savage, S.M., J. Osborn, J. Letey, and C. Heaton. 1972. Substances contributing to fire induced water repellency in soils. Soil Sci. Soc. Amer. Proc. 36:674–678.

Sawada, Y., L.A.G. Aylmore, and J.M. Hainsworth. 1989. Development of a soil water dispersion index (SOWADIN) for testing the effectiveness of soil wetting agents. Aust. J. Soil Res. 27:17–26.

Schantz, H.L., and R.L. Piemeisel. 1917. Fungus fairy rings in eastern Colorado and their effect on vegetation. J. Agr. Res. 11:191.

Schnitzer, M., and J.G. Desjardins. 1969. Chemical characteristics of a natural soil leachate from a humic podzol. Can. J. Soil Sci. 49:151–158.

Scholl, D.G. 1971. Soil wettability in Utah Juniper stands. Soil Sci. Soc. Amer. Proc. 35:344–345.

Scholl, D.G. 1975. Soil wettability and fire in Arizona Chaparral. Soil Sci. Soc. Amer. Proc. 39:356–361.

Schreiner, O., and E.C. Shorey. 1910. Chemical nature of soil organic matter. U.S. Dept. Agr. Bureau Soils Bull. 74:1–48.

Shiels, G. 1982. Dry patch—a new problem on Britain's fine turf. Turf Mgt. Oct.:18–19.

Singer, M.J., and F.C. Ugolini. 1976. Hydrophobicity in the soils of Findley Lake, Washington. Forest Sci. 22:54–58.

Stevenson, F.J. 1982. Humus Chemistry, Genesis, Composition, Reactions. John Wiley and Sons, New York.

Stevenson, F.J., and K.M. Goh. 1971. Infrared spectra of humic acid and related substances. Geoch. Cosmochim. Acta 35:534–557.

Taylor, D.H., and G.R. Blake. 1982. The effect of turfgrass thatch on water infiltration rates. Soil Sci. Soc. Amer. J. 46:616–619.

Templeton, A.R., and D.A. Rodriguez. 1988. Water infiltration using soil surfactants. Amer. Lawn Applicator/Maintenance Dec:68–70.

Tillman, R.W., D.R. Scotter, M.G. Wallis, and B.E. Clothier. 1989. Water repellency and its measurement using intrinsic sorptivity. Aust. J. Soil Res. 27(4):637–644.

Tucker, K.A., K.J. Karnok, D.E. Radcliffe, G.Jr. Landry, R.W. Roncadori, and K.H. Tan. 1990. Localised dry spots as caused by hydrophobic sands on bentgrass greens. Agron. J. 82:549–555.

Valoras, N., J. Letey, and J.F. Osborn. 1969. Adsorption of nonionic surfactants by soil materials. Soil Sci. Soc. Amer. Proc. 33:345–348.

van't Woudt, B.D. 1959. Particle coatings affecting the wettability of soils. J. Geophys. Res. 64(2):263–267.

van't Woudt, B.D. 1969. Resistance to wetting under tropical and subtropical conditions. Proc. Symp. Water Rep. Soils, Univ. Calif., Riverside.

Waddington, D.V. 1969. Observations of water repellency on turfgrass areas. Proc. Symp. Water Rep. Soils, Univ. Calif., Riverside.

Wallis, M.G., D.J. Horne, and K.W. McAuliffe. 1989a. A survey of dry patch and its management in New Zealand golf greens. 1. Questionnaire results. N.Z. Turf Mgt. J. 3(3):16–18.

Wallis, M.G., D.J. Horne, and K.W. McAuliffe. 1989b. A survey of dry patch and its management in New Zealand golf greens. 2. Soil core results and irrigation interaction. N.Z. Turf Mgt. J. 3(4):15–17.

Wallis, M.G., D.J. Horne, and K.W. McAuliffe. 1990a. A study of water repellency and its amelioration in a yellow brown sand. 1. Severity of water repellency and the effects of wetting and abrasion. N.Z. J. Agr. Res. 33: 139–144.

Wallis, M.G., D.J. Horne, and K.W. McAuliffe. 1990b. A study of water repellency and its amelioration in a yellow brown sand. 2. The use of wetting agents and their interaction with some aspects of irrigation. N.Z. J. Agr. Res. 33:145–150.

Wallis, M.G., D.R. Scotter, and D.J. Horne. 1991. An evaluation of the intrinsic sorptivity water repellency index on a range of New Zealand soils. Aust. J. Soil Res. 29:353–362.

Wander, I.W. 1949. An interpretation of the cause of water-repellent sandy soil found in citrus groves in Central Florida. Science 110:299–300.

Watson, C.L. 1969. Hydraulic conductivity of soils as influenced by surfactants. Proc. Symp. Water Rep. Soils, Univ. Calif., Riverside.

Watson, C.L., and J. Letey. 1970. Indices for characterising soil-water repellency based upon contact angle—surface tension relationships. Soil Sci. Soc. Amer. Proc. 34:841–844.

Watson, C.L., B.L. McNeal, and J. Letey. 1969. The effect of surfactants on the hydraulic conductivity of salt-affected soils. Soil Sci. 108(1):58–63.

Way, B. 1990. Irrigation system performance. N.Z. Turf Management J. 4(4):22–23.

Weil, J.R., R.E. Koos, W.M. Linfield, and N. Parris. 1979. Nonionic wetting agents. J. Amer. Oil Chemists Soc. 56:873–877.

Wessel, A.T. 1988. On using the effective contact angle and the water drop penetration time for classification of water repellency in dune soils. Earth Processes and Landforms 13:555–561.

Whitely, G., and N.A. Baldwin. 1990. Variations in soil strength in the zones of fairy rings formed by *Marasmius oreades*. J. Sports Turf Res. Inst. 66:109–114.

Wilkinson, J.F., and R.H. Miller. 1978. Investigation and treatment of localised dry spots on sand golf greens. Agron. J. 70:299–304.

Young, T. 1805. On the cohesion of fluids. Phil. Trans. Roy. Soc. 84.

Yuan, T.L., and L.C. Hammond. 1968. Evaluation of available methods for soil wettability with particular reference to soil-water contact angle determinations. Soil Crop Sci. Soc. Fla. Proc. 28:56–63.

Zisman, W.A. 1964. Relation of equilibrium contact angle to liquid and solid constitution. Contact angle, wettability and adhesion. pp. 1–51. In Advances in Chemistry Series 43. Amer. Chem. Soc., Washington, D.C.

Stem-Nodulating Legume-*Rhizobium* Symbiosis and Its Agronomic Use in Lowland Rice

J.K. Ladha, R.P. Pareek, and M. Becker

Advances in Soil Science, Volume 20

I. Introduction

In recent years there has been a great surge of interest and activity in the field of biological nitrogen fixation (BNF). The legume-*Rhizobium* symbiosis is estimated to account for 40% of the world's fixed nitrogen. Because the ultimate goal of BNF research is to increase crop yields on farmers' fields, researchers' attention must focus mainly on the understanding, improvement, and agronomic use of legume-*Rhizobium* symbiosis. Rightly so, since the increased cost of energy and the growing demand for N fertilizers are expected to cause dramatic increases in the price of this commodity in the years ahead. Rice (*Oryza sativa* L.) is the world's most important food crop as more than 40% of the world's population depend on rice as a major source of calories. Most of the rice in tropical developing countries is produced in lowland areas and by subsistence farmers whose resource base is so small and whose physical access to supply markets is so limited that they cannot take full advantage of technologies that are overdependent on purchased inputs. The BNF in legumes, used appropriately, reduces production constraints in low-input agricultural systems in the tropics. However, most traditionally used green manure (GM) species cannot tolerate prolonged periods of flooding that occur in most lowland rice-farming systems. A new and promising alternative may be the use of nitrogen-fixing, stem-nodulating legumes.

BNF by legumes is the result of a symbiotic relationship between the host plant and a microsymbiont, the *Rhizobium* which is located in a nodule. Until recently, only the roots were commonly known as sites for nodulation. However, with the discovery of *Sesbania rostrata* (Dreyfus and Dommergues, 1981) the attention of researchers was drawn to stem-nodulating legumes and their rhizobia. Since then, several species that form N_2-fixing nodules not only on roots but also on subepidermal primordia of the adventive roots on the stems have been described. (Alazard, 1985; Ladha et al., 1990).

Twenty-one species of the genus *Aeschynomene* (Alazard, 1985; Becker et al., 1988a; and Becker, unpublished), three species of the genus *Sesbania* (Dreyfus and Dommergues, 1981; Dreyfus et al., 1988; Ladha et al., 1990), and the species *Neptunia oleracea* (Schaede, 1940) have been reported to show stem nodulation. All these species grow in temporarily waterlogged soils, in swamps, in marshes, on riverbanks, or in the littoral region of freshwater lakes and rivers (Allen and Allen, 1981). Some appear as weeds in rice fields (Moody et al., 1984). They often dominate as pioneer plants in tropical freshwater and sometimes from floating islands with grasses and herbs in association (Berhaut, 1979). The natural habitat of these legumes leads to the hypothesis that stem nodulation may be a clever evolutionary adaptation to waterlogged conditions. Besides the trait of stem-nodule formation, some of the species develop a large number of parenchymatous cells, building spongy tissues that may store sufficient oxygen for various metabolic functions in water-saturated ecosystems.

The initial characterization of the microsymbionts of stem-nodulating legumes has led to the description of two new genera of rhizobial strains (Dreyfus et al., 1988; Eaglesham et al., 1990). As compared with root nodules, the anatomy and ecophysiology of stem nodules show distinct differences, and may partly explain the high BNF potential reported (Alazard and Duhoux, 1987; Adebayo et al., 1989; Ladha et al., 1990; Ladha et al., 1992). These and other features make the stem-nodulating legume-*Rhizobium* symbiosis a predominant area of basic research and will be discussed in this paper. Furthermore, high growth potential and waterlogging tolerance make stem-nodulating legumes promising crops for potential use as green manures in lowland rice.

Several exploratory reviews on stem nodulation have appeared (Subbao Rao and Yatazawa, 1984) and most of them dealt primarily with the agronomic aspects (Dreyfus et al., 1985; Becker et al., 1988a; Rinaudo et al., 1988). In recent years, this area of research has advanced considerably and a plethora of information on various aspects of the stem-nodule symbiosis and its agronomic use has been generated. This paper intends to review the current status of knowledge on stem-nodule symbiosis, with special emphasis on its development, ecology, potential, and its agronomic use in lowland rice-cropping systems.

II. Diversity of Stem-Nodulating Legumes

Nodule formation is the essential feature of the legume-*Rhizobium* symbiosis, as nodules are the sites where the energy-requiring reduction of molecular nitrogen takes place. Most species in the family Leguminosae form nodules on roots. However, only a few species exhibit nodule formation on above-ground stems and are therefore referred to as stem-nodulating legumes.

The phenomenon of nodulation on stems was first observed by Jaensch at the end of the last century on *A. elaphroxylon*, but its significance was not recognized until recently (Allen and Allen, 1981). The term "stem-nodule" first appeared in histological studies on *A. aspera* by Hagerup (1928). He first associated the formation of bacteria-filled host cells on stems with nitrogen assimilation. Further reports on stem-nodule formation followed for *A. paniculata* (Suessenguth and Beyerle, 1936), *Neptunia oleracea* (Schaede, 1940), *A. indica* (Arora, 1954), *A. elaphroxylon* (Jenik and Kubikova, 1969), *A. evenia*, and *A. villosa* (Barrios and Gonzales, 1971). But it was only in the 1980s when Dreyfus and Dommergues (1981) reported the profuse stem nodulation on fast-growing, subsahelien *S. rostrata* that the potential importance of this unique feature was recognized. The recognition of the agronomic potential of stem-nodulating legumes as green manure for lowland rice areas (Dreyfus et al., 1985) further gave an impetus to the research activity in this field.

Up to now, 26 different stem-nodulating legume species have been reported belonging exclusively to the genera *Aeschynomene*, *Neptunia*, and *Sesbania*. Botanically speaking, these legumes belong to the subfamilies Papilionoideae (*Aeschynomene* and *Sesbania*) and Mimosoideae (*Neptunia*) and the tribes Hedysareae (*Aeschynomene*), Robiniinae or Galegeae (*Sesbania*) and Adenantereae (*Neptunia*) (Allen and Allen, 1981). All of them are pioneer plants in tropical waterlogged ecosystems. They include annual and perennial herbs, shrubs, and small- to medium-size trees. The species, their geographic distribution, and phenotypic characterization are listed in Table 1.

Stem-nodulating *S. rostrata* and *S. punctata* are tall and fast-growing and show potential as green manure crops. However, in spite of their distinct phenotypic differences these species seem to be simply different ecotypes of *S. rostrata* (as identified by the Royal Botanical Garden, Kew, England).

Among the 21 *Aeschynomene* species only a few are fast-growing and therefore seem promising for agriculture. *A. crassicaulis* and *Neptunia oleracea* are slow-growing runners with stems often floating on the water surface, a characteristic that limits their effective use as green manure. Fast-growing, extensively nodulating, erect herbs and shrubs like *S. rostrata*, *A. afraspera*, and *A. nilotica* have been reported as most promising green manures for agricultural uses in lowland areas (Rinaudo et al., 1983; Alazard and Becker, 1987; Becker et al., 1990).

The most distinctive characteristic of stem-nodulating legumes is the presence of predetermined nodulation sites on the stems. The formation of these sites is independent of infection with rhizobia (Dreyfus et al., 1984). They comprise primordia of adventive roots that are able to grow out under waterlogged conditions (Tsien et al., 1983). These primordia can be distributed over the whole length of the stem or appear only on lower stem portions. They are arranged in straight vertical rows (*Sesbania*) or in spiral-shaped rows winding around the stem (*Aeschynomene*), giving the appearance of random distribution (Figure 1).

However, even though the presence of predetermined nodulation sites is a distinct feature of stem-nodulating legumes, it may not be sufficiently discriminative for classification purposes. Adventive root primordia can also be found on other legumes. Some cultivars of *Arachis hypogaea* bear nodules around the first node of the stem which is a site for adventitious root formation (Nambiar et al., 1982). Few *Vicea faba* cultivars (Fyson and Sprent, 1980) and tree species *Cassia tora* L. and *Parkinsonia aculeata* L. (Yatazawa et al., 1987) also form nodules on the lower stem portion. Eaglesham and Ayanaba (1984) reported that upon inundation *Glycine max* produced new roots on submerged stem portions. These can

Table 1. Stem-nodulating legumes, their geographic distribution, and phenotypic characterization

	Distribution	Habitat	Phenotype	Reference
Aeschynomene				
aspera	Africa, S. Asia	WM	A-Sp-Sh	Hagerup (1928)
afraspera	Africa	M	A-E-Sh	Alazard (1985)
ciliata	Africa, S. America	M	A-E-Sh	Alazard (1985)
crassicaulis	Africa	W	A-Sp-H	Alazard (1985)
cristata	Africa, Madagascar	M	AP-E-S	Becker (unpublished)
denticulata	S. America	M	P-E-H	Eaglesham and Szalay (1983)
elaphroxylon	Africa	WM	P-E-T	Jenik and Kubikova (1969)
evenia	S. America	M	P-Se-Sh	Barrios and Gonzales (1971)
fluminensis	S. America	M	P-E-T	Becker (unpublished)
indica	Pantropical	MG	AP-ESe-Sh	Arora (1954)
nilotica	Africa	WM	AP-E-Sh	Alazard (1985)
paniculata	S. America	Sv	P-E-Sh	Suessenguth and Beyerle (1936)
pfundii	Africa	WM	A-Se-Sh	Alazard (1985)
pratensis	S. America	M	P-E-H	Eaglesham and Szalay (1983)
rudis	S. America	MSv	AP-E-ShT	Eaglesham and Szalay (1983)
scabra	S. America	M	A-E-Sh	Eaglesham and Szalay (1983)
schimperi	Africa	WM	AP-E-Sh	Alazard (1985)
sensitiva	Pantropical	M	AP-Se-Sh	Eaglesham and Szalay (1983)
tambacoundensis	W. Africa	M	A-Sp-H	Alazard (1985)
uniflora	Africa	WM	AP-E-Sh	Becker (unpublished)
villosa	S. America	MF	A-Sp-H	Barrios and Gonzales (1971)
Sesbania				
punctata	Madagascar	M	A-Se-Sh	Dreyfus et al. (1988)
rostrata	West Africa	WM	A-E-Sh	Dreyfus and Dommergues (1981)
speciosa	Asia, W. Africa	M	P-E-Sh	Ladha, (unpublished)
javanica	S.E. Asia	M	AP-SeSp-Sh	Becker (unpublished)
Neptunia				
oleracea	Pantropical	W	AP-Sp-H	Schaede (1940)

W, water; M, marches; F, forest; Sv, savanna; A, annual; P, perennial; E, erect; Se, suberect; Sp, spreading; T, tree; S, shrub; Sh, shrubby herb; H, herb

Figure 1. Nodulated stems of 50-day-old *Sesbania rostrata* (A) and *Aeschynomene afraspera* (B).

be infected with rhizobia to form nodules that are functional even after drainage of water. Profuse nodulation on submerged stem portions of *S. speciosa* and *S. sesban* with increasing water level in the field has been observed (Ladha, unpublished). Furthermore, effective green nodules could be induced on aerial stem portions of young *S. speciosa* plants (Ladha et al., 1990). Stems of 30-day-old plants were wrapped in cotton wool, drenched with specific *Rhizobium* cell suspension, and kept moist. The dormancy of primordia was broken and adventive root growth took place. The formation of nodules on root primordia was recorded 7 to 10 days after inoculation. The induced nodules increased in number and formed an effective N_2-fixing mass of green nodules on the stem. This particular situation may not be specific to *S. speciosa* and *S. sesban*, and stem-nodule formation can probably be induced in other aquatic legumes. However, it was neither observed nor was it possible to induce stem nodules in *S. cannabina* (synonym. *S. aculeata*).

What appears important in the case of induced stem nodules is the existence of well-developed primordia that break through the epidermis at the point of forming an adventitious root. At this point specific *Rhizobium* infection can take place. What happens in fact is the rhizobial infection of an outgrowing adventive root and not of a primordium on the stem. Therefore, induced nodules on *S. sesban* and *S. speciosa* stems are, anatomically speaking, root nodules. A similar pattern of rhizobial infection may take place on the outgrowing primordia of submerged stem portions as observed in *A. elaphroxylon*.

However, it is important whether to consider these artificially induced structures as stem nodules. Likewise, it is also not clear whether the system or pattern of nodulation shown by *S. rostrata*, *A. afraspera*, and others that form nodules clearly on aerial portions of the stem should be regarded on a par with that of species such as *A. elaphroxylon* and others that form nodules only on submerged portions of the stem.

III. Working Definition of Stem Nodule

The discussion above leads to the need for a working definition of the stem nodule. Based on the appearance of the nodulation sites and the expression of the nodulation, legumes showing nodule formation on stems may be grouped into three major categories (Table 2): (1) legumes with visible primordia and profuse nodulation all over the stem; (2) legumes with visible primordia but scarce nodulation on aerial portions; (3) legumes with invisible primordia and nodulation occurring only in submerged stem portions. This grouping is somewhat similar to the one proposed by Dreyfus et al. (1984) and Alazard and Duhoux (1987), based on anatomical differences of the primordia on the stems of various *Aeschynomene* species. Taking into account the anatomy of the primordia and the physiological development of the nodulation sites during rhizobial infection, legumes belonging to group 3 (perhaps with the exception of *Aeschynomone* species) may not be considered as stem-nodulating species. However, stem-nodulating legumes of groups 1 and 2 also produce many adventive roots on submerged stem portions that bear profuse and highly active nodules. These nodules are related to root nodules. Therefore, "root nodules" are defined as nodules formed on below-ground roots, and on above-ground adventive roots. What will be referred to as "stem nodules" are those formed on aerial portions of the plant, whose location prevents contact with soil and whose formation is initiated on a primordium and not on an outgrowing adventive root.

IV. Classification of Stem-Nodulating Rhizobia

Nitrogen-fixing bacteria that can induce nodule formation on legume roots were classified, until recently, in the genera *Rhizobium* and *Bradyrhizobium* which include various subgroups and strains. This differentiation is based on host-specificity, nutritional/biochemical characteristics, nucleic acid hybridization, and 16s rRNA sequencing. Preliminary studies showed that stem-nodulating legumes cannot be easily

Table 2. Groupings of stem-nodulating legumes based on nodulation intensity

Group 1	Group 2	Group 3
	Nodulation	
Profuse nodulation all over the stem	Scarse nodulation on the aerial stem, but good nodulation on the submerged stem	No nodulation on the aerial stem, but good to scarse nodulation on the submerged stem
	Root primordia	
Root primordia well-developed and protrude out of epidermis	Root primordia less developed and embedded under epidermis; full development of primordia to be ready for infection requires high humidity in atmosphere	Root primordia hidden under cortical tissues and look like rudimentary primordia; inundation is required for an infection
	Examples	
S. rostrata	*A. aspera*	*A. fluminensis*
A. afraspera	*A. cilita*	*A. crassicaulis*
A. nilotica	*A. denticulata*	*A. pfundii*
S. punctata	*A. evenia*	*A. elaphroxylon*
	A. indica	*S. sesban*
	A. pratensis	*S. javanica*
	A. rudis	*S. speciosa*
	A. schimperi	*Neptunia oleracea*
	A. scabra	*Arachis hypogeae*
	A. sensitiva	
	A. tambacoundensis	*and others*
	A. uniflora	
	A. villosa	

Collated from Arora (1954), Tsien et al. (1983), Yatazawa et al. (1987) Dreyfus et al. (1984), Alazard and Duhoux (1987), and Ladha et al. (1990)

fitted into the established classification system (Stowers and Eaglesham, 1983; Dreyfus et al., 1984; Chakrabarti et al., 1986).

The following sections will summarize several studies undertaken to classify stem-nodulating rhizobia by cross-inoculation, host-specificity, nutritional and physiological characteristics, and DNA hybridizations.

A. Cross Inoculation and Host Specificity

The first step in rhizobial classification is the determination of cross-inoculation behavior and the host specificity. Few systematic studies characterize stem-nodulating rhizobia based on nodulation and cross-inoculation behavior. Apparently, rhizobia isolated either from root or stem nodules of *Sesbania* and *Aeschynomene* possess the ability to produce dual nodulation (Yatazawa and Yoshida, 1979; Legocki et al., 1983; Alazard, 1985; Subba Rao, 1988; Dreyfus et al., 1988). However, a specific root-nodulating *Rhizobium* of *S. rostrata* that is unable to nodulate stems has also been reported (Dreyfus et al., 1984). According to Subba Rao et al. (1980) rhizobia isolated from *A. indica* stem and root nodules failed to induce effective nodules on several legumes and vice versa. Alazard (1985) classified various stem-nodulating rhizobia of *Aeschynomene*, based on their cross-inoculation behavior, into three groups. Group 1 consists of rhizobia of *A. elaphroxylon*, *A. crassicaulis*, and *A. pfundii*. They were placed near the cowpea group because they form effective nodules on *Macroptilium atropurpureum*, the test plant for cowpea rhizobia. Group 2 consists of rhizobia of *A. ciliata*, *A. denticulata*, *A. evenia*, *A. indica*, *A. pratensis*, *A. rudis*, *A. scabra*, *A. sensitiva*, and *A. tambacoundensis*. Each isolate can effectively nodulate several species within this group. Group 3 consists only of rhizobia of *A. afraspera* and *A. nilotica* which seem to be highly host-specific. No effective cross inoculation between these strains and *S. rostrata* was found.

Dreyfus and Dommergues (1981) reported ineffective nodulation in *S. pachycarpa* and *S. bispinosa* by strain ORS 551, isolated from stem nodules of *S. rostrata*. *S. rostrata*, however, was unable to form nodules with rhizobia isolated from *S. pachycarpa*, *A. afraspera*, and *Vigna unguiculata*. The only effective cross inoculation reported on stems of *S. rostrata* was achieved with rhizobia of *S. cannabina* and *S. grandiflora* (Dreyfus et al., 1988). Aside from *S. rostrata*, *S. paludosa* seems to be the only species to produce effective root nodules upon inoculation with strain ORS 571 (Alazard et al., 1988). From these studies it can be concluded that rhizobia isolated from stem nodules of *S. rostrata* and some isolates from *Aeschynomene* seem to be highly host-specific, and with the exception of isolates from *A. crassicaulis*, *A. elaphroxylon*, and *A. pfundii* they apparently cannot be ascribed to the classical cross-inoculation groups.

B. Physiological Characterization

The biochemical and physiological characteristics of rhizobia of stem-nodulating legumes have been studied recently. Carroll (1934) placed *Aeschynomene* rhizobia in the slow-growing group, while Dreyfus et al. (1984) reported them to be both fast- and slow-growing types. Stowers and Eaglesham (1983) and Chakrabarti et al. (1986) characterized rhizobia strains from stem nodules of *A. indica*, *A. aspera*, and *S. rostrata*, respectively. These strains seem to be an intermediate type, sharing characteristics of both fast and slow growers. Based on phenotypic

Table 3. Types of symbiont of *S. rostrata*

Group 1	Group 2	Group 3	
Genus	*Azorhizobium*	*Rhizobium*	*Rhizobium*
Growth	Fast	Fast	Fast
Ex planta growth on N_2	Yes	No	No
Closest relative	*Xanthobacter*	*R. meliloti* and *R. leguminosarum*	*R. meliloti* and *R. leguminosarum*
Homologous nodulation (on *S. rostrata*)	Effective nodulation on root and stem	Effective nodulation on root and pseudo-nodulation on stem	Effective nodulation on root and stem
Heterologous nodulation (on other species)	Effective nodulation only on *S. paludosa*	Effective nodulation on many *Sesbania* species	Effective nodulation on many *Sesbania* species

Collated from Alazard et al. (1988) and Dreyfus et al. (1988).

relationships, Dreyfus et al. (1988) found *A. afraspera* isolate ORS 301 being clustered with *Bradyrhizobium*. However, a detailed taxonomic characterization of ORS 301 and rhizobia of other stem-nodulating *Aeschynomene* species is still needed. In *S. rostrata*, three types of strains have been identified which either produce effective nodules on the root only or on both root and stem (Table 3). All three types are fast growers and have recently been characterized in detail (Dreyfus et al., 1988). The strain isolated from a stem nodule of *S. rostrata* possesses the ability of free-living N_2 fixation and can grow with N_2 as sole N source (Dreyfus et al., 1983) (Table 3). Normally, rhizobia are obligate diazotrophs, i.e., they fix N_2 only with the host. However, a limited number of rhizobia (belonging to the genus *Bradyrhizobium*) show ex planta nitrogenase activity (Ramaswamy and Bal, 1986). These strains, however, require combined N for growth and N_2 fixation. *S. rostrata* isolate ORS 571 and *Rhizobium trifolii* isolate 0403 (Urban et al., 1986) are exceptions to this finding. Strain ORS 571 differs from 0403 in its fixation ability at a relatively high O_2 concentration (3%) and temperature (37°C). In a survey of a large number of rhizobia strains of stem-nodulating species, free-living growth and N_2 fixation have been shown for isolates of *A. indica*, *A. sensitiva*, *A. pratensis*, and *A. afraspera* (Ladha et al., 1992). The significance of free-living growth and N_2 fixation by stem nodule rhizobia is not clear. Ludwig (1988) has proposed a possible role of N_2 fixation under free-living conditions in the biphasic nature of N_2 fixation. Accordingly, in the proliferative phase during early nodulation, N_2

fixation is coupled with bacterial ammonium assimilation, thus allowing the growth of the invading rhizobia. In the subsequent altruistic phase, N_2 fixation is decoupled from rhizobial ammonium assimilation, thus, allowing ammonium export from bacteroids to the host plant. Recently, Ladha et al. (1989a) demonstrated the role of free-living, N_2-fixing rhizobia in rice-associative N_2 fixation. When *S. rostrata* is grown as a prerice GM crop and incorporated before transplanting, the rice rhizosphere harbors a large population of specific rhizobia which supports higher nitrogenase activity than that of a fallow-rice sequence. Sterile rice seedlings inoculated with specific rhizobia of *S. rostrata* showed nitrogenase activity comparable to seedlings inoculated with *Azospirillum lipoferum* 34H, a rice root isolate (Ladha et al., 1989a). However, rhizobia with no ability for free-living N_2-dependent growth and N_2 fixation did not support nitrogenase activity. The use of stem-nodulating legumes in rice may therefore not only improve the yield through the supply of large amounts of symbiotically fixed nitrogen, but may further contribute to N nutrition of the rice crop by enhancing associative N_2 fixation in the rhizosphere.

C. Azorhizobium

Dreyfus et al. (1988) performed whole-cell protein analysis, DNA-DNA, and DNA-rRNA hybridization of both root-nodulating and root- and stem-nodulating strains of *S. rostrata*, as well as of several reference strains of *Rhizobium* and *Bradyrhizobium*. Their study concluded that (a) the root-nodulating strains belong to the genus *Rhizobium* and (b) stem nodule isolates seem to belong to the rRNA branch of *Rhodopseudomonas palustris* but are quite distinct from *R. palustris*. A close relative of these strains is *Xanthobacter*. These features together with the ability to grow using N_2 under free-living conditions warranted the proposal of a new genus and species *Azorhizobium* (*Az.*) *caulinodans* for the stem-nodulating isolate ORS 571 of *S. rostrata*.

D. Photorhizobium

Subba Rao (1988) observed two types of endosymbionts or morphological variants in the bacteroids of *A. indica* stem nodules. One endophyte had a gram-negative cell wall with striking similarities to purple photosynthetic bacterium (PPSB) belonging to Rhodospirilliacea. Subsequently, Eaglesham et al. (1990) confirmed that the strain BTAil isolated from stem nodules of *A. indica* by Eaglesham and Szalay (1983) was able to produce bacteriochlorophyll "a" and a photosynthetic reaction center resembling that of photosynthetic bacteria. BTAil performs bacterial photosynthesis when grown aerobically ex planta.

In contrast with *Rhodopseudomonas rubrum*, BTAil does not grow on soybean trypton medium. It grows photosynthetically only under aerobic conditions and an intermittent light-dark cycle. The amount of bacteriochlorophyll "a" formed by BTAil (0.05 to 0.48 nmol/mg dry weight) was much less than that synthesized by anaerobically grown Rhodospirillaceae (11 to 13 μmol/mg dry weight). The bacteroids of *A. indica* could double the rate of acetylene reduction by using light at 730 μm, a wavelength which cannot be used by plants for photosynthesis (Eaglesham et al., 1990). Photosynthetically active strains isolated from stem nodules of *Aeschynomene* have been proposed to be assigned to the new genus *Photorhizobium* (Eaglesham et al., 1990).

Recently, rhizobia of *A. afraspera*, *A. denticulata*, *A. evenia*, *A. indica*, *A. nilotica*, *A. pratensis*, and *A. sensitiva* were also found to produce bacteriochlorophyll "a" (Ladha et al., 1990). The ability of bacteroids to use light in a wavelength region that cannot be absorbed by the plant may be of considerable importance to the efficiency of the symbiosis. To a certain degree, N_2 fixation might be independent from host plant energy supply. By generating ATP via photophosphorylation within the *Photorhizobium*, part of the chloroplast-produced photosynthates may be spared for production of plant biomass or seed yield.

V. Infection Process, Nodule Formation, and Fine Structure of Stem Nodules

Stem-nodulating rhizobia are true soil bacteria; without the appropriate macrosymbiont they thrive saprophytically on soil organic matter. Due to their unipolar flagellum they are highly mobile in the aquatic phase (Dreyfus, 1982). Besides the soil, the phyllosphere of the host plant seems to be an alternate ecological niche for nonsymbiotic growth of stem-nodule rhizobia. Adebayo et al. (1989) reported the presence of large *Az. caulinodans* populations on leaves of *S. rostrata*. Similarly, homologous rhizobia were also found on the leaves of *A. afraspera*, *A. pratensis*, and *A. sensitiva*. Inoculum for stem infection can therefore come from two sources: the soil and the phyllosphere. In both cases, rain seems to play a major role for stem inoculation. Rainwater can wash the epiphytically growing rhizobia down the stems and thereby infect nodulation sites. Furthermore, soil splash due to rain may bring rhizobia to the lower stem. This view is supported by the high correlation found between stem nodule numbers on *S. rostrata* and *A. afraspera* and the rainfall pattern (Becker et al., 1990). Dreyfus and Dommergues (1981) considered ants as a possible vector to inoculate primordia on the stem of *S. rostrata* under nonflooded conditions. Wind may be another important factor; often a high nodulation rate can be observed on stems of stem-nodulating legumes along dusty roads. Wind may transport contaminated soil particles on above-ground plant parts thus favoring stem inoculation.

After the rhizobia have reached the nodulation site, they penetrate into the cortical tissues of the stem. This depends on the anatomy of the root primordium. Three different situations can be found:

1. In *S. rostrata*, *A. afraspera*, and *A. nilotica* primordia are well developed and protrude distinctly out of the epidermis. Primordia are rod- or dome-shaped and are encircled by a 50 to 80 μm wide circular cavity, which seems to be the entry point for rhizobia (Tsien et al., 1983; Alazard and Duhoux, 1987).
2. In the second situation, a less-developed root primordium underlies the epidermis of the stem. The epidermal cells generally cover the root primordium without formation of an epidermal dome and with no circular cavity at the primordial base. The species belonging to this type are *A. indica*, *A. schimperi*, and *A. scabra*.
3. The least-developed type of primordium has been observed in *A. crassicaulis*, *A. elaphroxylon*, and *A. pfundii*. The root primordia, although clearly visible, remain embedded in the cortical tissues of the stem. External factors like waterlogging or mechanical scarification are needed to trigger primordial development and to allow the access of rhizobia (Alazard and Duhoux, 1987).

The anatomy of the adventive root primordia on the stem may explain why *S. rostrata*, *A. afraspera*, and *A. nilotica* show generally profuse nodulation, whereas other species nodulate scarcely or only on submerged stem portions.

Upon penetration of rhizobia in the intercellular space, infection pockets are formed into the basal layers of the root primordium, which then resume their meristematic activity (Alazard, 1985). As a result, cell division starts in the infective center and the nodule begins to form. From the infected basal cells of the primordium, intracellular infection threads are formed, as in the case of *Sesbania*. They are about two bacterial rod lengths wide and surrounded by electron-dense material (Duhoux and Dreyfus, 1982). In *Aeschynomene* spp. and *N. oleracea*, intercellular infection takes place without the formation of infection threads (Dreyfus et al., 1984; Schaede, 1940). It is most likely that the mode of infection is controlled by the host plant's genome as reported for root-nodulating legumes (Dart, 1977). At the infective stage, rhizobia change in shape and size. They become bacteroids and are surrounded by a membrane envelope. As the growing infective center forms the nodule, the majority of the cells in stem nodules are infected.

The formation of stem nodules becomes macroscopically visible within 5 to 7 days after inoculation and the nodules reach their full size in 15 to 20 days. In *S. rostrata*, stem nodules are spherical protrusions, 5 to 12 mm in diameter. They can easily be detached from the stem, as the basal portion of the nodule forms a narrow neck. In *Aeschynomene* flattened hemispheric nodules are formed. Because no neck formation is found, the nodules are hard to remove. Stem nodules of *N. oleracea* are 3 to 5 mm ovoid structures, largely protruding from the stem. Their length is 1.5 to 3 times the nodule width. In all cases stem nodules have a continuous epidermis with the stem.

The fine structure of stem nodules is similar to that found in root nodules (Yatazawa et al., 1987; De Bruijn, 1988). Cross sections show the presence of a red pigment (leghemoglobin), about 9 to 11 days after inoculation. Scanning electron micrographs show that most host cells are infected and densely packed with bacteroids, with five to ten of them being enveloped in a membrane (Legocki et al., 1983; De Bruijn, 1988).

There are two main differences between root and stem nodules: (a) the aerial location of stem nodules and (b) the presence of chloroplast-containing cortical cell layer around the stem nodules. Their specific location prevents the contact of stem nodules with soil solution or floodwater. A photosynthetically active tissue around the nodule allows the cyclic photophosphorylation as well as synthesis of carbohydrates and may therefore improve the energy supply for N_2 fixation. Photosynthetically active tissues generate oxygen. Because electron transport in nitrogenase enzyme complex is sensitive to oxygen, additional mechanisms to protect bacteroids from oxygen may be needed in stem nodules. Trinchant and Rigaud (1987) showed that a high tolerance of nitrogenase in bacteroids of *S. rostrata* towards increased oxygen concentrations was due to the presence of a highly efficient superoxide dismutase. Furthermore, *Az. caulinodans* ORS 571 had a high O_2 consumption in continuous culture. However, this O_2-tolerant, N_2-fixing system is not present if ORS 571 is grown at steady-state conditions (Bergersen et al., 1988).

Legocki et al. (1983) found a relatively high amount of leghemoglobin in the stem nodules of *A. scabra*. The molecular species LB2 was substantially more abundant in leghemoglobin of stem nodules than in that of root nodules. However, it is not known if this characteristic helps the stem nodule to deal with the problem of increased O_2 concentration.

VI. Biological Nitrogen Fixation and Nitrogen Accumulation by Stem-Nodulating Legumes

Data on amounts of N accumulation by different stem-nodulating legumes raised in diverse environments and soils after 40 to 55 days have been recently compiled by Becker et al. (1988a) and Buresh and De Datta (1990). The N-accumulation potential under flooded conditions ranges from 41 kg N/ha in *A. indica* (Crozat and Sangchyo-sawat, 1985) to more than 200 kg N/ha in *S. rostrata*, *A. afraspera*, and *A. nilotica* (Alazard and Becker, 1987; Becker et al., 1990; Rinaudo et al., 1983) and

Table 4. BNF contribution by stem-nodulating legumes under flooded conditions

Species	Growing season	Growth duration (days)	% Ndfa	Kg N fixed/ha	Method	Reference
S. rostrata	Long day	56	38	—	^{15}N dilution	Rinaudo and Moudiongui (1987)
S. rostrata	Long day	60	36–51	83–109	^{15}N dilution	N'doye and Dreyfus (1988)
S. rostrata	Long day	56	88	175	ARA/$^{15}N_2$	Becker et al. (1990)
	Short day	56	83	70		
A. afraspera	Long day	56	77	145		
	Short day	56	68	105		
S. rostrata	Long day	25	76	10	^{15}N dilution	Pareek et al. (1990)
		45	88	140		
		65	94	458		
	Short day	25	53	7		
		45	71	100		
		65	86	324		

is thereby considerably higher than in nonstem-nodulating *S. cannabina* and *S. sesban* (Buresh and De Datta, 1990).

Recently, ^{15}N dilution studies have been conducted to determine the fractions of plant N derived from air and soil in stem-nodulating species. The percent contribution of N derived from dinitrogen (Ndfa) by well-nodulated 45 to 55 days old *S. rostrata* plants was about 70% and increased further to more than 90% at 65 days (Table 4). There was good agreement between estimates obtained by ^{15}N dilution using enriched ^{15}N fertilizer or ^{15}N natural abundance (Yoneyama, 1990). Pareek et al. (1990) using ^{15}N stabilized soil compared percentage Ndfa of *S. rostrata* and *S. cannabina*. Although both species accumulated almost similar percentages of Ndfa, *S. rostrata* fixed higher total N.

Stem nodulation has been reported to be a more efficient N_2-fixing system than the root-nodulating system. Eaglesham and Szalay (1983) reported activities of over 160 µmol C_2H_2/h/plant of above-ground portions of *A. scabra*. Dreyfus et al. (1985) reported nitrogenase activity of about 600 µmol C_2H_2/(h·plant) in a 3-month-old *S. rostrata* in the field. Most of this nitrogenase activity was attributed to stem nodules and is the highest ever reported for any legume-*Rhizobium* symbiosis. The following discussion attempted to answer the questions whether (a) stem nodulation plays a more important role than root nodulation in plant N_2 fixation and (b) dual nodulation results in higher N_2 fixation than single nodulation.

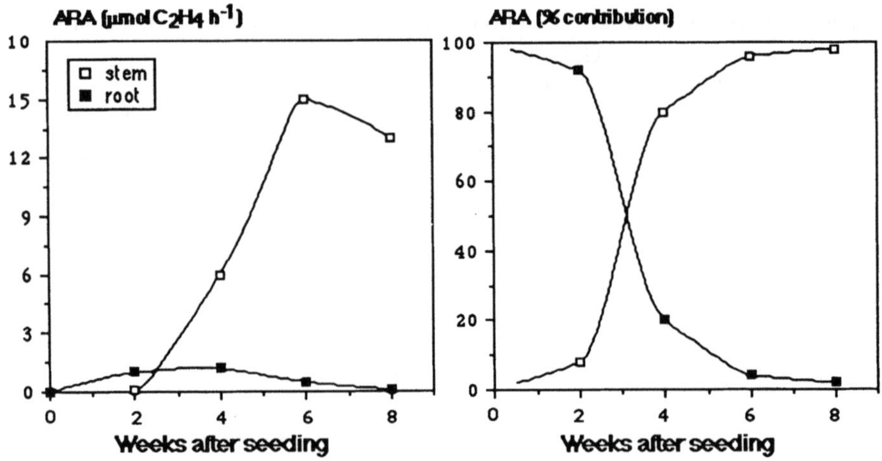

Figure 2. Trend and relative contribution of acetylene-reducing activity (ARA) by root and stem nodules of *Sesbania rostrata* (adapted from Becker et al., 1990).

A. Root vs Stem Nodulation

Rinaudo et al. (1988) in a large pot, outdoor experiment (four plants/pot) in Senegal, and Becker et al. (1990) in field experiments in the Philippines, reported negligible contributions of *S. rostrata* root nodules toward N_2 fixation as measured by acetylene-reducing activity (ARA). The stem nodules appeared at about 4 weeks of age and accounted for more than 80% of total ARA which further increased to almost 100% between 4 to 7 weeks (Figure 2). Greenhouse experiments conducted in the Philippines, however, showed that root nodules started functioning at a very early stage (1 week after germination) and were active until 40 to 50 days. The contribution to BNF by root nodules of 40- to 50-day-old *S. rostrata* was substantial even when plants were inoculated on the stem (Ladha et al., 1989b). A possible reason for this discrepancy might be differences in environmental conditions (field vs greenhouse). Compared with greenhouse conditions, field conditions are more congenial for the development of stem nodules.

Ladha et al. (1992) conducted several phytotron, greenhouse, and field experiments to assess the contribution of N_2 fixation (ARA) by stem nodules in the presence and absence of root nodules of *S. rostrata*. Root or stem nodulation was achieved by inoculating either the root or the stem. Furthermore, the effect of detaching already-formed stem nodules on restoration of root nodules and nitrogenase activity was examined. Stem inoculation to root-nodulating plants resulted in a sharp decline to less than 20% in root nodule dry weight and ARA. Root nodule dry weight, and ARA however, increased several-fold within 15 days of detachment of aerial stem nodules (Figure 3). These results suggest that

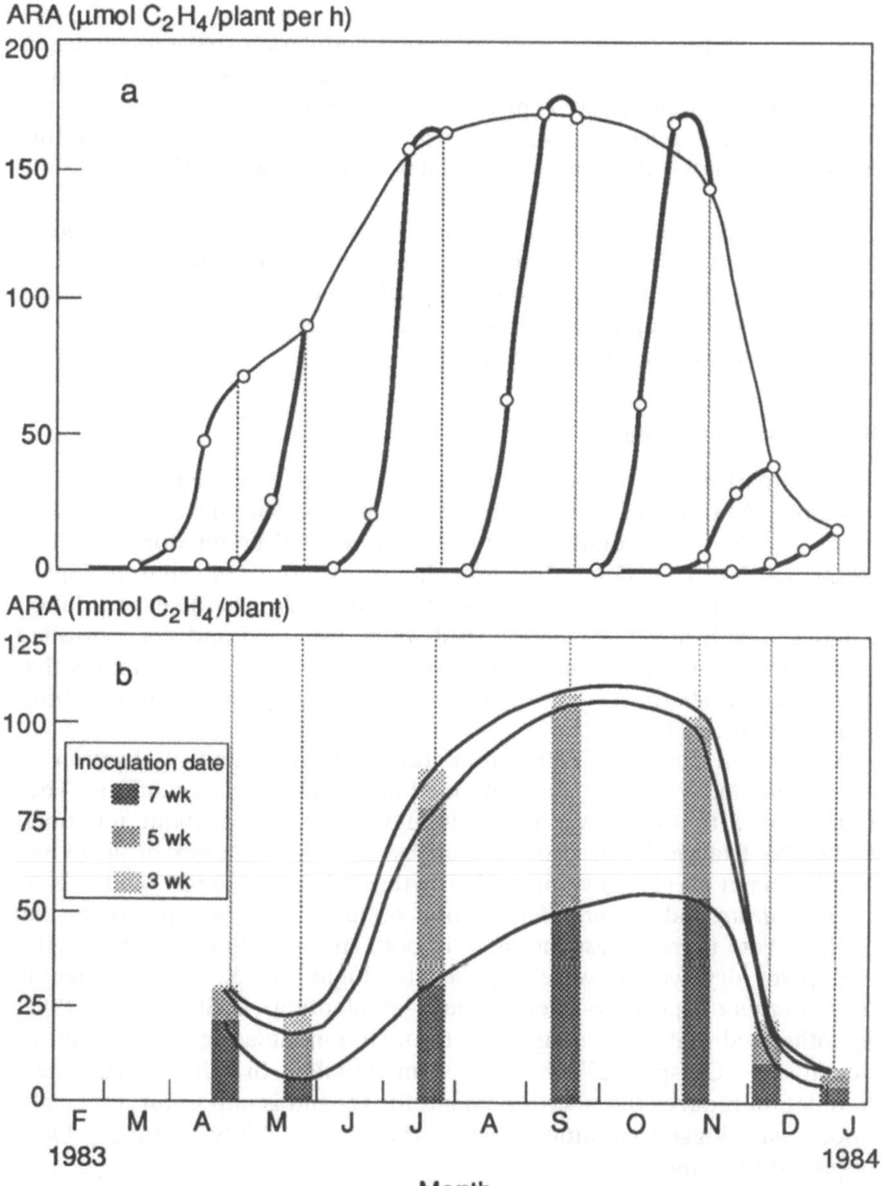

Figure 3. Acetylene-reducing activity (ARA) of field-grown *Sesbania rostrata* in Senegal. (a) ARA as a function of plant age, (b) contribution of stem portions inoculated 3, 5, and 7 weeks after seeding to total nitrogen-fixing activity of 9-week-old plants (adapted from Rinaudo et al., 1988).

in dual-nodulating legumes, root and stem nodulation strike a balance in competition for energy, which is controlled by stem nodulation. The controlling mechanism is probably the stem nodule's proximity to the photosynthetic apparatus, which permits stem nodules to control the supply of photosynthate to root nodules and determines their longevity and activity.

It can be concluded that in the early growth stage the contribution of root nodules to plant N is probably more important than that of stem nodules. However, the stem nodules which appear between 20 to 30 days after germination suppress the root nodulation. At this stage the major N contribution is from the stem nodules.

B. Dual vs Single Nodulation

Few comparisons of dual (root and stem) vs single (root) nodulations are available. As discussed earlier, stem-nodulating species have more nodule biomass than root-nodulating legumes due to dual nodulation (Dreyfus, 1982). Ladha et al. (1989b) in greenhouse experiments reported higher nodule biomass and N_2 fixation in *S. rostrata* due to dual nodulation compared with single nodulation. But when examined under a controlled environment, nodule biomass and ARA did not differ in treatments that inoculated either root, stem, or root plus single and multiple inoculations (Ladha et al., 1992).

Ndoye and Dreyfus (1988) compared nodulation and N_2 fixation by *S. rostrata* and *S. sesban* under flooded and nonflooded conditions. They found three to four times more nodule biomass and about ten times higher N_2 fixation by *S. rostrata* than by *S. sesban*. Pareek et al. (1990) and Ladha et al. (1992) compared growth, nodulation, ARA, and %Ndfa of *S. rostrata* and *S. cannabina*. Total nodule dry weight per plant and %Ndfa were more or less the same in both species. However, total ARA and plant dry weight were significantly higher in *S. rostrata* than in *S. cannabina* especially after 55 days of planting. Ladha et al. (1992) hypothesized that better plant growth of stem-nodulating plants could be due to an "O_2 sparing" effect of stem nodules. In other words, stem nodulation reduces the plant's demand for O_2 and is important, especially under waterlogged conditions, when the plant's ability to supply O_2 is impaired by aging.

VII. Factors Affecting Symbiosis

A. Survival of Rhizobia

Like root-nodule bacteria, stem-nodulating bacteria are true soil bacteria. A carbon source is critical for their survival in the soil. In vitro, *Az.*

caulinodans cannot use some of the sugars, aliphatic acids, alcohols, and amino acids which are used by rhizobia or bradyrhizobia for growth (Dreyfus et al., 1988). Information on the carbon sources for other stem-nodulating rhizobia (e.g., *Photorhizobium*) in soil is not available. Studies conducted by Ladha et al. (1989a) showed that the soil receiving inoculated *S. rostrata* as GM contained 3000 times more azorhizobia than a flooded fallow control treatment. Using the antibiotic tagged strain ORS 571, it was found that the inoculated strain declined during two *S. rostrata*/rice cycles. However, the population of native azorhizobia increased after the *S. rostrata* incorporation which could be due to the release of stem-nodule rhizobia together with the addition of organic matter.

1. Rice Rhizospheric Survival

The rhizosphere is the environment immediately under the influence of plant roots. It shows a high microbial activity that decreases with distance from roots. Survival of bacteria in the rhizosphere depends on their ability to compete with other microorganisms and the nature of plant exudates. Because of the submergence habit of rice, oxygen is also an important factor. Rhizobia, being fastidious in food habit, are generally poor competitors compared with other rhizospheric microorganisms; however, the rhizosphere of the homologous host is reported to have a positive effect on their survival (Van Egeraat, 1975; Bohlool et al., 1984). The effect of nonhost plants on the survival of rhizobia has not been adequately studied. Ladha et al. (1989a) reported that the population of stem-nodulating *Az. caulinodans* in the rhizosphere and histosphere of rice was maintained at the same level from tillering to the heading stage, whereas in nonrhizospheric soil a decline in population was recorded. The favorable effect of the rice rhizosphere and histosphere was evident even at the late-growth stage of thc ricc crop. This might be the effect of increased exudation (Bohlool et al., 1984) and oxygen supply through roots (Yoshida, 1981). High nitrogenase activity in rice rhizosphere inoculated with *Az. caulinodans* (Ladha et al., 1989a) signifies the importance of this niche for growth as well as for nonsymbiotic N_2 fixation of *Az. caulinodans*.

2. Phyllospheric or Epiphytic Survival

The phyllosphere is a particular kind of niche characterized by the presence of leaf diffusates or leachates which are reported to contain nutritional substances like amino acids and carbohydrates (Ruinen, 1961). The phyllosphere has been reported to harbor various kinds of micro-organisms including nitrogen-fixing bacteria. Recently Adebayo et al. (1989) reported the presence of a large population of *Az. caulinodans* on the leaves of *S. rostrata*. Rhizobia, although in lesser number, were also found on the flowers. *S. rostrata* and *A. afraspera* nodulating profusely on stems, harbored more of their specific rhizobia on above-ground plant parts than plants nodulating scarsely on stems. The high cell density of stem-nodulating bacteria on leaves and flowers of the host plants indicates that the phyllosphere may serve as an alternate adaptive niche for these organisms rather than their presence being only incidental. The phyllosphere may therefore play an important role in ensuring stem nodulation.

Ladha et al. (1988) found the spermatospheric presence of *Azorhizobium* in seeds of *S. rostrata*. They could not be eliminated by treatments normally used for seed sterilization or scarification (mercuric chloride, concentrated sulfuric acid). Azorhizobia in seeds may come from senescing stem nodules during flowering and seed formation. As nodule formation occurs on flowering branches, flowers may be contaminated with rhizobia; these may enter the seed through the hilum scar, pilar groove, and micropyle. It can be hypothesized that the mode of initial entry of azorhizobia into the soil is through the seeds (Ladha et al., 1990). This cyclic nature of symbiosis is a feature of *S. rostrata*; the bacterial symbiont is maintained epiphytically on the host plant, is carried into the flower, and is transmitted from one generation to the next enclosed in the seed. Because of the high survival of stem-nodulating rhizobia in soil, rhizosphere, and phyllosphere and the fact that rhizobia are carried through the seed, the question arises if inoculation of soil, seeds, or stems is necessary for maximum performance of stem-nodulating legumes.

B. Inoculation

There is little argument about the need for inoculation in grain legumes in most agricultural soils (Brockwell, 1982). In root-nodulating crops, inoculation is an established practice, and yield increases due to inoculation range from no response to significant responses (Gibson, 1977; Sekhon et al., 1978). Major constraints to effective inoculation in root-nodulating crops are strain competition and adverse soil environment (Eaglesham and Ayanaba, 1984; Dowling and Broughton, 1986). Stem nodulation, however, is a system where some of the serious problems of seed or soil inoculation may not be encountered. Unlike seed or soil inoculation, stem inoculation is less affected by the strain competition problem. Edaphic factors may have less effect on the efficiency of stem inoculation than of root inoculation. However, further work is required to confirm this hypothesis.

For stem inoculation, shoots can be sprayed with a suspension containing about 10^8 bacteria/ml, using either a liquid culture of rhizobia, a colloidal suspension obtained by mixing entrapped rhizobia (alginate polymer) in a phosphate buffer (0.06 m, pH 6.8), or a suspension of crushed stem nodules with water passed through a filter (Rinaudo et al., 1988).

Inoculation by spraying the appropriate rhizobia on stems of 2- to 7-week-old plants has been considered useful for high BNF by some authors. Alazard and Duhoux (1987) reported significant increases in plant dry weight, nodule number, and N_2-fixing activity (ARA) of *A. afraspera* after inoculation. Influence of stem inoculation on BNF of *S. rostrata* seems to depend on the time (plant age) of inoculation (Rinaudo et al., 1988). The positive effect of stem inoculation on growth and BNF of *S. rostrata* was reported by Ladha et al. (1989b). Adding 3% gum arabica can further increase the number of nodules as nodulation sites are exposed longer to the broth culture (Ladha et al., 1989b). Positive response to inoculation is usually found in pot/greenhouse experiments, while in the field an inoculation response seems limited to experiments conducted during the dry season and/or under upland conditions.

In greenhouse experiments the effect of single and dual inoculations on growth, nodulation, and N_2 fixation (ARA) by *S. rostrata* grown in *Sesbania*-rice cycle during both the wet and dry seasons was studied (Ladha et al., 1989b). Plants sprayed with ORS 571 inoculum showed profuse nodulation on the stem, while unsprayed plants did not produce stem nodules. Dual inoculation showed the maximum nodule dry weight and ARA. Uninoculated plants produced only root nodules and had significantly lower ARA. The inoculation effect was less pronounced in a second crop which may be due to (1) increased number of azorhizobia in control treatments as a result of *S. rostrata* incorporation and (2) poor plant growth due to short day length.

The response to inoculation under flooded field conditions in Senegal was poor which could be due to contamination of rhizobia from inoculated to noninoculated plots (Rinaudo et al., 1983). However, spraying the shoots of upland-grown *S. rostrata* with a suspension of 10^8 cells/ml remarkably increased nodule dry weight and ARA (IRRI, 1986). The first spray gave a fourfold increase and the second a 20-fold increase in stem-nodule dry weights. However, after several days, spontaneous stem nodulation, further enhanced by the onset of rains, was observed in the uninoculated plots.

Rinaudo et al. (1988) determined the optimum plant age for stem inoculation of *S. rostrata* grown at different times of the year. Plants were inoculated at 3, 5, and 7 weeks after seeding. ARA of 9-week-old plants was the sum of activities of the stem portion which was inoculated successively at 3, 5, and 7 weeks. Their results showed that the ARA was mainly due to nodules initiated 3 and 5 weeks after seeding (Figure 4). Stem nodules appearing at 3 weeks were probably due to spontaneous infection. It was concluded that the most suitable time for *S. rostrata* inoculation in Senegal was at a plant age of 5 weeks.

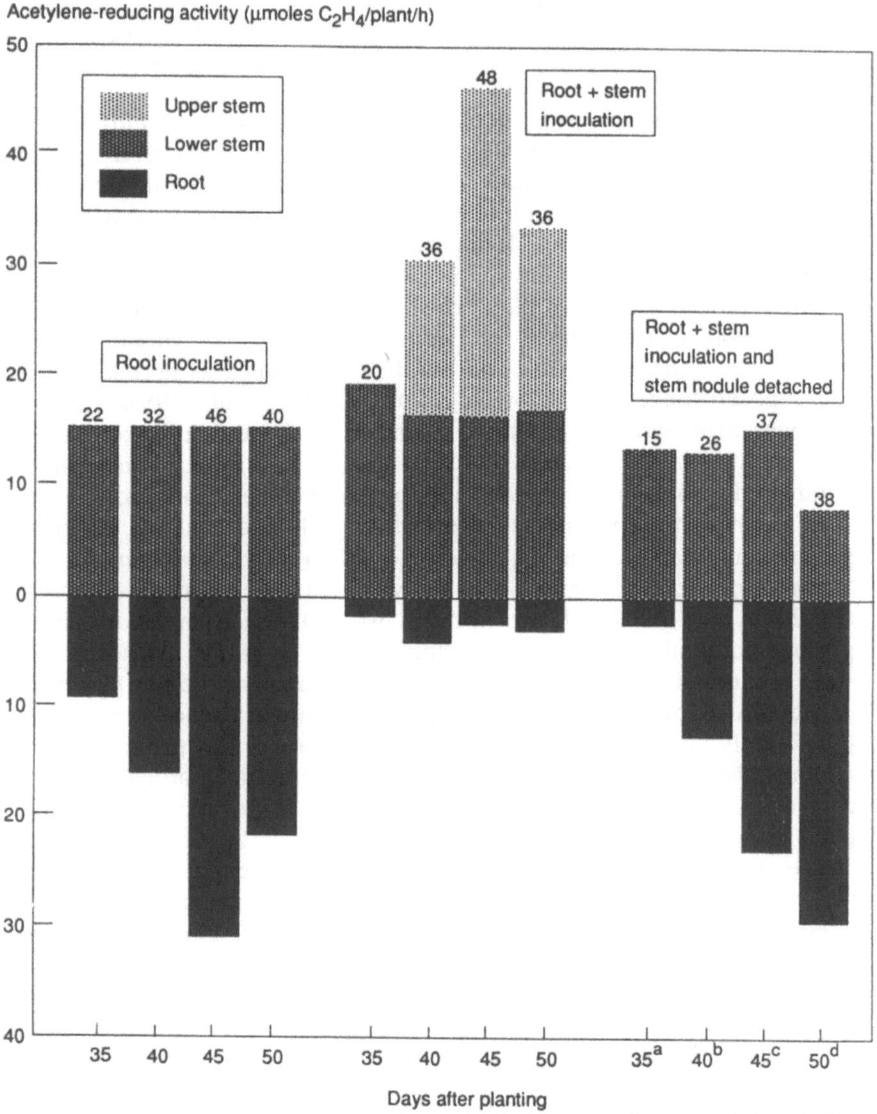

Figure 4. Acetylene-reducing activity (ARA) of of stem and root nodules of 35-, 40-, 45-, and 50-day-old *Sesbania rostrata* as influenced by inoculation and nodule detachment (from Ladha et al. (1992), International Rice Research Institute).

Although inoculation has improved BNF by *S. rostrata* and grain yield of the succeeding rice crop under pot conditions (Ladha et al., 1989b), no such effects have been reported from field studies. Spontaneous nodulation occurs in the field, and in most cases, is sufficient for high BNF. Even after long fallow periods, reinoculation of soil may not be needed.

C. Edaphic Factors Including Submergence

Voluminous work has been reported on edaphic factors affecting the different aspects of root nodule symbiosis (Bushby, 1982; Hodgson and Stacey, 1986). The optimum growth conditions for both host and symbiont partners have been reported to vary with crop, species, and bacterial strains. Edaphic factors mainly discussed in the literature are soil submergence, nutrient status of the soil (O'Hara et al., 1988), salinity (Arunin et al., 1988), extreme pH acidity and alkalinity (Munns, 1977), and available soil N or mineral N-fertilizer (Pate, 1977). Only limited information on the effect of these factors on stem-nodulating legumes is available and it will be discussed in the following paragraphs.

1. Soil Submergence

The tolerance of a green manure to flooding is an important trait for species used in lowland rice systems with poor water control. Species/ecotypes of aquatic plants differ in their abilities to tolerate flooding (Evans and Rotar, 1987; Saint Macary et al., 1984; Kretschmer and Bullock, 1980). This might be related to the extent of aerenchyma development and might explain the variability in nodulation and N_2 fixation among species. Assuming that root nodulation is an extra burden on the plant's need to supply oxygen under flooded conditions, the character of stem nodulation might have evolved as a result of adaptation to flooding. Although the inhibitory effect of waterlogging due to anoxia is well established in nonaquatic legumes, some degree of tolerance due to gradual adaptation has been reported in *Glycine max* (Hunter et al., 1980), *Vigna unguiculata* (Zablotowig and Foch, 1979), and *Vicia faba* (Gallacher and Sprent, 1978).

The influence of submergence on nodulation and N_2 fixation by root nodules of *Sesbania* and *Aeschynomene* has been poorly documented. Albrecht et al. (1981) and Queensberry et al. (1982) reported negligible effects of flooding on nodulation and N_2 fixation of *A. americana* and *A. villosa* grown in pots. Eaglesham and Szalay (1983) studied the effect of flooding on root and stem nodulation of *A. scabra* in pots containing sand and a nutrient solution. The number of root nodules decreased to almost zero compared with 1000 stem nodules per plant. Similarly, N'doye and Dreyfus (1988) reported the inhibitory effect of waterlogging on root nodulation and N_2 fixation by *S. rostrata* and *S. sesban*. Ladha et al. (1989b), however, found no difference in root nodule mass and ARA of 45-day-old *S. rostrata* grown in flooded or nonflooded pots.

2. Nutrient Status of the Soil, Acidity, and Salinity

Nutritional factors are responsible for the wide variation in the amount of N fixed by *Rhizobium* spp. in symbiotic association with various host legumes (O'Hara et al., 1988). Nutrient imbalances and other soil constraints may have a considerable impact on successful biofertilizer production at low-input agricultural practices.

Vigorous growth of *S. rostrata* and *A. afraspera* was reported on both productive and relatively poor sandy soils (Becker et al., 1990). However, mineral nutrients, especially phosphorus and potassium, seem to further increase biofertilizer production.

Inadequate available soil P is a limiting factor for growth of many legumes in the humid tropics, especially on acid soils (Lynd et al., 1984; Mokwunye et al., 1986). The P requirements for nodulation and maximum nodule activity are much greater than those for host plant growth (De Mooy et al., 1973). The impact of an insufficient P supply may be more pronounced in stem-nodulating legumes where the number of nodules is considerably higher than in root-nodulating legumes.

The favorable influence of adequate available soil K as a requisite for desirable nodulation and maintenance of high nitrogenase activity has been reported (Chamalet et al., 1987). Potassium is involved in the regulation of membrane permeability and in various synthetic processes particularly in photosynthesis, thus, providing the essential substrates for N_2 fixation (Bethlenfalvay et al., 1978).

Becker et al. (1991) showed that the application of mineral P and K fertilizers stimulated nodulation and ARA of *S. rostrata* and *A. afraspera*, which could lead to a 40% increase in N yield. Potassium alone mainly enhanced plant growth, whereas P stimulated BNF. A positive effect of PK fertilizers on biomass production of *S. rostrata* has also been reported by Meelu and Morris (1988). Becker et al. (1991) recommended the application of PK fertilizer to GM instead of rice. The availability of this organic-amended P and K to the rice crop may be higher as compared with the direct mineral application. Little is known about the influence of essential nutrients for N_2 fixation of stem-nodulating legumes.

Often, leguminous GM species are grown on problem soils such as saline, alkali, and acid soils to reclaim and stabilize such lands (Abrol and Palaniappan, 1988). Although species such as *S. cannabina*, *S. sesban*, *S. bispinosa*, and *S. exaltata* have been found to be fairly well adapted to saline, alkali, and acidic soils (Evans and Rotar, 1987), little is known about their nodulation and N_2-fixation performance. Qualitative observations like good or poor nodulation in some *Sesbania* species have been reported (Yadav and Vyas, 1971).

Recent studies reveal that *S. rostrata* performs satisfactorily in acid (Salam et al., 1989) and saline soils (Hansen and Munns, 1985; Arunin et al., 1988). In acid soils in northeastern Thailand, *S. rostrata* and *A. afraspera* performed better than traditional (root-nodulating) GM legumes (D.P. Garrity, personal communication). Further studies under varying soil and climate employing a wide range of genetic material are desirable.

3. Mineral Nitrogen

Normally, crop plants respond to fertilizer nitrogen in the field since most of the cultivated soils worldwide are deficient in nitrogen. For reasons of energy legumes prefer soil or fertilizer nitrogen rather than biologically fixed nitrogen (Houwaard, 1980). Mineral N can reduce both nodulation and N_2 fixation in root-nodulating legumes; the degree of inhibition varies with the compound and its concentration (Eaglesham and Szalay, 1983). The adverse effects of combined N on infection, nodule development, and N_2-fixing ability have been reviewed by Pate (1977). However, the N_2-fixing apparatus of stem-nodulating legumes is reported to be unaffected to some extent by mineral N (Dreyfus and Dommergues 1981, Eaglesham and Szalay 1983, Becker et al., 1986; Becker et al., 1989; and Becker et al., 1991).

Rinaudo et al. (1988) observed significant and complete inhibition respectively, of root nodule ARA in hydroponically grown *S. rostrata* where the NH_4NO_3 concentrations were constantly maintained at 1.5 and 3 m*M*. The concentration for a complete inhibition of stem nodule ARA was 6 m*M*. The specific and total ARA of *A. afraspera* root nodules were markedly reduced with 3 m*M* NO_3, but significant inhibition of stem nodule ARA required 6 m*M* NO_3–N. A similar trend was obtained by urea N in pots. Becker et al. (1986) reported that 10 ppm N reduced the number of nodules and ARA of roots, whereas the same concentration increased stem ARA. Stem nodule number and ARA were reduced at 40 ppm N. Becker et al. (1986) further studied the effects of mineral N on pot- and field-grown *A. afraspera*. Results showed positive effects on ARA and the biomass of stem nodules and negative effects on root nodules at 50 and 100 kg N/ha applied as urea. Nitrogen at 200 kg/ha inhibited both root and stem nodule parameters in the pot, but in the field, stem nodules were unaffected. In further field experiments, Becker et al. (1991) reported a gain of 30 kg N/ha by *S. rostrata* when 30 kg N/ha was applied as urea. Generally, a low N dose has a stimulatory effect on legume N_2 fixation and is referred to as starter dose. In a field study under flooded rice soil, Balasubramani and Kannaiyan (1990), observed that the dry weight of stem nodules and nitrogenase activity of root and stem nodules of *S. rostrata* were not affected adversely up to 90 kg N/ha; instead, 30 to 90 kg N/ha increased the fresh and dry weights of nodules. The lower inhibition of stem-nodule N_2 fixation may be explained by:

1. A general stimulation of plant growth by mineral N. Bigger plants can bear more stem nodules and provide a larger source of carbohydrates.
2. Anatomical separation of the sites of N uptake (roots) and the sites of N_2 fixation (stems) and therefore less contact with mineral N in the stem nodules.
3. Lower competition for carbohydrates between N_2 fixation in stem nodules and amino acid synthesis. Soil ammonium is converted into

amino acids in roots and transported to above-ground plant parts.
4. Unique regulation mechanisms in *nif* and *ntr* genes (Pawlowski et al., 1987) and for the glutamine pool (Donald and Ludwig, 1984) in stem nodule bacteria.

Despite the positive effect of mineral N on N accumulation of stem-nodulating GM crops, a starter N application is not practical for resource-poor farmers. However, cultivating stem-nodulating legumes on N-rich soils may enhance biological N accumulation and affect rice yield positively (Becker et al., 1991). Generally, waterlogging, high N content, extreme pH, and salinity which may adversely affect the formation and development of root nodules may not affect those of stem nodules. This aspect, however, needs further experimental proof.

D. Climatic Factors Including Photoperiod Sensitivity

Since stem nodules are formed on aerial parts of plants, atmospheric factors like temperature, relative humidity, and solar radiation assume more importance than edaphic factors.

The formation and functioning of stem nodules may be more affected by temperature changes than those of root nodules because the soil buffers temperature extremes. The temperature differences around roots and above-ground plant parts can be as high as 11°C (Thomas et al., 1975). Most stem-nodulating legumes are tropical plants and may therefore need a relatively high temperature for optimal growth. However, up to now, little is known about temperature requirements for germination, growth, and flowering. Rinaudo and Moudiongui (1987) reported temperature being the main source of variation for plant height and N yield by *S. rostrata* in northern Senegal. Besides mean temperature, the diurnal temperature amplitudes may be relevant for biomass production and nitrogenase activity. High average temperature during the day increases plant height, N_2-fixing activity, and N yield of root-nodulating legumes. Low night temperatures may slow down the respiration of photosynthates, thereby increasing net photosynthesis. This feature may explain why highest growth rates for *S. rostrata* were reported from subsahelian Senegal (Rinaudo et al., 1983), where diurnal temperature changes are as high as 15°C.

Humidity may not be as critical for plant growth as for the infection of the nodulation sites on the stems. High humidity during the rainy season is reported to break the dormancy of stem nodule primordia of *Aeschynomene* species such as *A. indica*, *A. ciliata*, and *A. rudis* (Alazard and Duhoux, 1987). Sometimes even waterlogging is required to have proper development of primordia for infection (Schaede, 1940). Alazard et al. (1988) have observed that in Senegal, primordia in *S. rostrata*, *A.*

afraspera, and *A. nilotica* are well developed and get readily infected. However, humidity further enhances primordial development and increases the success of infection. Becker et al. (1990) observed a correlation of the number of stem nodules in *S. rostrata* and *A. afraspera* with rainfall; it may have a dual role in enhancing stem nodulation. Rain may be the carrier of bacterial inoculum to the nodulation sites (see also Section V) and by increasing air humidity, rain may further trigger primordial development.

Light intensity is known to affect BNF through the photosynthetically produced reduced carbon compounds (Patterson and LaRue, 1983). Becker et al. (1990) showed that plant height, nitrogenase activity, and N yield of *S. rostrata* and *A. afraspera* are significantly related to solar radiation. At high light intensity, biomass production was high even on unproductive sandy soils.

Legumes are generally sensitive to changes in day length. The importance of photoperiod sensitivity on the performance of legume crops is well documented (Herridge and Pate, 1977). During the short-day season vegetative growth and BNF are cut short by early flowering. Flowers and pods represent a stronger sink for carbohydrates than the nodules do. As a result, biomass and N yield are poor. *S. rostrata* was found to be highly sensitive to photoperiod (Visperas et al., 1987). Dreyfus et al. (1985) observed poor plant growth in Senegal during short-day periods. Low biomass and N yield of *S. rostrata* in the Philippines between October and February have been ascribed to the day length of less than 12 h (Becker et al., 1990). Depending on day length, BNF in 8-week-old plants varied between 10 and 28 μmol C_2H_4 per h and plant. Dreyfus et al. (1985) have surmised that until photoperiod-insensitive varieties of *S. rostrata* are available, the use of this species in equatorial climates will remain limited. Although interspecies variability in photoperiod sensitivity of stem-nodulating legumes has been shown (Becker et al., 1990), it is not known if an intraspecies variability exists.

A. afraspera seems to be considerably less photoperiod sensitive than *S. rostrata*. December-planted *S. rostrata* flowered after 37 days, while *A. afraspera* flowered in 54 days. April-planted *S. rostrata* took 125 days to flower compared with 85 days in *A. afraspera* (Becker et al., 1990). Consequently, biomass and N accumulation in 8-week-old *A. afraspera* planted throughout the year were more or less constant, whereas *S. rostrata* showed large variations (3 to 15 t/ha) throughout the year (Figure 5). Nothing is known about the photoperiod sensitivity of other stem-nodulating legumes. It might be interesting to collect ecotypes at various distances from the equator (12-h day) and thereby identify a range of plants, varying widely in their sensitivity to photoperiod.

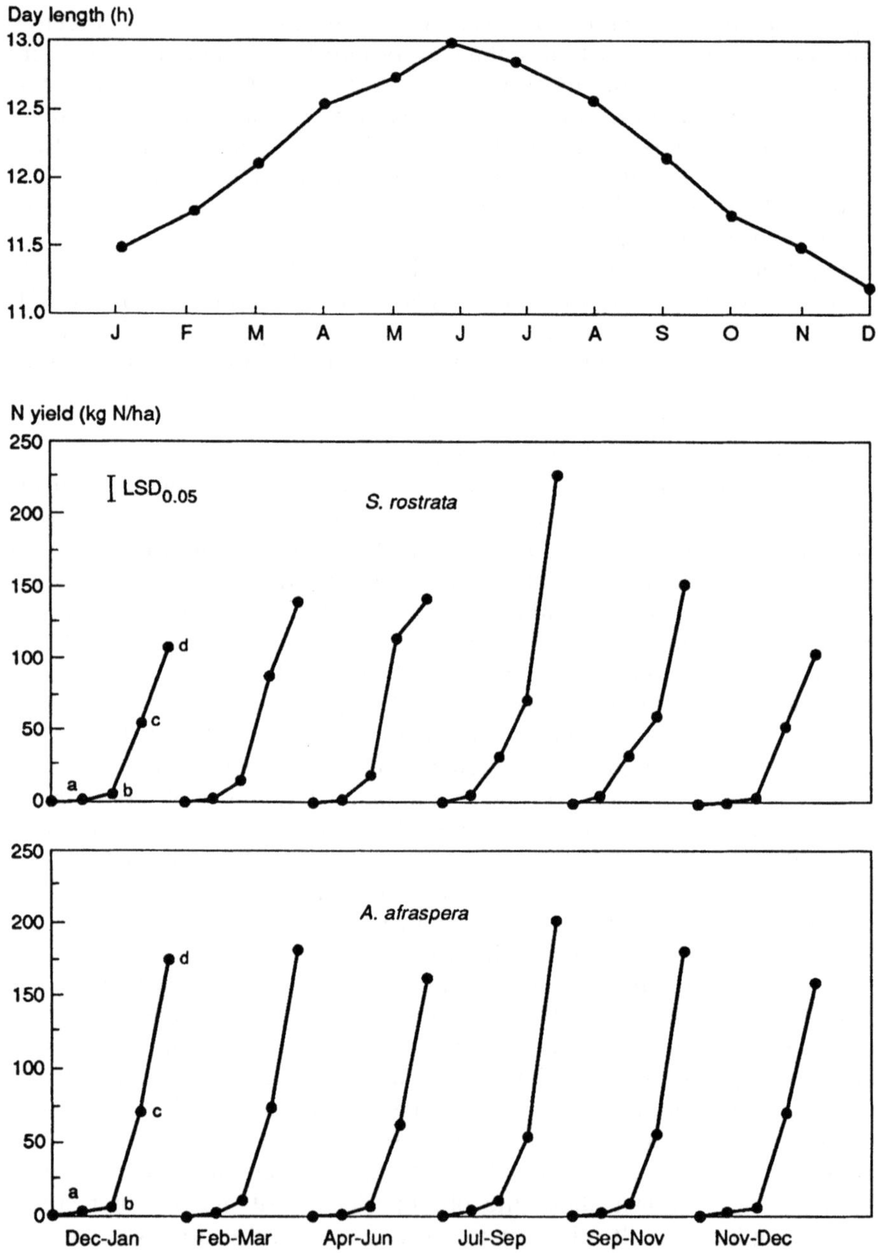

Figure 5. Pattern of N accumulation of field-grown *Sesbania rostrata* and *Aeschynomene afraspera* in a year-round planting in the Philippines. a, b, c, and d: measurements taken at plant ages of 2, 4, 6, and 8 weeks, respectively (adapted from Becker et al., 1990).

VIII. Agronomic Use of the Stem-Nodulating Legumes

A legume crop can be used as food, fodder, fuel, GM, a cover crop, a soil stabilizer, and a wind breaker. The work on the agronomic use of stem-nodulating legumes has concentrated mainly on the GM issue in lowland rice.

Leguminous GM crops are potential N sources with relatively high efficiency, but also have diverse non-N effects such as the improvement of soil properties and control of weeds, pests, and diseases. Positive effects on soil properties include (a) maintenance or improvement of soil organic matter, (b) conservation and recycling of nutrients, (c) increased mineralization, (d) increased biological activity, (e) increased cation exchange capacity, (f) reduction in soil erosion, and (g) reduction in fertilizer N losses (Jiao Bin, 1983; Hesse, 1984; Becker et al., 1988a).

Nonetheless, during the past few decades, GM use has declined in all rice-growing countries except in parts of China (Qi-Xiao Wen, 1984). The reasons for the decline and the various constraints associated with GM use have been reviewed by Roger and Watanabe (1986), Garrity and Flinn (1988), and Becker et al. (1988a). They include competition for time, labor, land, and water, substantial establishment cost, incorporation difficulties, and nonavailability of seeds. Furthermore, the increased availability of inorganic fertilizers and low emphasis on GM by researchers and extension services have contributed to the decline in GM use. However, emerging issues such as stagnation and decline of farm production and productivity, and the nonreliable future supply of N fertilizer, together with the discovery of fast-growing, stem-nodulating legumes, have revived interest in green manuring. In the following sections, the potential environments for GM use and efficient ways of GM management are discussed.

A. Potential Rice Environments for GM Use

Rice is grown in lowlands that may be irrigated or rainfed. The rainfed environment is estimated to cover more than 20 million hectares in Asia alone. Most bunded rainfed ricelands produce a single crop of transplanted rice a year, with a limited proportion planted to an upland crop preceding or following rice. The area is characterized by poor internal and surface drainage, and experiences unpredictable short-term waterlogging during the dry season–wet season transition. This is particularly common in ricelands on floodplains of major river systems such as those found in northern and eastern India and Bangladesh. Since the alternate possibilities are limited, flood-prone or waterlogged areas are the largest possible niche for stem-nodulating GM (Garrity and Flinn, 1988).

The irrigated lowland environment, characterized by an assured water supply, has a high cropping intensity. Rice is the main crop of irrigated lowlands and most areas produce two cereal crops (rice-rice or rice-wheat) a year, with a time gap before both wet season and dry season crops. The main problem of this environment is crop competition for time, space, and labor. In areas where the gap is very short or where early-season waterlogging occurs, fast-growing *Sesbania* or *Aeschynomene* would fit between the main crops. *S. rostrata* may be used as GM before wet season rice, during the long-day period (April/May) and *A. afraspera* before the dry season rice (November/December) during the short-day period (Becker et al., 1990).

B. GM Management in Rice-Based Cropping Systems

In most eco-socio-environments it is difficult to fit a GM into the established cropping sequence without raising the production costs. Therefore, there is a need to develop ways to grow GM crops as cheaply and conveniently as possible (Garrity and Flinn, 1988).

1. GM Establishment

A successful establishment of a GM crop with a minimum investment in seed, tillage, and labor is extremely important. Pradhan (1988) compared four methods of tillage in establishing *S. rostrata*, at different seeding rates in fallow riceland. His study concluded that if a duration of 60 days is available to grow and incorporate *S. rostrata* then zero tillage broadcast of 15 kg seed/ha is adequate to yield 95 kg N/ha. If the duration available is about 40 days, broadcasting 60 kg seed/ha on a saturated soil would yield 80 kg N/ha. These practices have recently been validated in farmer-managed trials in Central Luzon, Philippines (D.P. Garrity, personal communication).

An alternative to the use of seed is crop establishment by vegetative propagation. This practice seems suitable with stem-nodulating legumes, since the nodulation sites on the stem consist of primordia of adventive roots that can grow out under waterlogged conditions (Tsien et al., 1983). Liu (1988) suggested the use of stem cuttings to fill gaps in the canopy of interrow-seeded *A. indica* in China. Ratooning and/or the use of stem cuttings was reported to reduce growth duration requirements of *S. rostrata* by 2 weeks compared with seed use (Becker et al., 1988b, 1989). Vegetative propagation may have additional advantages since the GM crop can be grown on bunds and dikes or in some wasteland around the rice field and clipped when needed. This way, the GM will not compete for land with cash crops. Adoption of this system may result in (a) reduced time and labor requirements for land preparation, (b) availability of a continuous supply of organic N, and (c) establishment of seed production lots that will ensure seed supply (Becker et al., 1988a).

2. Growing GM as an Intercrop and on Bunds/Wastelands

To save time, land, and water, GM crops may be grown in association with rice (Vacchani and Murty, 1964). By changing the crop geometry, GM may be seeded in between rice rows either at the time of transplanting or about 3 weeks before harvest of the first rice crop (double wide-narrow row system). In the case of GM seeded into newly transplanted rice, the GM crop is machine-incorporated or simply trampled into the soil between rice rows of 30 to 40 days, thus coinciding with weeding operations. Incorporating 30- to 40-day-old *S. rostrata* between rice rows 40 cm apart resulted in a yield increase of rice grain equivalent to the application of 30 kg N/ha as urea (Torres, Pareek, unpublished). Hati (1987) defoliated 20-day-old *S. rostrata* intercropped with upland rice, by spraying 0.9% propanil. The incorporation of a fresh biomass of 4.3 to 6.3 tons per hectare resulted in a rice grain yield increase equivalent to 40 kg mineral N per hectare.

In rice-rice sequences it seems possible to seed GM into the standing rice 2 to 3 weeks before harvest. After the rice harvest, the established GM crop shows rapid growth. This postrice–mixed cropping or relay cropping system is known as *uttera* in India (Vacchani and Murty, 1964). Relay cropping shortens the interval between two rice crops by about 15 days, and may be useful in areas where the period between the two main crops is short. Furthermore, tillage costs as well as land and water requirements for the GM crop will be reduced. Becker et al. (1988a) suggested that the low C:N ratio of incorporated GM mixed with rice straw or stubble may slow down N released from decomposing organic matter, thus reducing N losses. Using *S. cannabina* for relay cropping, Liu (1988) reported fresh biomass yields of more than 8 tons per hectare which resulted in a rice grain yield increase of about 14% in the second rice crop. Palaniappan (personal communication) used 30-day-old *S. rostrata* seedlings transplanted between rice rows. After 30 days, GM was pruned and 8 to 10 tons of fresh biomass per hectare was incorporated. This resulted in a rice grain yield increase of about 1 ton. Stubbles were allowed to ratoon and were incorporated before the second rice crop.

Another intercropping or mixed cropping practice is found in large rainfed lowland rice areas in eastern India. *S. cannabina* is broadcast with rice on dry soil. The GM is incorporated 40 to 60 days after seeding by means of bushenning—an operation that involves plowing/cross-plowing and trampling of *Sesbania*. Rice plants are then redistributed. Rice grain yield increases of 35% were reported (Pillai, 1958). Recently at IRRI, this practice has been modified. Rice was dry-seeded in rows spaced alternately at 7- and 33-cm spacing. Simultaneously, *A. afraspera* was dry-seeded at the midline of the 33-cm spaces and was incorporated at about 40 days. This practice resulted in a rice grain yield increase of about

1 ton/ha as compared with that of traditional bushenning (Woodhead and Chaudhary, personal communication).

3. Optimum Time for GM Incorporation and Rice Transplanting

Depending on the species and plant age the amount of biomass and carbon, nitrogen, and lignin contents vary. These factors are important in relation to GM incorporation, time of rice transplanting, N use efficiency by rice, and N supply to the soil.

The recommended growth duration for fast-growing legumes ranges from 45 to 55 days (Ghai et al., 1988; IRRI, 1986). Optimal crop growth duration for desired N accumulation has been established for *S. rostrata* and *A. afraspera* grown under Philippine conditions (Becker et al., 1990). For instance, to accumulate 100 kg N/ha, *S. rostrata* should be grown for 55 days when seeded in December and for 42 days when seeded in May. *A. afraspera*, however, needs about 48 days of growth, regardless of the season. Depending on the species and the plant age, the lignin content in the dry matter also varies. Percentage lignin content in *S. rostrata* is generally higher than in *A. afraspera*, suggesting a more labor-intensive incorporation and a slower N release after incorporation of *S. rostrata* (Becker et al., 1990).

Besides the age of incorporated material, the interval from GM incorporation to rice transplanting is important to obtain optimum effects. The interval between incorporation and rice transplanting varies with the soil type. On nonacid tropical soils, transplanting immediately after incorporation gave higher yields as compared with delayed transplanting (Bhardwaj, 1982; Beri et al., 1989). Experiments conducted in the Philippines and Senegal with *S. rostrata* and *A. afraspera* also suggested that rice can be transplanted 1 to 2 days after incorporation (Becker et al., 1990; Rinaudo et al., 1988). However, in poorly buffered acidic soil, Garrity (personal communication), recommended an interval of 14 to 20 days between incorporation of *S. rostrata* and rice transplanting to obtain optimum yield. Diekmann et al. (1991) observed adverse effects on an early growth of rice when the rice was transplanted 1 day after incorporation of *S. rostrata* and *A. afraspera* in a clay soil. Organic acids and phenolic substances may accumulate during anaerobic decomposition and are known to affect root growth and nutrient uptake (Chou and Chiou, 1979; Cannell and Lynch, 1984).

4. Mineralization of GM-N and its Recovery by Rice

Unlike inorganic N fertilizers, GM biomass must decompose and undergo mineralization before its N becomes available to the rice plant. Apparently, decomposition depends on several factors such as duration of submergence, method of GM incorporation, soil type, soil temperature, and the amount and nature (C:N ratio and lignin content) of the applied material (Shi et al., 1978; Huang et al., 1981; Ito and Watanabe, 1985; Nagarajah, 1988; Ventura et al., 1987). Information on the pattern and amount of N release from GM-N and its recovery by rice is essential in developing management practices leading to the efficient use of GM in lowland rice.

Recently, Buresh and De Datta (1990) summarized the published data on GM-N mineralization as measured by the release of NH_4^+–N in flooded soils. The release of NH_4^+-N showed a typical and more or less similar pattern among different GM species; it increased sharply within 7 to 10 days of incorporation, maintained that level up to 20 days, and then gradually declined. Furoc and Morris (1989) studied the NH_4^+-N release of *S. rostrata* in a flooded soil. Significantly more NH_4^+–N was released from 60-day-old GM than 48-day-old GM treatment. It was generally higher in *S. rostrata* than in *S. cannabina* treatment. Becker et al. (1990) observed a higher carbon to nitrogen ratio and lignin content in *S. rostrata* than in *A. afraspera*. Based on these data, they assumed that the former may release NH_4^+–N more slowly than the latter species.

Data reported on the apparent N recovery of *S. rostrata* and *A. afraspera* by rice varies between 20 and 75% (Table 5) compared with 30 to 60% recovery of 50 to 60 kg of inorganic fertilizer N. Biswas (1988) and Diekmann et al. (1991) determined the N recovery in rice plant and soil using ^{15}N-labeled *S. rostrata*, *A. afraspera*, urea, and a combination of GM and urea (Table 6). Total fertilizer-N recovery ranged from 45 to 72% compared with 87 to 93% of GM-N. Fertilizer-N recovery improved slightly when combined with GM. Total GM-N recovery was lower in the study of Biswas (1988) with *S. rostrata* (71%), than the study of Diekmann et al. (1991) with *S. rostrata* and *A. afraspera* (87 to 93%). Both the studies showed a high percentage of leftover N in soil where GM-N was applied.

5. Rice Crop Response to GM and Residual Effects

The primary objective of green manuring is to increase the grain yield of the following crop. Rice grain yield increases due to use of stem-nodulating GM over a no GM control registered wide variations ranging from 45 to 130% depending on agroecological conditions and soil types.

Experiments conducted in Thailand (Arunin et al., 1988), Nigeria (Mulongoy and Nguu, 1988), Senegal (Rinaudo et al., 1988), and India (Kalidurai and Kannaiyan, 1989) showed substantial yield increases after incorporation of *S. rostrata*. In the Philippines, recent experiments with *S. rostrata* and *A. afraspera* showed rice yield increases ranging from 0.8 to 2.8 t/ha, with an average of 1.6 t/ha (Table 6). Generally, apparent GM-N recovery and agronomic GM-N use efficiency were similar to or higher than those of inorganic N fertilizers (Table 5). Recently, Becker (1990) and Diekmann (1991) have reported GM-N use efficiencies of up

Table 5. Effect of *S. rostrata* and *A. afraspera* as green manure on rice yield and N uptake/recovery utilization efficiency

Species/time of experiment	GM growth duration	GM−N applied (kg/ha) (A)	Rice grain yield (t/ha) W/out GM (B)	Rice grain yield (t/ha) W/ GM (C)	Rice N uptake (kg/ha) W/out GM (D)	Rice N uptake (kg/ha) W/ GM (E)	GM−N recovery (%) (E − D/A)	GM−N utilization efficiency (kg) grain increase/kg N applied (C − B/A)	Reference
S. rostrata									
Wet season (1985)	44	74	4.6	6.0	82	129	63 (30)	19 (18)	Ventura et al. (1987)
Dry season (1985–86)	57	64	4.0	5.6	66	114	75 (60)	25 (24)	
Wet season (1987)	47	79	4.9	6.3	83	131	61 (46)	18 (20)	
S. rostrata									
Dry season (1987)	45	90	4.8	6.3	73	118	50 (66)	17 (25)	Biswas (1988)
S. rostrata									
Wet season (1984)	48	79	2.1	3.0	35	45	73	14	Furoc and Morris (1989)
	60	162	2.1	3.6	35	74	24	9	
A. afraspera									
Wet season (1987)	(1) 49	115	4.9	6.5	92	124	28 (45)	14 (23)	Becker (1990)
	(2) 49	204	3.5	5.2	71	115	21 (50)	8 (19)	
Dry season (1987–88)	(1) 42	77	4.1	6.1	79	108	38 (43)	26 (24)	
	(2) 42	77	4.0	6.2	53	93	52 (54)	29 (30)	

S. rostrata									
Wet season (1987)	(1) 49	157	4.9	6.5	92	119	39 (45)	10 (23)	Becker (1990)
	(2) 49	194	3.5	4.9	71	116	23 (50)	7 (19)	
Dry season (1987–88)	(1) 42	55	4.1	6.0	79	105	47 (43)	38 (24)	
	(2) 42	48	4.0	5.9	53	91	78 (54)	40 (30)	
S. rostrata									
Wet season (1987)	(1) 49	150	4.9	6.5	92	154	41 (45)	11 (23)	Diekmann (1991)
	(2) 49	190	3.5	4.9	71	116	30 (52)	7 (18)	
	(3) 42	60	3.5	5.4	58	89	52 (48)	31 (26)	
Dry season (1988)	(3) 42	90	5.2	8.2	55	153	108 (90)	33 (38)	
A. afraspera									
Wet season (1988)	(1) 49	74	4.9	6.5	92	124	43 (45)	22 (23)	
	(2) 49	74	3.5	4.9	71	115	59 (52)	19 (18)	
	(3) 42	60	3.5	5.3	58	97	65 (48)	30 (26)	
Dry season (1988)	(3) 42	90	5.2	7.4	55	120	72 (90)	25 (38)	

Data in parentheses were obtained with inorganic fertilizer (50–60 kg N/ha/crop). Experiments conducted in Philippines: (1) IRRI experimental farm, Los Baños, (2) farmer's field in Floridablanca, Central Luzon, and (3) Maligaya Rice Research and Training Center, Nueva Ecija, Central Luzon

Table 6. GM and fertilizer-^{15}N recovery in rice plant and soil

Treatment	Season	Nitrogen applied (kg/ha) GM	IF	^{15}N recovery (%) Plant	Soil + root	Reference
S. rostrata	DS	90	0	32	39	Biswas (1988)
Prilled urea	DS	0	90	(51)	(21)	
S. rostrata + PU	DS	45	45	34	42	
				(60)	(29)	
S. rostrata	DS	90	0	49	44	Diekmann et al.
A. afraspera	DS	90	0	47	40	(1991)
Prilled urea	DS	0	60	(23)	(23)	
S. rostrata + PU	DS	30	60	(24)	(30)	
S. rostrata	WS	60	0	42	49	Diekmann et al.
A. afraspera	WS	60	0	40	44	(1991)
Prilled urea	WS	0	40	(31)	(38)	

Data in parentheses are ^{15}N recovery of inorganic fertilizer (IF) N; GM, green manure; DS, dry season; WS, wet season

to 40 kg grain/kg N with *S. rostrata* and *A. afraspera*. The lower efficiency levels (<15 kg grain/kg N) were generally found in experiments where the amount of GM-N exceeded 100 kg/ha. Similar results were also found by Morris et al. (1989) and Crozat and Sanchyo-sawat (1985).

The fertilizer N equivalent of a GM crop has been summarized by Ladha et al. (1988) and Becker et al. (1988a). Data suggest that incorporating a 45- to 60-day GM crop will result in rice yield increases equivalent to those obtained with 50 to 100 kg fertilizer N/ha. If GM incorporation is combined with fertilizer N application, rice yields are higher than those obtained with an equivalent amount of fertilizer N alone. A conceptual framework for interpreting the effects of GM-N on succeeding crop yields has been proposed by Bouldin (1988). On the one hand, the GM acts as a replacement of fertilizer N to produce a given yield. On the other hand, the GM has additional beneficial effects on yield that cannot be ascribed to the N effect alone, suggesting that GM crops are more than just a source of N.

Yield increases in subsequent crops due to a residual effect from green manuring have been reported in both upland (such as in LGM/rice/wheat system) and lowland crops (such as in LGM/rice/rice system) (Buresh and De Datta, 1990). Responses ranged from none to high and were normally related to the amount of GM-N incorporated. Morris et al. (1986) did not detect a residual response for GM-N in a second crop after 4 years of GM incorporation. Diekmann et al. (1991) reported yield increase of 1.6 t/ha of a second rice crop due to the residual effect of incoporation of *S. rostrata* (equivalent to 120 kg N/ha). Using ^{15}N-labeled material,

Diekmann et al. (1991) quantified the amount of GM-N left in the soil after the first crop and its recovery in the subsequent rice crop. Thirty -five to 42% ^{15}N GM-N was recovered in the soil (Table 6), but only 3 to 5% was recovered in the subsequent rice crop. This represents about 14% recovery of the total residual ^{15}N present in the soil at harvest of the first rice crop. Watanabe (unpublished) observed a small residual effect after seven crop cycles of *S. rostrata*/rice where an average 73 kg GM-N/ha/crop was incorporated in situ. A total soil-N analysis made at the end of the seventh crop showed an increase in total N per gram of dry soil.

The extent of a residual effect depends on the amount of N added, soil texture, cation exchange capacity, nutrient status, C:N ratio, organic matter content, dry/wet cycles, and the depth of the plow layer. Furthermore, the decomposition pattern of applied GM may be important. Lignified GM releases N slowly; thus, recovery by the first crop may be relatively low, thereby increasing the residual effects on the second crop (Bouldin, 1988). Application of legume GM in combination with an appreciable amount of straw may slow down decomposition and thereby increase residual effects.

IX. Conclusion and Agenda for Future Research

The review of 10 years of research shows that the phenomenon of stem nodulation, the stem-nodulating bacteria, and the agronomic use of stem-nodulating legumes show various unique features, making this relatively new field a rewarding area for both basic and applied research.

Formation of nodules on stems has been reported for 25 species of the genera *Sesbania*, *Aeschynomene*, and *Neptunia*, and is a clever evolutionary adaptation to flooded conditions. It implies the specific location of these structures on above-ground plant parts and therefore their proximity to the energy-providing photosynthetic apparatus. Furthermore, the stem-nodule cortex is green and its photosynthetic activity may provide additional substrates for energy-requiring BNF. The number of potential nodulation sites increases with plant growth and they are perhaps preferentially supplied with carbohydrates. This may explain the higher specific and total BNF observed in stem nodules than in root nodules.

Many rhizobia that form nodules on above-ground plant parts are unique with regard to host specificity and various physiological characteristics. They are able to grow and fix N_2 ex planta under free-living conditions. Rhizobia isolated from several stem-nodulating species of *Aeschynomene* produce bacteriochlorophyll "a." The ability of the stem nodule bacteroid to produce bacteriochlorophyll "a" and use light in a wavelength that is not used by the plant could be an advantage, diminishing competition for energy for C and N reduction. The two new genera

Azorhizobium and *Photorhizobium* have been proposed for stem nodule-forming bacteria. Because the bacteria can be carried through the seeds, the symbiosis is sustained from one plant generation to the next. This cyclic nature of the symbiosis has not been reported for other than stem-nodulating legumes. Besides flooding tolerance, other features contributed to the increasing interest in stem-nodulating legumes as green manure for lowland rice. Stem nodules tolerate increased levels of soil or mineral nitrogen, while adverse soil conditions affect the BNF of stem nodules only indirectly. Fast growth and high nitrogen-fixing potential make some species (e.g., *S. rostrata*, *A. afraspera*, *A. nilotica*) appear promising for use as green manure in lowland rice especially in rainfed areas.

However, up to now, systematic studies are limited and concentrate mainly on *S. rostrata* and *A. afraspera*. A detailed characterization of the legumes and the bacteria as well as studies on their technical and socioeconomic implications of their agronomic use are needed. The following proposed agenda for future research is not to be limiting and pinpoints only some particular research issues considered important:

The collection and evaluation of species/ecotypes of stem-nodulating legumes should continue to receive priority.

A detailed rhizobia characterization and screening with regard to free-living N_2 fixation and photopigment production is needed. Little is known about the survival of stem-nodule rhizobia in different soils and under flooded conditions and the mechanisms by which these rhizobia cope with high O_2 concentrations in the tissue that arise as a consequence of photosynthetic activity. The contribution of the photosynthetic activity of both the stem-nodule cortex and the bacteria to nitrogenase activity should be assessed.

For a better knowledge of potential agronomic use, the legumes have to be screened in progression for N accumulation, photoperiod insensitivity, resistance to pests and diseases, and adaptability to marginal and problem soils. Elite species or ecotypes must then be tested within local cropping patterns.

The next step would be to conceive and test production techniques and systems with a view to reduced time, labor, land, and water requirements, and to assess the economic feasibility of the agronomic use of stem-nodulating legumes. Demonstration trials are an important tool of agricultural extension that could help introduce these new species as green manure to farmers.

Finally, the genetic base and functions for stem-nodule formation need to be explored with regard to transferring this characteristic to other legumes.

References

Abrol, I.P., and S.P. Palaniappan. 1988. Green manure crops in irrigated and rainfed lowland rice-based cropping systems in South Asia. In Sustainable Agriculture. Green Manure Rice Farming. pp. 71–82. International Rice Research Institute, P.O. Box 933, Manila, Philippines.

Adebayo, A., I. Watanabe, and J.K. Ladha. 1989. Epiphytic occurrence of *Azorhizobium caulinodans* and other rhizobia on the host and non host legumes. Appl. and Environ. Microbiol. 55:2407–2409.

Alazard, D. 1985. Stem and root nodulation in *Aeschynomene* spp. Appl. and Environ. Microbiol. 50:732–734.

Alazard, D., and M. Becker. 1987. *Aeschynomene* as green manure for rice. Plant Soil 101:141–143.

Alazard, D., and E. Duhoux. 1987. Diversity of stem nodulation sites in *Aeschynomene* spp. J. Plant Physiol. 132:123–125.

Alazard, D., I. Ndoye, and B. Dreyfus. 1988. *Sesbania rostrata* and other stem nodulation legumes. In Bothe, F.J. De Bruijn and Newton (eds.). Nitrogen fixation hundred years after. pp. 765–1269. Gustav Fischer, Stuttgart, New York.

Albrecht, S.L., J.M. Bennett, and K.M. Queensberry. 1981. Growth and nitrogen fixation of Aeschynomene under water stress conditions. Plant Soil 60:309–315.

Allen, O.N., and E.K. Allen. 1981. In The leguminosae, a source book of characteristics, uses and nodulation. pp. 604–607. The University of Wisconsin Press, Madison.

Arora, N. 1954. Morphological development of the root and stem nodules of *Aeschynomene indica*. L. Phytomorphology 4:211–216.

Arunin, S., C. Dissataporn, Y. Anuluxtipan, and D. Nana. 1988. Potential of Sesbania as a green manure in *Sa*line rice soils in Thailand. In Sustainable Agriculture. Green manure in rice farming. pp. 83–95. International Rice Research Institute, P.O. Box 933, Manila, Philippines.

Balasubramani, G., and S. Kannaiyan. 1990. Studies on the effect of fertilizer nitrogen on nodulation nitrogen fixation, and biomass production in *Sesbania rostrata*. Presented at INSURF Meeting held at IRRI, Los Banos, on 11–13 Sept. 1990.

Barrios, S., and V. Gonzales. 1971. Rhizobial symbiosis on Venezuelan savannas. Plant Soil 34:707–719.

Becker, M. 1990. Potential use of the stem-nodulating legumes *Sesbania rostrata* and *Aeschynomene afraspera* as green manure for lowland rice. Ph.D. thesis. Justus-Liebig University, Giessen, F.R.G.

Becker, M., D. Alazard, and J.C.G. Ottow. 1986. Mineral nitrogen effect on nodulation and nitrogen fixation of the stem-nodulating legume *Aeschynomene afraspera*. Z. pflanzenernaehr. Bodenk. 149:485–491.

Becker, M., J.K. Ladha, and J.C. Ottow. 1988a. Stem-nodulating legumes as green manure for lowland rice. Phil. J. Crop Sci. 13(3):121–127.

Becker, M., J.K. Ladha, I. Watanabe, and J.C.G. Ottow. 1988b. Seeding vs. vegetative propagations of the stem-nodulating green manure. *Sesbania rostrata* Biol. Fertil. Soils. 6:279–281.

Becker, M., R.P. Pareek., J.K. Ladha, and J.C.G. Ottow. 1989. Biofertilizer production of stem-cut planted and seeded *Sesbania rostrata*. Int. Rice Res. Newsl. 14(2):30–31.

Becker, M., J.K. Ladha, and J.C. Ottow. 1990. Growth and N_2 fixation of two stem-nodulating legumes and their effect as green manure on lowland rice. Soil Biol. Biochem. 22:1109–1119.

Becker, M., K.H. Diekmann, J.K. Ladha, S.K. De Datta, and J.C.G. Ottow. 1991. Effect of N.P.K. on growth and nitrogen fixation of *Sesbania rostrata* as green manure for lowland rice (*Oryza sativa* L.). Plant Soil 132:149–158.

Bergersen, F.J., G.L. Turner, D. Bogusz, and C.A. Appleby. 1988. Fixation of N_2 by bacteroids from stem nodules of *Sesbania rostrata*. J. Gen. Microbiol. 134:1807–1810.

Berhaut, J. 1979. Flore illustrée du Sénégal Tome V., ed. Clairafrique, Dakar, Senegal.

Beri, V., O.P. Meelu, and C.S. Khind. 1989. Biomass production, N accumulation, symbiotic effectiveness and mineralization of green manures in relation to yield of wetland rice. Trop. Agric. (Trinidad) 66:11–16.

Bethlenfalvay, G.J., S.S. Abu-Sharkra, K. Fishbeck, and D.A. Philipps. 1978. The effect of source-sink manipulations on nitrogen fixation in pea. Physiol. Plant. 43:31–34.

Bhardwaj, K.K.R. 1982. Effect of the age and decomposition period of dhaincha in the yield of rice. Indian J. Agron. 27:284–285.

Biswas, T.K. 1988. Nitrogen dynamics and ^{15}N balance in lowland rice as affected by green manure and urea application. Ph.D., Thesis. Indian Agric. Res. Inst., New Delhi, India.

Bohlool, B.B., R. Kosslak, and R. Woolfenden. 1984. The ecology of *Rhizobium* in the rhizosphere: Survival, growth and competition. In C. Verger and W.E. Newton (eds.). Advances in Nitrogen Fixation Research. pp. 287–293. Martinus Nijhoff/Junk Publishers, The Hague.

Bouldin, D.R. 1988. Effect of green manure on soil organic matter content and nitrogen availability. In Sustainable agriculture. Green manure in rice farming. pp. 152–163. International Rice Research Institute, P.O. Box 933, Manila, Philippines.

Brockwell, J. 1982. Inoculation methods for field experiments and farmers In J.M. Vincent (ed.). Nitrogen fixation in legumes. pp. 211–227. Academic Press, Australia.

Buresh, R.J., and S.K. De Datta. 1990. Nitrogen dynamics and management in rice-legume cropping systems. Advances in Agronomy. 45:1–59.

Bushby, H.V.A. 1982. Ecology. In W.J. Broughton (ed.). Nitrogen fixation. Vol. 2. *Rhizobium*. pp. 36–75. Clarendon Press. Oxford.

Cannell, R.Q., and J.M. Lynch. 1984. Possible adverse effects of decomposing crop residues on plant growth. In Organic Matter and Rice. pp. 455–475. International Rice Research Institute, P.O. Box 933, Manila, Philippines.

Carroll, W.R. 1934. A study of Rhizobium species in relation to nodule formation on the roots of Florida Legumes I. Soil Sci. 37:117–135.

Chakrabarti, S.K., A.K. Mishra, and P.K. Chakrabaritty. 1986. A note on physiological characteristics and genetic relatedness of a fast growing *Rhizobium* from stem nodules of *Aeschynomene aspera*. L. J. Appl. Bacteriol. 60:463–468.

Chamalet, A., J.M. Audergon, J.P. Maitre, and A.M. Domenach. 1987. Study of influence of potassium on the *Trifolium pratense* fixation by delta ^{15}N method. Plant Soil 98:347–352.

Chou, C.H., and S.J. Chiou. 1979. Auto intoxification mechanism of *Oryza sativa*. II. Effect of culture treatments on the chemical nature of paddy soil and rice productivity. J. Chem. Ecol. 5:839–859.

Crozat, I., and C. Sangchyo-sawat. 1985. Evaluation of different green manures on rice yield in Sonkla Laka Basin. Sonklanadarin J. Sci. Technol. 7:391–397.

Dart. P.J. 1977. Infection and development of leguminous nodule. In R.W.F. Mardes and W.S. Silver (eds.). A Treatise on Dinitrogen Fixation. Sec. III. Biology. pp. 367–672. John Wiley, New York.

De Bruijn, F.J. 1988. The unusual symbiosis between diazotrophic stem-nodulating bacterium *Azorhizobium caulinodans* ORS571 and its host, the tropical legume *Sesbania rostrata*. In T. Kasuge and E.W. Nester (eds.). Plant Microbe Interactions Molecular and Genetic Perspectives. Vol. 3. McGraw-Hill Publishing Company.

De Mooy, J.C., J. Pesek, and E. Spaldon. 1973. Mineral nutrition. In B.E. Caldwell (ed.). Soybean improvement, production and uses. pp. 267–352. American Society of Agronomy, Madison, W.

Diekmann, K.H. 1991. Nitrogen transformation processes involving ^{15}N-labeled *Sesbania rostrata* and *Aeschynomene afraspera* green manure in lowland rice (*Oryza sativa* L.). Ph.D. Thesis. Justus-Leibig-University, Giessen, F.R.G.

Diekmann, K.H., S.K. De Datta, and J.C.G. Ottow. 1991. Nitrogen-15 balance in lowland rice involving various soil N fractions as affected by green manure and urea amendment. Plant Soil (in press).

Donald, R.G.K., and R.A. Ludwig. 1984. *Rhizobium* sp. strain ORS571 ammonium assimilation and nitrogen fixation. J. Bacteriol. 158:1144–1151.

Dowling, D.N., and W.J. Broughton. 1986. Competition for nodulation of legume. Ann. Rev. Microbiol. 40:131–157.

Dreyfus, B.L. 1982. La symbiose entre *Rhizobium* et *Sesbania rostrata* une legumineuse a nodules caulinaires. Ph.D. Thesis, University of Paris.

Dreyfus, B.L., and Y.R. Dommergues. 1981. Nitrogen-fixing nodules induced by *Rhizobium* on the stem of the tropical legume *Sesbania rostrata*. FEMS Microbiol. Letter. 10:313–317.

Dreyfus, B., J.L. Garcia, and M. Gillis. 1988. Characterization of *Azorhizobium caulinodans*. Gen. Nov. sp. Nov, a stem-nodulating nitrogen-fixing bacterium isolated from *Sesbania rostrata*. Int. Systematic Bacteriol. 38:89–98.

Dreyfus, B.L., C. Elmerich, and Y.R. Dommergues. 1983. Free-living *Rhizobium* strain able to grow on N_2 as the sole nitrogen source. Appl. Environ. Microbial. 45:711–713.

Dreyfus, B.L., D. Alazard, and Y.R. Dommergues. 1984. Stem-nodulating rhizobia. In M.G. Klug, and C.E. Reddy (eds.). Current Perspectives of Microbial Ecology. pp. 161–169. American Society of Microbiology, Washington, D.C.

Dreyfus, B., G. Rinaudo, and Dommergues, Y. 1985. Observation on the use of *Sesbania rostrata* as green manure in paddy fields. MIRCEN Journal 1:711–721.

Duhoux, E., and B. Dreyfus. 1982. Nature des site d'infection par le *Rhizobium* de la tige de la legumineuse *Sesbania rostrata* Brem. C.R. Acad. Sci. Paris. 249:407–411.

Eaglesham, A.R.J., and A. Ayanaba. 1984. Tropical Stress Ecology of *Rhizobium* root nodulation and legume N_2 fixation. In N.S. Subba Rao (ed.). Current development in biological nitrogen fixation. pp. 1–35. Oxford and IBH Publishing Co., New Delhi.

Eaglesham A.R.J., and A.A. Szalay. 1983. Aerial Stem nodules on *Aeschynomene* spp. Plant Sci. Letters. 29:265–272.

Eaglesham, A.R.J., J.M. Ellis, W.R. Evans, D.E. Feischman, M. Hungaria, and R.W.F. Hardy. 1990. The first photosynthetic N_2-fixing *Rhizobium*: characteristics. In P.M. Gresshof, L.E. Roth, G. Stacey and W.L. Newton (eds.). N_2 fixation achievements and objectives. Chapman and Hall, New York.

Evans, D.O., and P.P. Rotar. 1987. *Sesbania* in Agriculture. Westview Press, Boulder, CO.

Furoc, R.C., and R.A. Morris. 1989. Apparent recovery and physiological efficiency of nitrogen in *Sesbania* incorporated before rice. Agron. J. 81:797–802.

Fyson, A., and J.I. Sprent. 1980. A light and scanning electron microscope study of stem nodules in *Vicia faba* L. J. Exptl. Bot. 31:1101–1106.

Gallacher, A.E., and J.I. Sprent. 1978. The effect of different water regimes on growth and nodule development of greenhouse grown *Vicia faba*. J. Exp. Bot. 29:413–423.

Garrity, D.P., and J.C. Flinn. 1988. Farm level management systems for green manure crops in Asian rice environments. In Sustainable agriculture. Green Manure in Rice Farming. pp. 112–129. International Rice Research Institute, P.O. Box 933, Manila, Philippines.

Ghai, S.K., D.L.N. Rao, and L. Batra. 1988. Nitrogen contribution to wet land rice by green-manuring with *Sesbania* spp. in an alkaline soil. Biol. Fert. Soils. 6:22–25.

Gibson, A.H. 1977. The influence of the environment and managerial practice on the legume-*Rhizobium* symbiosis. In R.W.F. Hardy and A.M. Gibson (eds.). A Treatise on Dinitrogen Fixation. pp. 393–450. Sec. IV. John Wiley & Sons, New York.

Hagerup, O. 1928. En hygrofil Baelgplante (*Aescyhnomene aspera* L.) med bakterieknolde paa staengelen. Dan Bot. 5:19.

Hansen, E.H., and D.N. Munns. 1985. Screening of *Sesbania* species for NaCL tolerance. Nitrogen-Fixing Tree Research Reports 3:60–61.

Hati, N. 1987. Effect of combining chemical N and *Sesbania aculeata* in upland rice. Int. Rice Res. Newsletter. 12:44–45.

Herridge, D.F., and J.S. Pate. 1977. Utilization of net photosynthate for nitrogen fixation and protein production in an annual legume. Plant Physiol. 60:759–764.

Hesse, P.R. 1984. Potential of organic materials for soil improvement. In Organic matter and rice. pp. 35–44. International Rice Research Institute. P.O. Box 933, Manila, Philippines.

Hodgson, D.I.M., and G. Stacey. 1986. Potential for *Rhizobium* improvement CRC critical Rev. in Biotechnol. 3:1–74.

Houwaard, F. 1980. Influence of ammonia and nitrate nitrogen on nitrogenase activity of pea plants as affected by light intensity and sugar addition. Plant Soil. 54:271–282.

Huang, D.M., J.H. Gao, and P.L. Zhu. 1981. The transformation on nitrogenase activity of pea plants as affected by light intensity and sugar additiona. Plant Soil. 54:271–282.

Hunter, M.N., P.L.M. de Jabrun, and D.E. Byth. 1980. Response of nine soybean lines to soil moisture condition close to saturation. J. Exptl. Agric. Amim. Hus. 20:339–345.

IRRI. 1986. Effect of Rhizobium inoculation method on nodulation and nitrogen fixation by *Sesbania rostrata*. Annual Report. International Rice Research Institute. P.O. Box 933, Manila, Philippines.

Ito, O., and I. Watanabe. 1985. Availability to rice plants of nitrogen fixed by *Azolla*. Soil Sci. Plant Nutr. 31:91–104.

Jenik, J., and J. Kubikova. 1969. Root systems of tropical trees. Peslia, Prague 14:59–65.

Jiao, Bin. 1983. Utilization of green manure for raising soil fertility in China. Soil Sci. 135:65–69.

Kalidurai, M., and S. Kannaiyan. 1989. Effect of *Sesbania rostrata* on N uptake and yield of lowland rice. J. Agro. Crop. Sci. 163:284–288.

Kretschsmer, A.E., Jr., and R. Bullock. 1980. *Aeschynomene* spp. Distribution and potential use. Soil Crop Sci. Soc. of Florida. 39:145.

Ladha, J.K., I. Watanabe, and S. Soano. 1988. Nitrogen fixation by leguminous green manure and practice. In Sustainable Agriculture. Green Manure in Rice Farming. pp. 165–183. International Rice Research Institute, P.O. Box 933, Manila, Philippines.

Ladha, J.K., M. Garcia, S. Miyan, A.T. Padre, and I. Watanabe. 1989a. Survival of Azorhizobium caulinodans in the soil, rhizosphere of wetland rice under *Sesbania rostrata* rice rotation. Applied and Environ. Microbiol. 55:454–460.

Ladha, J.K., Garcia, M., Pareek, R.P., and G. Rarivoson. (1992). Relative contributions to nitrogenanse (acetylene-reducing) activity of stem and root nodules in *Sesbania rostrata*. Canadian J. Microbiol. 38:00–00. (In Press)

Ladha, J.K., S. Miyan, and M. Garcia. 1989b. *Sesbania rostrata* green manure for lowland rice: Growth, N fixation *Azorhizobium* spp. inoculation and effects on suceeding crop yields and nitrogen balance. Biol. Fert. Soils. 7:191–197.

Ladha, J.K., R.P. Pareek, R. So, and M. Becker. 1990. Stem nodule symbiosis and its unusual properties. In P.M. Gresshoff, L.E. Roth, C. Stacy and W.L. Newton (eds.). N_2 fixation: Achievements and Objectives. pp. 633–640. Chapman & Hall. New York.

Legocki, R.P., A.R.J. Eaglesham, and A.A. Szalay. 1983. Stem nodulation in Aeschynomene: A model system for Bacterium-Plant interaction. In A. Puhler (ed.). Molecular Genetics of the Bacteria-Plant Interaction. pp. 210–219. Springer Verlag, Berlin, Heidelberg.

Liu, Chungchu. 1988. Integrated use of green manure in rice fields in South China. In Sustainable agriculture. Green manure rice farming. pp. 319–331. International Rice Research Institute, P.O. Box 933, Manila, Philippines.

Ludwig, R.A. 1988. Regulation of *Azorhizobium caulinodans* ORS 571 N_2 fixation. In Proceedings of an International Symposium on *S. rostrata*, Dakar, Senegal.

Lynd, L.Q., E.A. Hanlon, Jr., and G.V. Odell, Jr. 1984. Nodulation and nitrogen fixation by arrowleaf clover. The effect of phosphorus and potassium. Soil Biol. Biochem. 16:589–594.

Meelu, O.P., and R.A. Morris. 1988. Green manure management in rice based cropping systems. In Sustainable agriculture. Green manure rice farming. pp. 210–222. International Rice Research Institute, P.O. Box 933, Manila, Philippines.

Mokwunye, A., S.H. Chien, and E. Rhodes. (1986). Phosphate reactions with tropical soils. In A. Mokwunye and P.L.G. Vlek (eds.). Management of nitrogen and phosphorus fertilizers in sub-sharan Africa. pp. 253–281. Nijhoff/ Junk Publishers, The Hague, Netherlands.

Moody, K., C.E. Munroe, R.T. Lubigan, and E.C. Paller, Jr. 1984. Major weeds of the Philippines at Los Banos, Laguna, Philippines. pp. 151–152.

Morris, R.A., R.E. Furoc, and M.A. Dizon. 1986. Response to a short duration green manure I. Grain Yield Agron. J. 78:409–412.

Morris, R.A., R.E. Furoc, N.K. Rajbhandri, E.P. Marqueses, and M.A. Dizon. 1989. Rice response to waterlog tolerant green manures. Agron. J. 81:803–809.

Mulongoy, K., and N.V. Nguu. 1988. Use of *Sesbania rostrata* in alley cropping systems. Presented at the workshop on "*Sesbania rostrata* Fundamental Research and Agricultural Applications" held at Dakar, Senegal on Jan. 4–8, 1988.

Munns, D.N. 1977. Mineral nutrition and legume symbiosis. In R.W.F. Hardy and A.H. Gibson (eds.). A Treatise on Dinitrogen Fixation. Sec. IV. Agron. 4. Ecol. Wiley, New York.

Nagarajah, S. 1988. Transformation of green manure nitrogen in lowland rice soils. In Sustainable agriculture. Green manure in rice farming. pp. 193–208. International Rice Research Institute. P.O. Box 933, Manila, Philippines.

Nambiar, P.T.C., P.J. Dart, B. Srinivasa Rao, and V. Ramanatha Rao. 1982. Nodulation in the hypocotyl region of ground nut (*Arachis hypogaea*) Exptl. Agric. 18:203–207.

N doye, I., and Dreyfus, B. 1988. N_2 fixation by *Sesbania rostrata* and *Sesbania sesban* estimated using ^{15}N and total N difference methods. Soil Biol. Biochem 20:209–213.

O'Hara, G.W., N. Boonderd, and M.J. Dilworth. 1988. Mineral constraints to nitrogen fixation. Plant Soil 108:93–110.

Pareek, R.P., J.K. Ladha, and I. Watanabe 1990. Estimating N_2 fixation by *Sesbania rostrata* and *S. cannabina* (Syn. *S. aculeata*) in lowland rice soil by ^{15}N dilution method. Biol. Fertil. Soils. 10:77–88.

Pate, J.S. 1977. Functional biology of dinitrogen fixation by legumes. In R.W. Hardy and W.F. Silver (eds.). A Treatise on Dinitrogen Fixation. Sec. III. pp. 473–517. Wiley, New York.

Patterson, T.G., and T.A. LaRue. 1983. N_2 fixation (C_2H_2) and ureide content of soybeans: Environmental effect and source-ink manipulations. Crop Sci. 23:819–824.

Pawlowski, K., P. Ratet, J. Schell, and F.J. De Bruijn. 1987. Cloning and characterization of *nif* A and *ntr* C genes of the stem-nodulating bacterium ORS 571, the nitrogen-fixing symbiont of *S. rostrata*. Mol. Gen. Genet. 206:207–219.

Pillai, P.S. 1958. Cultural trials and practices of rice in India. Ind. Council of Agric. Res., Monograph No. 27. New Delhi.

Pradhan, P. 1988. The establishment requirements of Sesbania rostrata as pre-rice green manure. Ph.D. Thesis, University of the Philippines, Los Banos, Laguna, Philippines.

Qi-Xiao, Wen. 1984. Utilization of organic materials in rice production in China. In Organic matter and rice. pp. 45–55. International Rice Research Institute, P.O. Box 933, Manila, Philippines.

Queensberry, K.H., S.L. Abrecht, and J.M. Bennett. 1982. Nitrogen fixation and forage characterization of *Aeschynomene* spp. in a subtropical climate. In P.H. Graham and S. Harris (eds.). BNF Technology for Tropical Agriculture. pp. 347-360. Centro Int. de Agricultural Tropical, Cali, Colombia.

Ramaswamy, P., and A.K. Bal. 1986. Media-induced changes in the assymbiotic nitrogen fixation bacteroids of *Bradyrhizobium* sp. 32H1. Curr. Microbiol. 14:81–185.

Rinaudo, G., and A. Moudiongui. 1987. Fixation d'azote par *Sesbania rostrata* son utilization comme engrais vert. Bulletin de Recherche Agronomiques Gembloux.

Rinaudo, G., B. Dreyfus, and Y. Dommergues. 1983. Sesbania green manure and the nitrogen content of rice crop and soil. Soil Biol. Biochem. 15:111–113.

Rinaudo, G., D. Alazard, and M. Moudiongui. 1988. Stem-nodulating legumes as green manure for rice in West Africa. In Sustainable Agriculture. Green manure rice farming. pp. 98–109. International Rice Research Institute. P.O. Box 933, Manila, Philippines.

Roger, P.A., and I. Watanabe. 1986. Technologies for using biological nitrogen fixation in wetland rice: Potentialities, current stage and limiting factors. Fertil. Res. 9:39–77.

Ruinen, J. 1961. The phyllosphere, An ecological neglected milion. Plant Soil 15:81–109.

Saint Macary, H., E.P. Marqueses, R.O. Torres, and R.A. Morris. 1984. Effect of flooding on growth and nitrogen fixation of two *Sesbania* species. Paper presented at the 15th Annual Scientific Meeting of the Crop Science Society of the Philippines, 16–18 May, 1984. Batac, Philippines.

Salam, M.A., S.M.S. Hameed, P. Sivaprasad, E. Tajuddin, and Y. Thomas. 1989. Performance of *Sesbania rostrata* in acid soils. Int. Rice Res. Newsl. 14:33–34.

Schaede, R. 1940. Die Knollchen der adventiven Wasser Wurzeln von *Neptunia oleracea* und ihre Bakteriensymbiose. Planta. 31:1–21.

Sekhon, H.S., J.N. Kaul, and B.S. Dahiya. 1978. Response of lentil to *Rhizobium* inoculation and N fertilizer. J. Agric. Sci. 90:325–327.

Shi, L.S., J.N. Kaul, and B.S. Dahiya. 1978. The availability of nitrogen of green manures in relation to their chemical composition. Acta Pedol. Sin. 17:240–246.

Stowers, M.D., and A.R.J. Eaglesham. 1983. A stem-nodulating *Rhizobium* with physiological characteristics of both fast and slow growers. J. Gen. Microbiol. 129:3651–3655.

Subba Rao, N.S. 1988. Microbiological aspects of green manure in lowland rice soil. In Sustainable Agriculture. Green manure rice farming. pp. 132–149. International Rice Research Institute. P.O. Box 933, Manila, Philippines.

Subba Rao, N.S., and M. Yatazawa. 1984. Stem nodules. In N.S. Subba Rao (ed.). Current developments in biological nitrogen fixation. pp. 101–110. Oxford and IBM Publishing Co., New Delhi.

Subba Rao, N.S., K.B.V.R. Tilak, and S. Singh, Sc. 1980. Root nodulation studies on *Aeschynomene aspera*. Plant Soil. 56:491–494.

Suessenguth, K., and R. Beyerle. 1936. Über Bakterienknöllchen am Sproß von *Aeschynomene paniculata*. Wild. Hedwigia 75:234–237.

Thomas, M., S.L. Ranson, and J.A. Richardson. 1975. Plant physiology. 5th ed. pp. 299–301. Longman Group Ltd. London.

Trichant, J.C., and J. Rigaud. 1987. Acetylene-reduction by bacteroids isolated from stem nodules of *Sesbania rostrata*. Specific role of lactate as an energy-yielding substrate. J. Gen. Microbiol. 133:37–43.

Tsien, H.C., B.L. Dreyfus, and E.L. Schmitt. 1983. Initial stages in the morphogensis of nitrogen-fixing stem nodules of *Sesbania rostrata*. J. Bacteriol. 150:888–897.

Urban, J.S., L.C. Davis, and S.J. Brown. 1986. *Rhizobium trifolii* 0403 is capable of growth in the absence of combined nitrogen. Appl. Environ. Microbiol. 52:1060–1067.

Vacchani, M.V., and K.S. Murty. 1964. Green-manuring for rice. Research Report Series No. 17. ICAR.

Van Egeraat, A.W.S.M. 1975. The possible role of homoserine in the development of *Rhizobium leguminosarum* in the rhizosphere of pea seedlings. Plant Soi. 42:381–386.

Ventura, W., G.B. Mascarinia, R.E. Furoc, and I. Watanabe. 1987. *Azolla* and *Sesbania* as biofertilizers for lowland rice. Philipp. J. Crop Sci. 12:61–69.

Visperas, R.M., R.E. Furoc, R.A. Morris, B.S. Vergara, and G. Patena. 1987. Flowering response of *Sesbania rostrata* to photoperiod. Philipp. J. Crop. Sci. 12:147–149.

Yadav, N.K., and S.R. Vyas. 1971. Response of root nodule rhizobia to saline alkaline and acidic conditions. Indian J. Agric. Sci. 41:875–881.

Yatazawa, M., and S. Yoshida. 1979. Stem nodules in *Aeschynomene indica* and their capacity of nitrogen fixation. Plant Physiol. 45:293–295.

Yatazawa, M., G.G. Hambali, M. Wiriadinata, M. Susilo, and F. Uchino. 1987. Nitrogen fixing stem nodules and stem-warts of tropical plants. Biotrop. Spec. Publ. No. 31:191–205.

Yoneyama, T. 1990. Use of natural ^{15}N abundance for evaluation of N_2 fixation by legumes and *Azolla* in remote fields. In Proceedings of 14th International Congress of Soil Science, Commission IV. pp. 126–130.

Yoshida, S. 1981. In Fundamentals of crop science. International Rice Research Institute, P.O. Box 933, Manila, Philippines.

Zablotowig, R.M., and D.D. Foch. 1979. Denitrification and anaerobic nitrate dependent acetylene reduction in cowpea *Rhizobium* J. Gen. Microbiol. 3:445–448.

Analysis of Lognormal Data

T.B. Parkin and J.A. Robinson

1992 by Springer-Verlag New York, Inc.
Advances in Soil Science, Volume 20

I. Introduction

The field of statistics exists for the purpose of improving the quality of decisions made from data, and further, to improve the quality of data obtained through experimentation. The field of statistics can be divided roughly into two areas: estimation and comparison. In estimation, the goal is to construct an estimate for a certain attribute (e.g., the rate of microbial respiration for a particular soil) along with some measure of the uncertainty of that estimate. Classic examples include the mean and standard error of the mean and a confidence interval. Testing differences between soils with respect to the rate of denitrification is an example of comparison. The need to make comparisons between batches of numbers (implying populations of attributes) provided much of the impetus for the development of hypothesis-testing methodology currently used by practitioners of statistics. While differing in mechanics, both of these branches of statistics serve to incorporate the impact of uncertainty on decisions made regarding a measured variable. The purpose of this chapter is to delineate methods for estimation and making comparisons between soil processes and properties that exhibit a particular frequency distribution, namely, the two-parameter lognormal distribution. This chapter is not a general review of classical statistical methods and data analytic approaches. There are several monographs and review papers that have effectively accomplished this task already. Rather, we have focused on one area of statistics, namely, the estimation and comparison problems where the observed data appear to conform to the two-parameter lognormal family of distributions. This skewed distribution approximates many of the soils variables that have been reported to be highly and positively skewed. Of course, other positively skewed distributions have been described and employed for data analysis purposes. However, the two-parameter lognormal family has been widely used by biologists and soil scientists, and it is this audience for which this chapter was intended. Indeed, Johnson and Kotz (1970) have stated that "It is quite likely that the lognormal distribution will be one of the most widely applied distributions in practical statistical work in the near future." Given the practical utility of this family of distributions we feel that a methods chapter describing the mechanics of estimating lognormal parameters and comparing batches of lognormally distributed data fills a need for environmental scientists in general and soil scientists in particular.

A. Lognormally Distributed Variables in Nature

Environmental data often exhibit a high degree of variability. This high variability is not "unnatural" but is a characteristic of natural processes. Nature is heterogeneous, especially at the spatial scale at which microorganisms are operating. Atlas and Bartha (1987) have succinctly emphasized this point by stating that statistics is essential "in the field of microbial ecology because of the high degree of variability in natural ecosystems and the heterogeneity of distribution of microorganisms within habitats." The high variability of natural processes translates into high uncertainty with regard to the statistical estimation and inference processes. In more specific terms, high uncertainty results in wide confidence intervals for point estimators of location (e.g., the sample mean or median) and scale (e.g., the standard deviation or variance), which results in low power associated with hypothesis-testing procedures. This suggests that better quantification of variability, or at least the recognition that high variability exists, is necessary to optimize the statistical and experimental design.

In many instances, the high variability exhibited by natural processes is manifested in positively skewed distributions. Positively skewed distributions are a consequence of the fact that many environmental variables cannot take on negative values and are, hence, constrained by zero. Several skewed probability density models have been used to describe environmental data, including the Poisson, negative binomial, Weibull, gamma, exponential, and lognormal. Of these distributions, the lognormal has been most commonly applied. Table 1 shows an abbreviated list of natural variables that can be approximated by a lognormal distribution. Because of the relatively widespread application of the lognormal distribution to environmental variables, the following discussion will focus on the analysis of lognormally distributed data.

B. Characteristics of the Lognormal Distribution

The properties of the lognormal distribution have been extensively investigated by mathematical statisticians (Finney, 1941; Sichel, 1952; Aitchinson and Brown, 1957; Johnson and Kotz, 1970). One reason for this is that a simple transformation will convert a lognormally distributed variable to a Gaussian distribution, for which the statistical properties are well known. Discussions of the application of lognormal distributions to environmental data have also recently appeared (Gilbert, 1987; Crow and Shimizu, 1988).

A variable is considered to be lognormally distributed if the logarithm of the variable is normally distributed. In practical terms, it does not make any difference whether $\log_{10}$ or $\log_e$ is used since log values to different bases are linearly related. For all of the examples presented in this chapter the natural logarithm (base e) has been used.

1. The Lognormal Probability Density Function

The equation describing the probability density function for the three parameter lognormal distribution is presented in Eq. 1.

Table 1. Abbreviated listing of abiotic and biological variables that have been observed to be lognormally distributed

Variable	References
Environmental pollution	Mage (1981)
	Georgopoulos and Seinfeld (1982)
	Kushner (1976)
Soil aggregate size	Allmaras et al. (1965)
Soil water flux	Jones and Wagenet (1984)
Soil hydraulic conductivity	Warrick et al. (1977)
Soil nitrate	Cameron and Wild (1984)
	Ruess et al. (1977)
	Tabor et al. (1985)
	White et al. (1987)
Total soil nitrogen	Sells et al. (1986)
Soil pore CO_2 concentrations	Focht et al. (1979)
Soil pore N_2O concentrations	Focht et al. (1979)
Soil respiration	Parkin and Robinson (1989)
Soil denitrification	Folorunso and Rolston (1984)
	Parkin et al. (1985)
	Parkin et al. (1987)
	Christensen et al. (1990)
Bacterial numbers	
Plant leaves	Hirano et al. (1982)
Rhizosphere	Loper et al. (1984)
Aquatic environments	Standard Methods (1985)
	Christian and Pipes (1983)
Nemetodes and Cestodes	Malhotra (1984)
Species numbers	Bonsdorff and Koivisto (1982)
Species size	Ugland and Gray (1982)
	Lewin (1987)

$$p[x] = \frac{1}{(x-\theta)\sigma\sqrt{2x}} \exp\left[-\frac{1}{2\sigma^2}(\ln(x-\theta)-\mu)^2\right] \tag{1}$$

where μ is the mean of the ln-transformed variable, σ^2 is the variance of the ln-transformed variable, and θ is the location shift parameter. In Eq. 1 the location shift parameter (θ) has no influence on the shape of the probability density curve, but simply influences where the curve is located along the x-axis. Figure 1 shows the influence of changing θ on the density function.

When θ is equal to 0 the three parameter lognormal distribution reduces to the two-parameter distribution (Eq. 2). In the following material only the two-parameter model will be discussed.

$$p[x] = \frac{1}{x\sigma\sqrt{2x}} \exp\left[-\frac{1}{2\sigma^2}(\ln x-\mu)^2\right] \tag{2}$$

where μ and σ^2 are as described above.

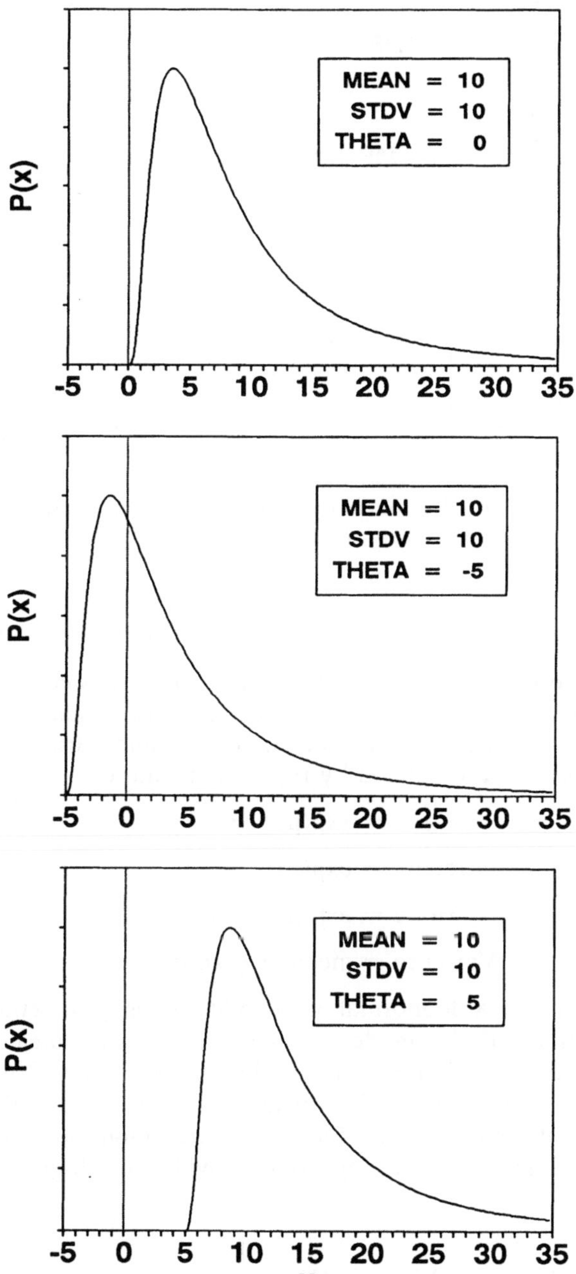

Figure 1. The three-parameter lognormal distribution.

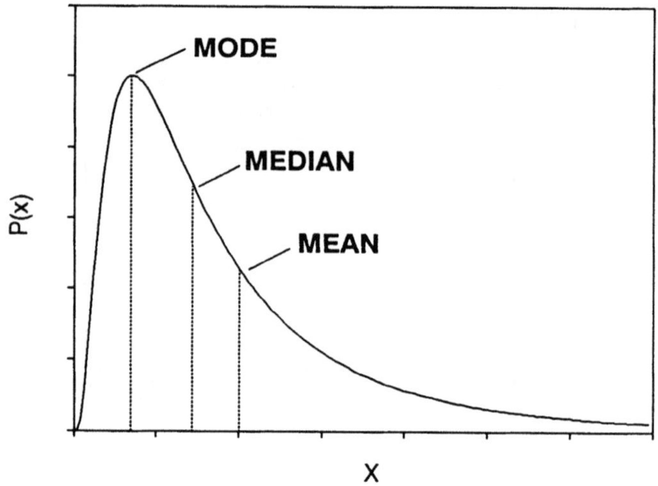

Figure 2. The lognormal probability density function showing the relative locations of the population mean, median, and mode.

2. Location and Scale Parameters (Mean, Median, Mode, Variance)

Unlike symmetric distributions, because of the skewed nature of the lognormal distribution, the mean, median, and mode have different values. The population mean, median, mode, and variance of the lognormal distribution are defined by Eq. 3, 4, 5, and 6.

$$\text{Mean} = \exp(\mu + \sigma^2/2) \tag{3}$$

$$\text{Median} = \exp(\mu) \tag{4}$$

$$\text{Mode} = \exp(\mu + \sigma^2) \tag{5}$$

$$\text{Variance} = \text{mean}^2 * (\exp(\sigma^2) - 1) \tag{6}$$

Figure 2 shows a lognormal probability density function with the relative positions of the mode, median, and mean indicated. In most instances the two location parameters that are of interest are the median and the mean. From Figure 2 it is evident that the value of the median is less than that of the mean. This fact has several implications in parameter estimation and hypothesis testing, some of which are discussed in Section IV.

C. Genesis of Lognormal Distributions

It is important to consider the underlying factors resulting in the observed distribution of the variable under study. The rational for this view is presented by Kapteyn (1903): "Knowledge of the connection between the

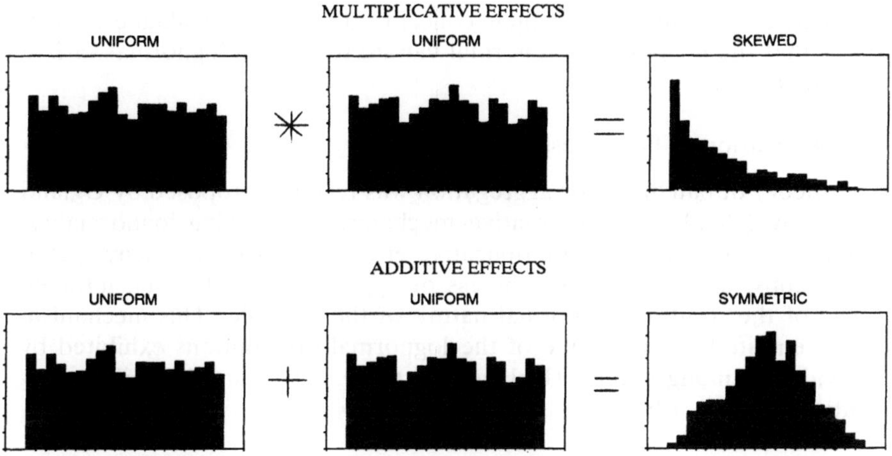

Figure 3. Multiplicative and additive effects on distributions.

form of the curves and the causes to which the form is due may lead us, in some cases, to precious indications about the nature of the true causes."

There have been three mechanisms proposed as the source of lognormal distributions. These mechanisms are: (1) breakage, (2) multiplicative effects, and (3) aggregation. These three mechanisms are discussed briefly below.

1. Breakage

Aitchinson and Brown (1957) present a detailed mathematical discussion of the theories underlying breakage. This mechanism has its foundations based on the Law of Proportionate Effect, which states that "A variate subject to a process of change is said to obey the law of proportionate effect if the change at any step of the process is a random proportion of the previous value of the variate" (Aitchinson and Brown, 1957). This mechanism of Breakage has been evoked as the source of the lognormal distributions displayed by geological deposits.

2. Multiplicative Effects

Multiplicative effects as a source of lognormal distributions has theoretical foundations based on the Central Limit Theorem (discussed by Aitchinson and Brown, 1957). Briefly stated, this mechanism predicts that a variable defined as the product of a number of elementary variables tends to be lognormally distributed. A simple example of this can be observed by considering Figure 3. The product of randomly selected, uniformly distributed variables yields a skewed distribution. Conversely, the sum of randomly selected, uniformly distributed variables yields a symmetric distribution. A similar concept advanced by Haldane (1942), states that the nth power of a normal variate tends to a lognormal variate as n increases.

3. Aggregation or Patchiness

The theory of patchiness or aggregation was recently proposed by Ugland and Gray (1982) as an alternative mechanism underlying lognormality. Briefly stated it says that multiplicative effects give rise to patchy spatial dispersions, and it is this patchiness or aggregation that is the ultimate cause of the resulting lognormal nature of the variable. This mechanism has been cited as the source of the lognormal distributions exhibited by individuals among species (Ugland and Gray, 1982) and by soil denitrification (Parkin, 1987).

II. Diagnosing Lognormality

A. Sample Statistics as Indicators of Asymmetry

A key characteristic of the lognormal probability model is its asymmetry. Several indications or clues of asymmetry in a given data set can be gleaned from inspection of the descriptive sample statistics. This can be illustrated by considering sample data (Table 2). In the first column of Table 2 are presented 16 measurements of ponded infiltration and in column 2 the logarithms of the sample values are presented. Also shown in Table 2 are the sample mean ($\bar{x}$), sample standard deviation (s), coefficient of variation (cv), and the sample geometric mean which is the antilogarithm of the average of the ln-transformed values ($\exp(\bar{y})$). The wide difference between the sample mean (62.83) and the sample geometric mean (31.44) indicates that asymmetry exists, and a lognormal model may be appropriate. Another clue suggesting that asymmetry may exist is the high coefficient of variation.

B. The Sample Histogram

Asymmetry in a data set can also be visualized by graphical techniques. One such technique is to construct a sample histogram, overlaid by a plot of the probability density function (Eq. 2) using $\bar{y}$ and $\hat{\sigma}^2$ in place of μ and σ^2. Figure 4 shows a sample histogram of the untransformed infiltration data (column 1, Table 2). From Figure 4 it is evident that a high proportion of samples have low values, but a few samples exhibit extreme values. If the data are lognormally distributed then the histogram of the ln-transformed values will exhibit a Gaussian distribution (Figure 5). Plotting the sample histogram is useful in visualizing skewness but does not necessarily confirm that the sample data are indeed from a lognormal

Table 2. Ponded infiltrations rates measured at 16 locations in a corn/soybean field in central Iowa (from unpublished data of S. Logsdon)

Infiltration (μm/s)	ln(infiltration)
5.9	1.775
7.0	1.946
12.5	2.526
12.7	2.542
12.9	2.557
15.5	2.741
21.6	3.073
22.8	3.127
28.7	3.357
29.1	3.371
56.2	4.029
62.3	4.132
67.2	4.208
113	4.727
179	5.187
359	5.883
$\bar{x} = 62.8$	$\bar{y} = 3.448$
$s^2 = 8370$	$\hat{\sigma} = 1.353$
$s = 91.5$	$\exp(\bar{y}) = 31.4$
$\%cv = 146$	

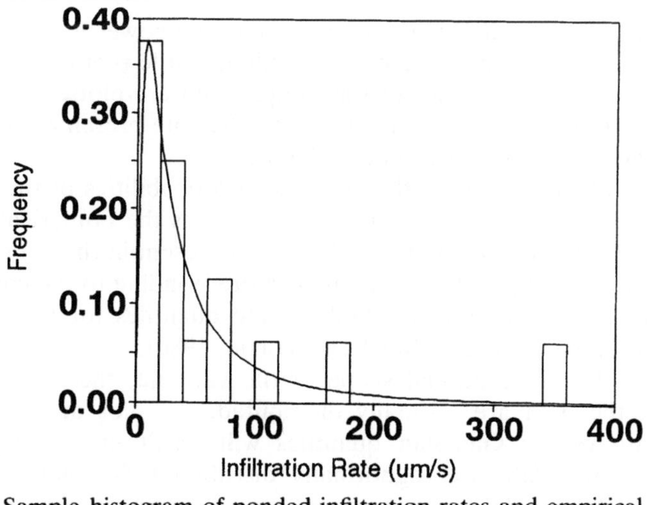

Figure 4. Sample histogram of ponded infiltration rates and empirical lognormal density function.

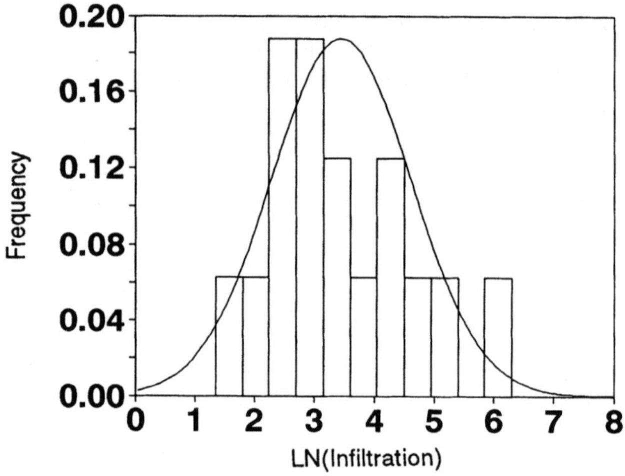

Figure 5. Sample histogram of ln-transformed infiltration rate data and empirical normal density function.

distribution, as the shape of the histogram is influenced, in part, by how the data are distributed into the different size classes during construction of the histogram. Computer programs that compute sample histograms and plot lognormal probability density functions are available (USDA Technology Transfer Documents NSTL91-4 and NSTL91-5).

C. Probit Analysis

Another graphical method with greater diagnostic power is a probit plot. Probit analysis is a special case of a quantile-quantile plot (Miller, 1986), where the empirical quantiles of the sample data are plotted against the expected quantiles for a given distribution. The data presented in Table 3 illustrate the application of probit analysis.

Column 3 of Table 3 gives the cumulative probabilities of the ordered data that are calculated by $(i - 0.5)/n$, where i is the numerical rank of the ith sample and n is the number of samples present in the data set. The theoretical quantiles (Table 3, column 4) corresponding to the cumulative probabilities can be found in a table of the quantiles for the Gaussian distribution (e.g., Table A4; Steel and Torrie, 1980).

To assess the distributional shape of the variable, the data are then plotted. If the data are normally distributed, the plot of the untransformed data vs the Gaussian quantiles will result in a straight line. However, if the data are lognormally distributed the plot of the ln-transformed data vs the Gaussian quantiles will be linear. Figure 6 shows a probit plot of the untransformed and ln-transformed data presented in

Table 3. Untransformed and ln-transformed infiltration data of Table 2 used in constructing probability plot

Infiltration (μm/s)	ln(infiltration)	Cumulative probability	Gaussian quantile
5.900	1.775	0.031	−1.862
7.000	1.946	0.094	−1.318
12.500	2.526	0.156	−1.010
12.700	2.542	0.219	−0.776
12.900	2.557	0.281	−0.579
15.500	2.741	0.344	−0.402
21.600	3.073	0.406	−0.237
22.800	3.127	0.469	−0.078
28.700	3.357	0.531	0.078
29.100	3.371	0.594	0.237
56.200	4.029	0.656	0.402
62.300	4.132	0.719	0.579
67.200	4.208	0.781	0.776
113.000	4.727	0.844	1.010
179.000	5.187	0.906	1.318
359.000	5.883	0.969	1.862

Table 3. The linearity of the relationship between the ln-transformed infiltration data of Table 3 and the Gaussian quantiles indicates that the infiltration data are well approximated by a lognormal distribution (Figure 6). A computer program to calculate quantiles for probit analysis is available from the authors (USDA Technology Transfer Document NSTL91-2).

D. Goodness-of-Fit Tests

Graphical methods such as sample histograms and probit plots only provide indications of lognormality. There are several statistical methods available for testing normality, and when these tests are performed on ln-transformed data they can be used as tests of lognormality.

The chi-square goodness-of-fit test is perhaps the most widely applied; however, this test is only applicable to discrete distributions. For this and other reasons the chi-square test is not as appropriate for testing lognormality as other available tests.

The Kolomogorov-Smirnov test (Daniel, 1978) was developed for use with continuous distributions. Two forms of this test exist: a one-sample goodness-of-fit test and a two-sample goodness-of-fit test. In the one-sample test a hypothesized cumulative distribution is compared with the sample distribution. The two-sample test compares the observed distributions from two samples. These tests are more powerful than the chi-

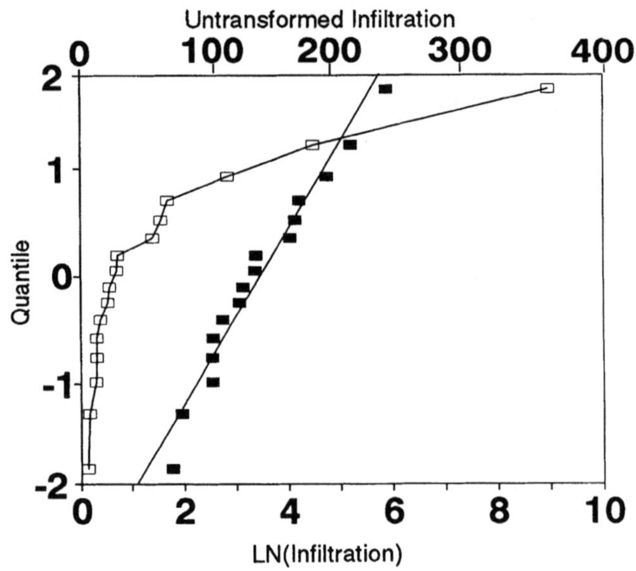

Figure 6. Probit plot of infiltration data from Table 3.

square test. However, the one-sample test is not valid if the hypothesized distribution is estimated from the sample data (Conover, 1980). For this reason, the Kolomogorov-Smirnov one-sample test is not very useful for diagnosing lognormality. The two-sample Kolomogorov-Smirnov test, however, is a powerful procedure for detecting differences between two distributions. It is considered an omnibus test since it is sensitive to all types of differences that may exist.

For testing normality (or lognormality) Gilbert (1987) recommends the use of the *W* test developed by Shapiro and Wilk (1965). This test has good power for detecting departures from normality; however, tables have only been developed for sample sizes ≤ 50. The D'Agostino test (D'Agostino, 1971) is similar to the *W* test, and tables of the test statistic have been developed for $50 < n < 1000$. A description of the implementation of these tests is presented below. In addition, a computer program that implements these two tests is available (USDA Technology Transfer Document NSTL91-2).

E. Shapiro-Wilk's *W* Test

The ln-transformed infiltration data presented in Table 2 are used as an example in the description of this test.

Step 1: Compute the denominator of the *W* statistic.

$$d = \sum_{i=1}^{n} x_i^2 - \frac{1}{n}\left[\sum_{i=1}^{n} x_i\right]^2$$

$d = 20.255$ for the ln-transformed infiltration data of Table 2

Step 2: Sort the data from smallest to largest (this has already been done in Table 2).

Step 3: Compute k.

$$k = \frac{n}{2}, \qquad \text{if } n \text{ is even}$$

$$k = \frac{(n-1)}{2}, \qquad \text{if } n \text{ is odd}$$

For the data in Table 2:

$$k = \frac{16}{2} = 8$$

Step 4: Find the coefficients $a_1, a_2, \ldots a_k$, corresponding to the $n = 16$ samples available (Table A1). For the example data the coefficients are:

$$a_1 = .5056 \; a_2 = .3290 \; a_3 = .2521 \; a_4 = .1939$$
$$a_5 = .1447 \; a_6 = .1005 \; a_7 = .0593 \; a_8 = .0193$$

Step 5: Compute the W statistic.

$$W = \frac{1}{d}\left[\sum_{i=1}^{k} a_i * (x_{n-(i+1)} - x_i)\right]^2$$

For the example data this equation becomes:

$$\begin{aligned} W = \frac{1}{20.255} & [(.5056)(5.883 - 1.775) + (.329)(5.187 - 1.946) \\ & + (.2521)(4.727 - 2.526) + (.1939)(4.208 - 2.542) \\ & + (.1447)(4.132 - 2.557) + (.1005)(4.029 - 2.741) \\ & + (.0593)(3.371 - 3.073) + (0.0193)(3.357 - 3.127)]^2 \end{aligned}$$

$$W = 0.9563$$

Step 6: Reject the null hypothesis if W is less than the quantile given in Table A2 for the given sample n and α level.

H_o = the population is normally distributed.
H_A = the population is not normally distributed.

From Table A2 the quantile for $\alpha = 0.05$ and $n = 16$ is 0.887. Since the W for the ln-transformed infiltration data (0.9563) is greater than 0.887 we cannot reject the null hypothesis and conclude that the ln-transformed data are normally distributed (i.e., the untransformed data are lognormal).

F. D'Agostino's Test

This test compliments the Shapiro-Wilk test and is used for $n \geqslant 50$ (up to $n = 1000$).

H_o = the underlying population is normally distributed.
H_A = the underlying population is not normally distributed.

Step 1: Order the n data from smallest to largest

Step 2: Compute the D statistic.

$$D = \frac{\sum_{i=1}^{n} [i - 1/2\,(n + 1)]x_i}{n^2 s}$$

where,

$$s = \left[\frac{1}{n}\sum_{i=1}^{n} (x_i - \bar{x})^2\right]^{1/2}$$

Step 3: Transform the D statistic to the Y statistic.

$$Y = \frac{D - 0.28209479}{0.02998598/\sqrt{n}}$$

Step 4: Reject the null hypothesis if Y is less than the $\alpha/2$ quantile or greater than the $(1 - \alpha/2)$ quantile for the given n from Table A3.

III. Estimating Population Parameters from Sample Data

A. General Considerations

In most environmental studies it is impossible to sample the entire population of the variable of interest. Thus, we are forced to estimate the parameters of the underlying population from sample data. Several statistical procedures for estimating these population parameters have been applied to lognormally distributed data. However, a statistically complete evaluation has not been summarized in a source accessible to most soil scientists. Even in cases where the observed distribution of a given variable appears to be lognormal, confusion exists on how to best estimate the population mean, median, and variance from sample data (e.g., Warrick and Nielsen, 1980; Koch and Link, 1970).

An example of this confusion can be found in a popular statistics textbook (Sokal and Rohlf, 1969). In this text it is suggested that the sample mean of a lognormally distributed data set should be calculated by taking the antilogarithm of the average of the log-transformed sample values. From the preceding discussion of the general characteristics of the lognormal distribution, it is evident that this is the sample geometric mean which is an estimator of the population median, not the sample mean (see Eqs. 3 and 4).

The following discussion considers some of the statistical properties of commonly applied estimators of the population mean, median, and variance from lognormally distributed sample data. This discussion serves two purposes: (1) to allow researchers to decide themselves what the trade-offs are between ease of calculation, bias, and efficiency of a given estimation method for analyzing data, as well as (2) to allow researchers and administrators to evaluate past studies with regard to the bias and efficiency of the statistical estimation methods used.

1. Statistical Properties of Estimators, Bias, and Efficiency

In the evaluation of statistical estimators two criteria are commonly used, bias and efficiency (Barnett, 1973). Bias refers to the systematic under- or overestimation of the population parameter. However, bias alone is not a sufficient criterion to evaluate a statistical estimator as the variance associated with the estimator may be so large as to preclude its use. For example, while the arithmetic average of sample values is an unbiased estimate of the population mean regardless of the shape of the underlying distribution, for the lognormal distribution other unbiased estimators exist that have lower associated variance. The variance associated with an estimator is known as the efficiency. Ideally, the statistical estimator of choice is the estimator with zero bias and low variance (high efficiency). However, a situation often occurs when a choice must be made between estimators that are unbiased but inefficient or, conversely, biased yet efficient. In these situations another criterion that can be used in distinguishing estimators is mean square error (MSE). The MSE is a combination of the bias and efficiency and is equal to the $\text{Bias}^2 + \text{Efficiency}$.

Recently Parkin et al. (1988) investigated the statistical efficacy of three classes of estimators for estimating the population mean, variance, and coefficient of variation of lognormal data. A description of these estimation methods and a brief discussion of their relative merits follows.

B. Commonly Applied Estimators

Generally, three estimators have been most commonly used. These are: (1) the method of moments (MM), (2) the maximum likelihood method (ML), and (3) the uniformly minimum variance unbiased estimators (UMVUE). A comparison of the bias and relative efficiencies of the MM, ML, and UMVUE estimators of the mean follows.

1. Mean and Variance Estimators

a. Method 1: The Method of Moments

The method of moments is typically applied when a normal frequency distribution is observed or when the sample size is too small to adequately determine the true population mean, but a normal distribution is assumed. In this method the sample mean and variance are taken to be estimators of the population mean and variance. The sample mean and variance are calculated using untransformed data as described by Eqs. 7 and 8, respectively. The variance estimate that appears in Eq. 7 is not strictly a method of moments estimator as the term $n - 1$ rather than n appears in the denominator. The term $n - 1$ is used here because of the known bias that exists for low sample numbers when n is used (Snedecor and Cochran, 1967).

$$m_1 = \frac{1}{n} \sum_{i=1}^{n} x_i \tag{7}$$

$$s_1^2 = \frac{1}{n-1} \sum_{i=1}^{n} (x_i - m_1)^2 \tag{8}$$

where x_i is the untransformed ith observation, n the number of observations, m the estimate of the population mean, and s_1^2 = the estimate of the population variance. Subscripts refer to method 1.

b. Method 2: The Maximum Likelihood Method

The maximum likelihood method has been recommended when the observations are thought to be lognormally distributed (Warrick and Nielsen, 1980). Estimates of the population mean and variance (Aitchinson and Brown, 1957) are presented in Eqs. 9 and 10.

$$m_2 = \exp(\bar{y} + \hat{\sigma}^2/2) \tag{9}$$

$$s_2^2 = m_2^2(\exp(\hat{\sigma}^2) - 1) \tag{10}$$

where $\bar{y}$ and $\hat{\sigma}^2$ are the mean and variance of the ln-transformed sample data given by Eqs. 11 and 12.

$$\bar{y} = \frac{1}{n} \sum_{i=1}^{n} \ln(x_i) \tag{11}$$

and

$$\hat{\sigma}^2 = \frac{1}{n-1} \sum_{i=1}^{n} (\ln(x_i) - \bar{y})^2 \tag{12}$$

c. Method 3: Uniformly Minimum Variance Unbiased Estimators

The uniformly minimum variance unbiased estimators (UMVUE) were developed independently by Finney (1941) and Sichel (1952) and have been typically applied to the analysis of geological data (Krige, 1981; Koch and Link, 1970). Estimators of the population mean and variance are given by Eqs. 13 and 14, respectively.

$$m_2 = \exp(\bar{y})\psi_n(\hat{\sigma}^2/2) \tag{13}$$

$$s_3^2 = \exp(2\bar{y})\ (\psi_n(2\hat{\sigma}^2) - \psi_n\left(\frac{(n-2)}{(n-1)}\hat{\sigma}^2\right) \tag{14}$$

where ψ_n is the power series given in Eq. (15).

$$\psi_n(t) = 1 + \frac{t(n-1)}{n} + \frac{t^2(n-1)^3}{n^2(n+1)2!} + \frac{t^3(n-1)^5}{n^3(n+1)(n+3)3!} + \frac{t^4(n-1)^7}{n^4(n+1)(n+3)(n+5)4!} + \ldots \tag{15}$$

where n is the number of samples.

Depending on n and s^2, 6 to 10 terms of Eq. 15 must be evaluated to achieve stability. A computer program for evaluating this function to achieve a user specified precision is available (USDA Technology Transfer Document NSTL91-3). A brief description of results of the MM, ML, and UMVU estimators of the mean follows. A more detailed discussion is presented by Parkin et al. (1988).

2. Comparison of Estimators

a. Estimating the Mean

In estimating the mean of a lognormal population, the ML estimator (m_2, Eq. 9) has been recommended as the method of choice (Warrick and Nielsen, 1980); however, this estimator is biased unless the number of samples is large. The arithmetic sample average (m_1) is an unbiased estimate of the population mean regardless of the shape of the underlying population distribution. For a lognormal population the UMVU estimator (m_3) is preferred, since it is unbiased and has minimum variance.

b. Estimating the Variance

In a manner similar to the mean, the MM, ML, and UMVU estimators of the variance can also be evaluated by comparing bias and relative efficiencies over a range of sample numbers and populations. The ML estimator of the variance (s_2^2) is asymptotically unbiased but exhibits bias at low sample numbers. The high bias of the ML estimator precludes its use for sample sizes of $n < 100$ when the population exceeds 100%. The MM (s_1^2) and UMVU (s_3^2) estimators are unbiased; however, the UMVU estimator is preferred since it has the highest efficiency (lowest variance).

3. Recommendations

A table of recommendations regarding the application of the three estimators for the population mean is presented (Table 4). These recommendations are based, primarily, on the relative magnitudes of the

Table 4. Summary recommendations of estimation methods for three sampling intensity ranges from three lognormal populations

Lognormal population (%cv)	Sample size (n)	Recommended method[a]	
		Mean	Variance
	4–20	MM, UMVU	MM, UMVU
50	20–40	MM, UMVU	UMVU
	40–100	MM, UMVU, ML	UMVU, ML
	4–20	UMVU, MM	UMVU
100	20–40	UMVU, MM	UMVU
	40–100	UMVU, MM, ML	UMVU
	4–20	UMVU	UMVU
200	20–40	UMVU	UMVU
	40–100	UMVU, ML	UMVU

[a] When more than one method is recommended, the methods are presented in order of most to least preferable.

MSEs of the estimators for the given populations and sample ranges tested. When no substantial differences in the MSEs were apparent, two secondary criteria, (1) bias of the estimator and (2) ease of implementation were used to rank the methods.

4. Estimating the Median

In addition to the mean, the median is a location parameter often used in the analysis of skewed data. Thus, consideration must be given to selection of the optimum estimator of the population median. Two common estimators of the population median are the sample geometric mean (Eq. 16) and the sample median (Eqs. 17 and 18).

a. The Sample Geometric Mean

The sample geometric mean (GM) is the nth root of the product of the n samples. This is equivalent to the antilogarithm of the average of the log-transformed sample values (y) as given in Eq. 16.

$$GM = \exp(\bar{y}) \tag{16}$$

b. The Sample Median

The sample median (SM) is the center value of the ranked (sorted) sample data. If an odd number of samples is available then the median is the sample corresponding to the $(n + 1)/2$ rank. If n is even then the median is calculated as the average of the two center values of the ranked data.

$$SM = x_{(n\,+\,1/2)}, \qquad n = \text{odd} \tag{17}$$

$$SM = [x_{(n/2)} + x_{(n\,+\,2/2)}]/2, \quad n = \text{even} \tag{18}$$

Both the GM and the SM are asymptotically unbiased estimators of the population median but at small n bias exists (Kleijnen, 1987; Landwehr, 1978). Monte Carlo studies reveal that both the *GM* and the *SM* estimators yield positively biased estimates of the true population median and that the extent of bias is dependant on the skewness of the underlying population (Parkin and Robinson, 1992). These studies also report on two alternative estimators of the population median, the UMVU estimator and the bias-corrected geometric mean (BCGM). Descriptions of the implementation of these estimators are presented below.

c. The UMVU Estimator

The UMVU estimator of the median (UM) was developed by Bradu and Mundlak (1970), based on Finney's work (1941), and is calculated according to Eq. 19.

$$UM = \exp(\bar{y})\psi_n(-\hat{\sigma}^2/(2 * (n - 1))) \tag{19}$$

where $\bar{y}$ is the average of the ln-transformed samples and $\hat{\sigma}^2$ is the variance of the ln-transformed samples, and ψ_n is the power function presented in Eq. 15.

d. Bias Corrected Geometric Mean

Gilbert reports that the BCGM can be used as an estimator of the population median if σ^2 is known a priori (Gilbert, 1987). He suggests that the variance of the ln-transformed sample data can be substituted if σ^2 is not known. The BCGM is calculated according to Eq. 20.

$$BCGM = \exp(\bar{y} - \hat{\sigma}^2/2n) \tag{20}$$

where $\hat{\sigma}^2$ is the variance of the ln-transformed sample data.

Both the BCGM and the UMVU estimator are unbiased estimators of the population median and over the range of sample sizes $n = 4$ to 100 the mean square errors associated with these estimators were comparable (Parkin and Robinson, 1992). Thus, the use of these two estimators of the population median is recommended over the GM and the SM.

C. Confidence Intervals About the Mean

It was illustrated in the preceding discussion that knowledge of underlying distribution allows for the selection of better estimators for the mean, median, and variance. However, the real value of indentifying the underlying distribution is in selection of the appropriate method for calculating confidence intervals about the mean.

Historically, confusion has existed concerning the calculation of confidence limits about the mean of a lognormally distributed variable. The primary difficulty with confidence limit calculations is that the mean and the variance of the lognormally distributed variable are not independent. As a result, there have been several recommendations for the calculation of confidence intervals (Koch and Link, 1970). To illustrate this problem a comparison of the three confidence interval methods, as applied to the infiltration data (Table 2), is presented below.

1. Method 1

This is the asymptotic method and is typically applied when the population normally is distributed or the sample size is too small to determine the actual population distribution and a normal distribution is assumed (Snedecor and Cochran, 1967).

$$CL_1 = \bar{x} \pm s * \mathrm{t}_{(1-\alpha/2)}/\sqrt{n} \tag{21}$$

2. Method 2

This is the method recommended by Sokal and Rohlf (1969) and is actually a confidence interval about the geometric mean.

$$CL_2 = \exp(\bar{y} \pm \hat{\sigma} * t_{(1-\alpha/2)}/\sqrt{n}) \tag{22}$$

3. Method 3: Land's Exact Method

A method for the calculation of nearly exact confidence intervals was developed by Sichel (1966). A refinement of this approach was later developed by Land (1973). This method is based on the distribution function of a test statistic (C). Abbreviated tables of Land's (1973) constants for the upper and lower 95% confidence level are presented in Tables A4 and A5 of the Appendix.

$$LCL_3 = \exp(\bar{y} + \hat{\sigma}^2/2 + \hat{\sigma} * C_1/\sqrt{n-1} \tag{23}$$

$$UCL_3 = \exp(y + \hat{\sigma}^2/2 + \hat{\sigma} * C_u/\sqrt{n-1} \tag{24}$$

where $\bar{y}$ is the mean of the ln-transformed sample values, $\hat{\sigma}^2$ is the variance of the ln-transformed sample values, C_1 is the tabulated constant (Table A4) for the lower 95% confidence limit, and C_u is the tabulated constant (Table A5) for the upper 95% confidence limit. When applied to the infiltration data in Table 2 the results in Table 5 are obtained.

Table 5. Upper and lower 95% confidence limits of the mean as calculated by three methods

Method	Lower CL	Upper CL
1	22.7	102.9
2	37.1	102.9
3	36.2	156.2

It is evident from Table 5 that the three methods differ with regard to the calculated upper and lower confidence limits. However, it is impossible to determine which method is providing the correct coverage by simply observing the results from sample data.

In a recent study several methods of confidence interval calculation were evaluated using Monte Carlo simulation (Parkin et al., 1990). The influence of sample number on the efficacy was evaluated by comparing the coverages of the different methods for sample sizes from 6 to 100. These evaluations illustrate the utility of Land's method (Method 3 here) as an exact method.

D. Application of the UMVU Estimators

At first glance, implementation of these estimators may appear to be complex, as in all cases the power series given in Eq. 15 must be evaluated. However, this equation has been converted to machine form (USDA Technology Transfer Document NSTL91-2), which makes the implementation of the UMVU estimators straightforward. Below are presented sample calculations of the UMVU estimators of the mean, variance, and median (Eqs. 13, 14, and 19, respectively), as well as calculations of Land's confidence limit method. These calculations are performed using the sample denitrification rate data shown in Table 6.

1. Calculating the UMVUE Estimator of the Mean

Application of Eq. 13 to calculate the UMVU estimator of the mean first requires calculating the mean and variance of the ln-transformed values using Eqs. 11 and 12:

$$\bar{y} = 2.8 \quad \text{and} \quad \hat{\sigma}^2 = 3.51$$

Next the power series (ψ_n) given in Eq. 15 must be evaluated for $\hat{\sigma}^2/2$, with $n = 36$:

$$\hat{\sigma}^2/2 = 1.755, \quad \text{thus} \quad \psi_n(1.755) = 5.13$$

Now solving for Eq. 13 yields,

$$\exp(\bar{y})\psi_n(\hat{\sigma}^2/2) = (16.4) * (5.13) = 83.6$$

Note that the UMVUE estimate of the mean (418) is different than the arithmetic average of untransformed sample values ($\bar{x} = 56.5$) given in Table 6.

2. Calculating the UMVU Estimate of the Variance

To estimate the population variance Eq. 14 is used. This requires that the ψ function (Eq. 15) be evaluated for both the quantities $2\hat{\sigma}^2$ and $[\hat{\sigma}^2 * (n - 2)/(n - 1)]$:

Table 6. Sample denitrification rate data

Denitrification rate ng-N/(g·day)	ln(denitrification) ln(ng-N/(g·day))
0.06	−2.805
0.87	−0.143
1.04	0.039
1.47	0.385
1.61	0.476
1.87	0.626
2.30	0.833
4.56	1.517
5.56	1.716
7.07	1.956
9.42	2.243
10.00	2.303
11.10	2.407
12.90	2.557
12.90	2.557
13.20	2.580
14.30	2.660
16.80	2.821
19.10	2.950
20.40	3.016
22.70	3.122
37.20	3.616
41.00	3.714
43.30	3.768
48.20	3.875
51.20	3.936
54.10	3.991
55.80	4.022
70.90	4.261
77.70	4.353
106.00	4.663
114.00	4.736
126.00	4.836
226.00	5.421
358.00	5.881
435.00	6.075

$\bar{x} = 56.5$; $\bar{y} = 2.80$; $\hat{\sigma}^2 = 3.51$

$$\psi_n(2\hat{\sigma}^2) = \psi_n(7.02) = 378 \quad \text{and}$$

$$\psi_n[\hat{\sigma}^2 * (n-2)/(n-1)] = \psi_n(3.41) = 21.5$$

Subsequent application of Eq. 14 results in:

$\exp(2\bar{y}) * (378 - 21.5) = (270) * (357) = 96390$ or a standard deviation of $\sqrt{96390} = 31$

The coefficient of variation is then calculated by dividing the standard deviation by the mean:

$$311/83.6 = 3.72, \text{ or } 372\%$$

3. Calculating the UMVU Estimate of the Median

The UMVU estimator of the median is given by Eq. 19, which requires the power function ψ_n to be evaluated for the quantity $(-\hat{\sigma}^2/(2 * (n - 1)))$:

$$(-\hat{\sigma}^2/(2 * (n - 1))) = \psi_n(-.051) = 0.950$$

The estimate of the median from Eq. 19 is then:

$$\exp(\bar{y}) * (.924) = (16.4) * (.950) = 15.6$$

4. Calculating 95% Confidence Limits About the Mean

From simulation studies it has been determined that the method of Land (Method 3) is an exact method and yields confidence intervals which are equal to the intended limits. An illustration of the implementation of Land's (1973) method is presented below.

Using Eq. 23 and 24 to calculate the lower and upper confidence limits about the mean requires $\bar{y}$, $\hat{\sigma}^2$ and the confidence limit factors from Tables A4 and A5 given in the Appendix. Plugging in these values from the denitrification data (Table 6) yields:

$$LCL = \exp(2.80 + 3.51/2 + 1.95 * C_l/\sqrt{(36 - 1)})$$
$$UCL = \exp(2.80 + 3.51/2 + 1.95 * C_u/\sqrt{(36 - 1)})$$

Values of C_l and C_u are interpolated from Tables A4 and A5. For $n = 36$ and $\hat{\sigma} = 1.95$ these correspond to $C_u = 3.622$ and $C_l = -2.255$. Plugging these into the above equations gives:

$$LCL = \exp(2.80 + 3.51/2 + 1.95 * - 2.255/\sqrt{(36 - 1)}) = \exp(3.812) = 45.2$$

$$UCL = \exp(2.80 + 3.51/2 + 1.95 * 3.622/\sqrt{(36 - 1)}) = \exp(5.749) = 314$$

Software to perform these statistical analyses is available (USDA Technology Transfer Document NSTL91-3).

E. Sample Number Reguirements

The preceding sections discussed some of the effects of sample number on bias and efficiency of statistical estimators and subsequent sections will address sample number effects on the power of existing hypothesis-testing machinery. However, the question "How many samples do I need to accurately estimate the mean?" has not yet been directly addressed. This question and questions like it are the most frequently asked statistical questions raised by experimentalists.

The formulas for computing the number of observations needed in the one-sample and unpaired two-sample problems appear in many statistics texts (e.g., Gilbert, 1987; Green, 1979). Of course, it should be emphasized that these formulas rely on the Gaussian distribution being correct in the limit as the number of observations increases. When the underlying probability distribution is highly skewed application of the asymptotic formula may suggest that more samples are needed to accurately estimate the mean at a prespecified precision and a given Type I error rate (α) than are actually necessary. This is a direct result of the fact that highly skewed samples can be the norm for the lognormal family but not for the Gaussian distribution. The number of values needed to estimate the mean of a lognormal distribution, with a given probability level of having correctly done so, can be computed from the confidence interval estimation procedure used to estimate the mean. This is accomplished by iteratively searching for the value n in Eqs. 23 and 24. In this process α equals the probability level desired and the upper and lower confidence limits are set equal to the estimate of the mean plus or minus the user-specified degree of precision. This number can then be rounded up to the nearest integer to afford some protection against replicating too few times. It should be noted that, since Eqs. 23 and 24 yield an asymmetric confidence interval, the sample number requirement for obtaining an estimate of the mean at the lower precision boundry will differ from the number of samples required at the upper limit of precision.

F. Robustness of Estimators

In sampling we rarely have the luxury of knowing, precisely, the underlying distribution function. In many instances the lognormal model has been shown to apply. If the data are from a true lognormal distribution the UMVU estimators of the mean, variance, and median are the estimators of choice, as they are both unbiased and have minimum variance. Robustness of these UMVU estimators to deviations from true lognormality has only recently been addressed (Sichel, 1987; Meyers and Pepin, 1990). Meyers and Pepin (1990) reported that the UMVU estimators, when applied to data from contaminated lognormal distributions, are inefficient relative to the MM estimators. They recommend that unless it is reasonably assured that sample data are from a lognormal distribution, the UMVU estimators should not be applied.

The methods described in this chapter are appropriate and optimal when the data set or sets being analyzed are truly lognormal. If the investigator is unwilling to make the assumption that his or her data are lognormally distributed, then using the traditional method of moments approach to estimating the mean and variance. With regard to calculation of confidence limits of the mean, the asymptotic method (Eq. 21) cannot be recommended for use with skewed distribution. Application of this method to skewed data will result in a symmetric confidence interval about the sample mean. The resulting upper confidence limit will have very poor coverage relative to the nominal α level at which the confidence interval is applied. Land's lognormal confidence interval method provides exact coverage for true lognormal data; however, the efficacy of this procedure when applied to "contaminated" lognormal data is unknown and needs to be investigated.

As an alternative to parametric methods, nonparametric methods applicable to the analysis of skewed data have been recently developed. Horn (1990) reports on robust quantile estimators for skewed data, and a pseudo-nonparametric confidence interval estimation procedure has also been described (Parkin et al., 1990). Again it must be cautioned that the efficiency of these estimators has not been evaluated. Because of the lack of studies on robustness, it is not possible to make these kinds of general recommendations beyond this point.

G. Monte Carlo Simulation

Computer simulation is a powerful tool that can be used to evaluate the efficacy of statistical estimation procedures as well as to test hypotheses (Carmer and Swanson, 1971; Parkin et al., 1990). This section describes in greater detail what Monte Carlo simulations are and how then they can be used in evaluating statistical methods. Additional discussion of the role of computer simulation in hypothesis testing is presented by Green (1979).

"Monte Carlo" is a term that simply describes the procedure by which random variates with known properties can be obtained. The utility of Monte Carlo simulation when applied to the problem of statistical methods evaluation is that variates from known distributions can be obtained. Only when the underlying population distribution is known can be inferences drawn from application of a given statistical method to sample data be evaluated. A Monte Carlo procedure for obtaining random variates from normal and lognormal distributions is described below.

Several methods exist for generating random variates from a normal distribution (Abramowitz and Stegun, 1964). Two of these are described below.

1. Sum of Uniform Deviates

A random variate from a unit normal population (mean = 0, variance = 1) can be generated by summing variates from a uniform distribution on the interval (0, 1) according to the following equation:

$$R = \left(\sum_{i=1}^{n} U_i - \frac{n}{2}\right)\left(\frac{n}{12}\right)$$

where R is a random unit normal varite and U_i random variate from a uniform distribution. This method is approximate and little gain is achieved for $n > 12$.

2. The Direct Method

This is an exact method that generates a pair of random unit normal variates from a pair of uniform variates on the interval (0, 1).

$$R_1 = \sqrt{(-2\ln(U_1))}\cos(2\pi U_2)$$
$$R_2 = \sqrt{(-2\ln(U_1))}\sin(2\pi U_2)$$

The random variates (R) generated by either method 1 or 2 are from a unit normal distribution (mean = 0, variance = 1). Variates from normal distributions other than the unit normal are obtained by multiplying the unit normal variate by the scale parameter (standard deviation) and adding the mean. For example, the following equation will generate a random variate from a normal distribution having mean = 10 and standard deviation = 2:

$$R = (R_1 * 2) + 10$$

Since, ln-transformed variates from a lognormal distribution are normally distributed, the procedure described above can be modified to obtain random variates from a lognormal distribution. The following equation details this procedure.

$$L = \exp[(R_1 * \sigma) + \mu]$$

where μ and σ are the mean and variance of the ln-transformed population.

By specifying the mean and variance on the untransformed scale, Eqs. 3 and 6 can be solved to obtain μ and σ. A computer program that generates random variates from a user-defined lognormal distribution is available in USDA Technology Transfer Document NSTL91-4.

IV. Hypothesis Testing

A. Selecting the Appropriate Location Parameter: The Mean vs the Median

If the sample data are lognormally distributed choices exist concerning the location parameter that is to be used as the summary statistic of the data. The two summary location parameters which are most often used are the population mean and the median. For a symmetric distribution these two parameters will have the same value; however, for an asymmetric distribution, such as the lognormal distribution, these parameters have different expected values. The mean and the median of a lognormal distribution actually convey different information about the distribution. The mean is the center of gravity of the distribution, while the median is the center of probability of the distribution. The choice of the appropriate location parameter is critical, as it can affect the conclusions drawn from the data. Few guidelines exist concerning the validity of focusing on the mean or median as a summary statistic. Often the median or the geometric mean is arbitrarily chosen over the mean because of the resistance of these two parameters to the influence of extreme values often observed with lognormal sample data. However, this rationale is not always valid.

It has been recognized that bacterial numbers in nature can be approximated by lognormal distributions, and it has been recommended that "the preferred statistic for summarizing microbiological data is the geometric mean" (Standard Methods for the Examination of Water and Wastewater, 1985). This recommendation is based on the assumption that the sample geometric mean is the best indicator of the central tendency of the lognormal distribution (for lognormal distributions the geometric mean is an estimator of the median). However, the mean and median both indicate central tendency (i.e., the mean is the central tendency of mass and the median is the central tendency of the individuals in the population). The recommendation given above, that the geometric mean is the best estimator of central tendency for bacterial numbers is somewhat arbitrary, as the definition of "best" is dependent on the exact nature of the problem being investigated and the objective of the study. Careful consideration of the nature of the problem is required before a selection can be made. A major factor which must be considered is the influence of sample volume on the median.

B. Sample Volume Effects on the Median

The central limit theorem predicts that, regardless of the form of the underlying population, the distribution of sample mean approaches normality as the number of samples used in computing the means increases. To illustrate this effect, Figure 7 (top) presents a histogram of 1000 samples randomly drawn from a lognormal population with a mean of 400, median of 179, and a standard deviation of 800. The arithmetic average, standard deviation, and geometric mean of these samples are 391, 675, and 179, respectively. When samples from the top panel are randomly bulked or averaged in groups of two we obtain a histogram of 500 samples (midde panel). These samples have a mean of 391, a standard deviation of 479, and a geometric mean of 246. Bulking the samples in groups of ten results in a more symmetric histogram with mean

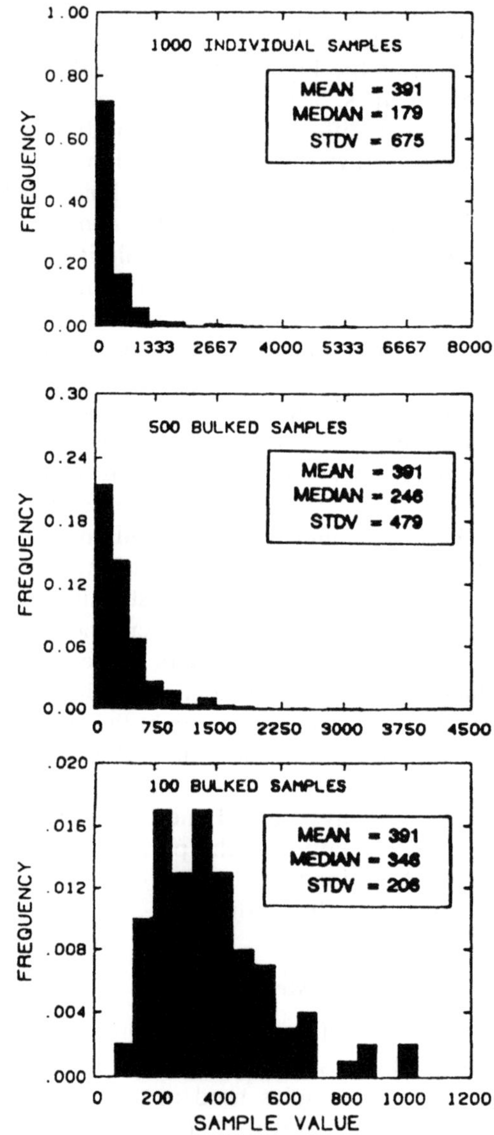

Figure 7. Influence of sample bulking on the shape of the sample histogram.

of 391, standard deviation of 206, and geometric mean of 346 (bottom panel).

Through the process of averaging, the population of bulked samples approaches symmetry and the value of the median approaches the value of the mean. In natural systems if the variable of interest is randomly dispersed, collecting large samples has the same effect as bulking or pooling small samples. Thus, the value of the population median is functionally dependent upon the sample volume collected. Because of this effect focus on the median or geometric mean may only be appropriate when the samples themselves have some inherent identity and significance. This implies that in an aquatic or terrestrial system, where the sample volume is arbitrarily defined by the size of the bucket or shovel available to the investigator, the median (or geometric mean) may not be the appropriate population parameter to estimate.

In many soils studies the size of the sample support has no inherent significance and is, in effect, arbitrarily defined. In such cases it may be inappropriate to describe the data in terms of the median (or geometric mean) unless the number and volume of samples are also specified. The following discussion more clearly illustrates when the median and mean are to be emphasized as the location parameter of choice.

C. The Median as the Location Parameter of Choice

When it is appropriate to use the median? A classical example illustrating a valid use of the median as a summary variable exists in the field of economics. Personal income data are skewed and have been approximated by a lognormal distribution (Aitchinson and Brown, 1957). Use of the median as the summary parameter for income data is desirable because the samples themselves (the individuals) have identity and significance. The median income level allows the individuals to gauge themselves against other individuals (samples) of the population.

In the environmental sciences an excellent example of the appropriate use of the median is presented by Hirano et al. (1982). These researchers analyzed data sets of 24 to 36 individual leaves and found that total epiphytic bacteria, ice nucleation active strains, and a strain of an inoculated bacterium were all lognormally distributed. According to our criterion stated above, the appropriate use of the median as the summary parameter requires that the samples have identity and significance. The significance of considering the bacterial populations on individual leaves is given by Hirano et al. (1982). They state, "The quantitative variability of epiphytic bacterial populations on individual leaves may be an expression of the uniqueness of each leaf as an ecosystem with one or more environmental or biological characteristics significantly different from that of a neighboring leaf." In addition Hirano et al. conclude that "Since foliar plant diseases occur on individual leaves within a given plant canopy, the mean pathogen population for that canopy is of less importance than the pathogenic population on each leaf." It was also observed in this study that the practice of bulking individual leaves results in overestimates of the population median for individual leaves. From a study on bacterial populations in the rhizosphere, Loper et al. (1984)

reached similar conclusion concerning the effects of sample bulking on estimates of the median.

D. The Mean as the Location Parameter of Choice

There have been several reports of environmental pollution exhibiting positively skewed distributions (e.g., Table 1). Suppose we are interested in the total mass of a pollutant or some other chemical species present at a particular site. Gilbert (1987) defines the total mass of pollutant at a site as the "inventory" of the pollutant. If the inventory of the pollutant is the desired summary variable, then the median is the wrong estimator of location since it will systematically underestimate the total mass of material at the contaminated site.

In the area of soil science, often what is desired is an estimate of the magnitude of a given microbial process in a particular ecosystem. For example, soil denitrification in agricultural systems may be an important mechanism of fertilizer N loss. Measurements of denitrification exhibit highly skewed frequency distributions. Since the individual soil samples have no inherent significance, the median, which estimates the center of probability of the distribution, is not the location parameter of choice. Rather, what is desired is an estimate of N loss from a given field, thus the mean, which is an estimator of the center of mass if the distribution, is the appropriate location parameter (Parkin, 1990).

For situations where an estimate of the total mass of a variable (either a chemical constituent or a process) is required, the mean should be the location estimator of choice.

E. Power of Hypothesis-Testing Procedures

In previous sections optimum methods for computing summary statistics of lognormally distributed data have been discussed. However, a typical objective of many studies extends beyond estimation of population parameters from sample data. In many cases sampling is conducted to evaluate treatment effects. Application of standard analysis of variance procedures requires several assumptions concerning the underlying error structure of the data, and among these is the assumption of normality. The effects of violations of the assumption of normality on the efficacy of parametric statistical tests, such as the t test, have long been known (Hey, 1938; Cochran, 1947). Nonnormality will influence the ability of a statistical test to perform at the stated α level. Nonnormality will also affect the power of a statistical test to detect differences when real differences in the underlying populations actually exist. Two common procedures have been recommended when the normality assumption has been violated. These are: (1) first transform for normality or (2) apply nonparametric statistical methods (Snedecor and Cochran, 1967). However, the consequences of implementing these two approaches are not typically considered.

In the preceding section it was observed that with lognormally distributed variables a choice of location parameters exists and that the choice of the appropriate location parameter must be consistent with the objectives and methodologies of the problem under study. After the choice of the appropriate location parameter has been made, consideration must be given to the statistical methods used at the hypothesis-testing stage. It is imperative that the experimenter who has to analyze positively skewed data understand what is being compared when ln-transformed data or nonparametric procedures are used. This concern is illustrated in the following discussion of the efficacy of five hypothesis-testing techniques applied to samples drawn from two different populations.

The statistical tests evaluated in this analysis are: (1) the unpaired, two-tailed t test (equal sized batches) on untransformed sample data; (2) the unpaired two-tailed t test applied to the batches of log-transformed data; (3) the Mann-Whitney form of the Wilcoxin test (Miller, 1986); (4) a test based on the overlap of 95% confidence limits of the median; and (5) a test based on the overlap of 95% confidence limits of the mean. Tests 1, 2, and 3 were applied at an α level = 0.05. Monte Carlo simulation was used to differentiate size samples (n = 10 to 100) from two different lognormal populations. The sample data were then compared using the tests listed above. This process was repeated 10,000 times for each sample size and the power of each statistical test was calculated. Here, power refers to the frequency with which a given test detects a difference between the batches of samples.

In the first example the two lognormal populations compared have identical medians but means that differ by a factor of 2 (Figure 8A). Results of the simulations (Figure 8B) indicate that the nonparametric Mann-Whitney method (NONPARA), the t test on ln-transformed data (TTEST-LN), and the median confidence limit overlap method (CL-MED) are all insensitive to underlying differences in population means. The probabilities associated with detecting differences between samples are not substantially influenced by sample size. Probability levels observed for CL-MED, TTEST-LN, and NONPARA, are 1, 5, and 7%, respectively. In contrast, the t test on untransformed data (TTEST) and the mean confidence limit overlap method (CL-MEAN) are sensitive to differences in samples drawn from the two populations. The power of the TTEST and CL-MEAN to detect differences is poor for small sample sizes, but increases with increasing sample sizes. It is also observed that at any given sample size ($n > 4$) the CL-MEAN test has greater power to detect differences that TTEST.

An entirely different pattern is observed when these tests are applied to sample data from populations with equal means, but different medians

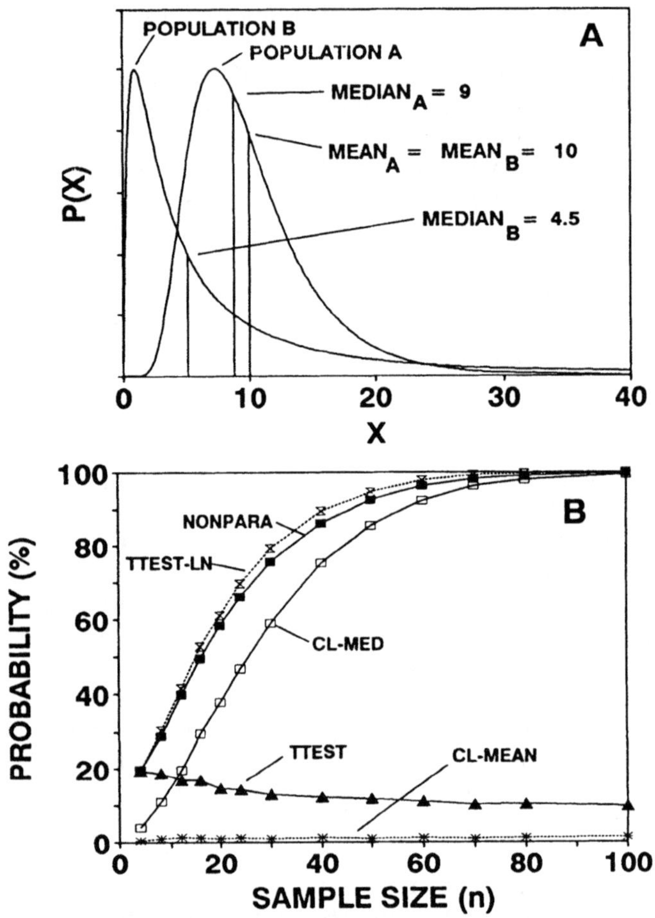

Figure 8. Comparison of hypothesis tests on samples drawn from populations with equal medians but unequal means. Probability (panel B) refers to the percentage of occurrences a given test detects differences between batches of samples drawn from the two lognormal populations shown in panel A.

(Figure 9). In this example, population A has a mean of 10 and a median of 9, while population B has a mean of 10 and a median of 4.5 (Figure 9A). When samples drawn from these two populations were compared using the five statistical tests it is observed that the TTEST and the CL-MEAN methods lack power to detect differences. The probability of detecting differences associated with the TTEST decreases from 20% at $n = 4$ to approximately 10% at $n = 100$. The CL-MEAN method operated at the 1% level over the range of sample sizes evaluated. In contrast, the TTEST-LN, NONPARA, and CL-MED methods are sensitive to differences between samples from populations with differing

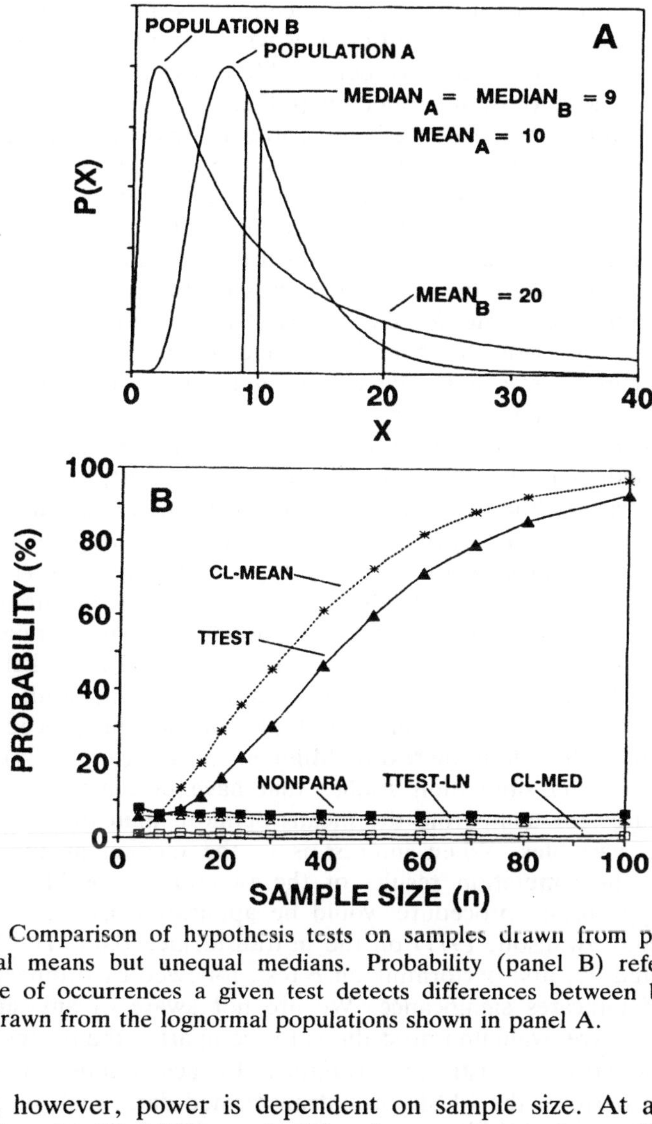

Figure 9. Comparison of hypothesis tests on samples drawn from populations with equal means but unequal medians. Probability (panel B) refers to the percentage of occurrences a given test detects differences between batches of samples drawn from the lognormal populations shown in panel A.

medians; however, power is dependent on sample size. At any given sample size the CL-MED method has less power to detect differences than either the NONPARA or TTEST-LN methods. These latter two methods performed at comparable levels over the range of sample sizes studied.

These results indicate that the nonparametric Mann-Whitney test, and the *t* test on ln-transformed data and the median confidence limit overlap technique are all sensitive to sample differences when the underlying populations differ with regard to medians. Conversely, the *t* test on untransformed data and mean confidence limit overlap method are sensitive to differences in population means. It is also apparent that all the tests lack power to detect differences when sample numbers are low. High replication is needed to detect differences between lognormal populations whose location parameters (either the mean or median) differ by only a factor of two. Alternatively, power to detect differences can be increased by increasing the α level at which the test is applied.

Based on simulation studies conducted over a range of population variances, the recommended approach to solving the two-sample problem when the mean is of interest involves constructing confidence intervals for the two batches and comparing them with one another (Parkin, 1992). If the 90% confidence limits for the means of two lognormally distributed batches overlap, then the hypothesis that the two lognormal means are equal can be rejected. The precise α level, or Type I error rate is dependent, to some degree, upon the variance of the underlying population, however, is usually less than 0.1. If it has been determined that the median is the location parameter of interest, then either the nonparametric method or the t test on ln-transformed data can be used.

While the above approach is acceptable for comparing two samples, extending the approach to make multiple pair-wise comparisons of several batches of lognormally distributed data is not encouraged. The reason for this stems from the fact that if enough pair-wise comparisons at the 5 or 10% level of significance are performed, then it is likely that at least one of the comparisons would produce a significant result. This is a commonly encountered problem in multiple comparisons procedures and is succinctly dealt with in the text by Milliken and Johnson (1984). We are not aware of any simulation studies that have evaluated a procedure or procedures for comparing the means of more than two lognormally distributed variables. When interest is in the median as the location estimator, the simulation results of the two-sample problem indicate that an appropriate procedure would be application of the FLSD test (Carmer and Swanson, 1971) on the ln-transformed data. The FLSD is implemented by, first performing an F test, followed by ordinary LSD if the F test indicates significance. We are not aware of any simulation studies that have evaluated procedures for comparing the means of more than two lognormally distributed variables. In such situations one option may be to perform the FLSD on untransformed data, realizing that the power to detect real differences may be poor. Alternatively, in the spirit of exploratory data analysis (Hoaglin et al., 1983), one could compute a plot of means and confidence intervals and visually observe overlapping ranges between treatments. Due to the general lack of information on this topic, it is not possible to make specific recommendations beyond this point.

V. Concluding Remarks

There are many situations where the mean and not the sample median is the summary parameter of interest for positively skewed data. In the authors' opinions, it is easier to construct examples of when the mean is important to soils studies than when the median should be the location estimator of choice. But lest undue emphasis is placed on the mean, it must be recognized that there may even be cases where a higher or lower quantile may be the location parameter of choice. Indeed there are many location parameters of interest when the data are asymmetric, the choice of which depends on the specific problem at hand. When data are symmetric, less debate occurs about the location estimator of greatest interest, it is the population mean. But when environmental data are asymmetric serious thought should be given to the location estimator of greatest interest: Should it be the population mean, the median, or some other quantile? In the end it is probably in the best interests of the data analyst and the subject matter specialist to not rely on a single summary parameter for location, when the data are positively skewed.

When the data are lognormally distributed, then methods developed assuming this distribution should be used in estimating the population mean and variance from sample data. Methods optimal for Gaussian distributed data will be suboptimal for the lognormal parametric family. The lack of power which exists when traditional approaches are applied to positively skewed data must be recognized. Why perform a costly experiment and then collect insufficient samples or use suboptimal statistical procedures for data analysis?

As a cautionary note, it should be stated that the consequences of applying parametric analysis methods to positively skewed data that are not truly lognormally distributed are unknown. But assuming a parametric family that is skewed is at least better than assuming a distribution that is symmetric. What is really needed is for the gap that Miller (1986) refers to, regarding the absence of robust statistical theory for asymmetric distributions, be filled. When these methods are in place, then the soil scientist with asymmetric data to analyze will have tools for analyzing positively skewed data that may equal or just approximate the lognormal probability density family. The correctness of the inferences drawn from these studies will depend on the statistical methods used to summarize the data.

References

Abramowitz, M., and I.A. Stegun. 1964. Handbook of mathematical functions with formulas, graphs, and mathematical tables. National Bureau of Standards, Applied Mathematics Series 55.

Aitchinson, J., and J.A.C. Brown. 1957. The Lognormal Distribution. Cambridge University Press, London.

Allmaras, R.R., R.E. Burwell, W.B. Voorhees, and W.E. Larson. 1965. Aggregate size distribution in the row zone of tillage experiments. Soil Sci. Soc. Am. Proc. 29:645–650.

Atlas, R.M., and R. Bartha. 1987. Microbiol Ecology; Fundamentals and Applications. Benjamin/Cummings Publishing Co., Menlo Park, CA.

Barnett, V. 1973. Comparative Statistical Inference. Wiley, New York.

Bonsdorff, E., and V. Koivisto. 1982. The use of the log-normal distribution of individuals among species in monitoring zoobenthos in the Northern Baltic Archipelago. Marine Pollut. Bull. 13:324–327.

Bradu, D., and Y. Mundlak. 1970. Estimation in lognormal linear models. J. Am. Stats. Assoc. 65:198–211.

Cameron, K.C., and A. Wild. 1984. Potential aquifer pollution leaching following the plowing of temporary grassland. J. Environ. Qual. 13:274–277.

Carmer, S.G., and M.R. Swanson. 1971. Detection of differences between means: A Monte Carlo study of five pairwise multiple comparison procedures. Agron. J. 63:940–945.

Christensen, S., S. Simkins, and J.M. Tiedje. 1990. Spatial variation in denitrification: Dependency of activity centers on the soil environment. Soil Sci. Soc. Am. J. 54:1608–1613.

Christian, R.R., and W.O. Pipes. 1983. Frequency distribution of coliforms in water distribution systems. Appl. Environ. Microbiol. 45:603–609.

Cochran, W.G. 1947. Some consequences when the assumptions for the analysis of variance are not satisfied. Biometrics. 3:22–38.

Conover, W.J. 1980. Practical Nonparametric Statistics, 2nd ed. Wiley, New York.

Crow E.L., and K. Shimizu. 1988. Lognormal Distributions. Marcel Dekker, New York.

D'Agostino, R.B. 1971. An omnibus test of normality for moderate and large size samples. Biometrika. 58:341–348.

Daniel, W.W. 1978. Applied Nonparametric Statistics. Houghton Mifflin, Boston.

Finney, D.J. 1941. On the distribution of a variate whose logarithm is normally distributed. J. Royal Statis. Soc. Suppl. 7:144–161.

Focht, D.C., L.H. Stolzy, and B.D. Meek. 1979. Sequential reduction of nitrous oxide under field conditions as brought about by organic amendments and irrigation management. Soil Biol. Biochem. 11:37–46.

Folorunso, O.A., and D.E. Rolston. 1984. Spatial variability of field-measured denitrification gas flux. Soil Sci. Soc. Am. J. 48:1214–1219.

Georgopoulos, P.G., and J.H. Seinfeld. 1982. Statistical distributions of air pollutant concentrations. Environ. Sci. Technol. 16:401–416.

Gilbert, R.O. 1987. Statistical Methods for Environmental Pollution Monitoring. Van Nostrand Reinhold, New York.

Green, R.H. 1979. Sampling design and statistical methods for environmental biologists. Wiley, New York.

Haldane, J.B.S. 1942. Moments of the distributions of powers and products of normal variates. Biometrika. 32:226–246.

Hey, G.B. 1938. A new method of experimental sampling illustrated on certain non-normal populations. Biometrika. 30:68–80.

Hirano, S.S., E.V. Nordheim, D.C. Arny, and C.D. Upper. 1982. Lognormal distribution of epiphytic bacterial populations on leaf surfaces. Appl. and Environ. Microbiol. 44:695–700.

Hoaglin, D.C., F. Mosteller, and J.W. Tuekey. 1983. Understanding Robust and Exploratory Data Analysis. Wiley, New York.

Horn, P.S. 1990. Robust quantile estimators of skewed populations. Biometrika. 77:631–636.

Johnson, N.L., and S. Kotz. 1970. Continuous Univariate Distributions-1. Houghton Mifflin, Boston.

Jones, A.J., and R.J. Wagenet. 1984. In situ estimation of hydraulic conductivity using simplified methods. Water Resour. Res. 20:1620–1626.

Kapteyn, J.C. 1903. Skew frequency curves in Biology and Statistics Astronomical Lab., Noordhoff, Groningen.

Kleijnen, J.P.C. 1987. Statistical Tools for Simulation Practitioners. Marcel Dekker, New York.

Koch, G.S., and R.F. Link. 1970. Statistical Analysis of Geological Data. Wiley, New York.

Krige, D.G. 1981. Lognormal-de Wijsian Geostatistics for Ore Evaluation. South African Institute of Mining and Metallurgy (PB1.) Johannesburg, South Africa.

Kushner, E.J. 1976. On determining the statistical parameters for pollution concentrations from a truncated data set. Atmospheric Environm. 10:975–979.

Land, C.E. 1973. Standard confidence limits for linear functions of the normal mean and variance. J. Am. Stat. Assoc. 68:960–963.

Landwehr, J.M. 1978. Some properties of the geometric mean and its use in water quality standards. Water Resources Res. 14:467–473.

Lewin, R. 1987. How big can a species be? Science. 236:117.

Loper, J.E., T.V. Suslow, and N.M. Schroth. 1984. Lognormal distribution of bacterial populations in the rhizosphere. Phytopath. 74:1454–1460.

Mage, D.T. 1981. A review of the application of probability models for describing aerometric data. In Environmetrics 81: Selected papers, SIAM-SIMS Conference Series 8. pp. 42–51. Society for Industrial and Appl. Mathematics, Phildelphia.

Malhotra, S.K. 1984. Log-normal distributions in parasitology. Proc. Indian Acad. Sci. 93:591–598.

Meyers, R.A., and P. Pepin. 1990. The robustness of lognormal-based estimators of abundance. Biometrics 46:1185–1192.

Miller, R.G. 1986. Beyond ANOVA, Basics of Applied Statistics. Wiley, New York.

Milliken, G.A., and D.E. Johnson. 1984. Analysis of Messy Data, Vol. 1 Designed Experiments. Von Nostrand Reinhold, New York.

Parkin, T.B. 1987. Soil microsites as a source of denitrification variability. Soil Sci. Soc. Am. J. 51:1194–1199.

Parkin, T.B. 1990. Characterizing the variability of denitrification. In N.P. Revsbech and J. Sorensen (eds.) Denitrification in soil and sediment. pp. 213–228. Plenum Press, New York.

Parkin, T.B. 1992. Evaluation of statistical methods for determining differences between lognormal populations. Agron. J. (in press).

Parkin, T.B., and J.A. Robinson. 1989. Stochastic models of soil denitrification. Appl. Environ. Microbiol. 55:72–77.

Parkin, T.B., and J.A. Robinson. 1992. Statistical evaluation of median estimators for lognormally distributed variables. Soil Sci. Soc. Am. J. (in press).

Parkin, T.B., A.J. Sexstone, and J.M. Tiedje. 1985. Comparison of field denitrification rates determined by acetylene-based soil core and nitrogen-15 methods. Soil Sci. Soc. Am. J. 49:94–99.

Parkin, T.B., J.L. Starr, and J.M. Meisinger. 1987. Influence of sample size on measurement of soil denitrification. Soil Sci. Soc. Am. J. 51:1492–1501.

Parkin, T.B., J.J. Meisinger, S.T. Chester, J.L. Starr, and J.A. Robinson. 1988. Evaluation of statistical methods for lognormally distributed variables. Soil Sci. Soc. Am. J. 52:323–329.

Parkin, T.B., S.T. Chester, and J.A. Robinson. 1990. Evaluation of methods for calculating confidence intervals about the mean of a lognormally distributed variable. Soil Sci. Soc. Am. J. 54:321–326.

Ruess, J.O., P.N. Soltanpour, and A.E. Ludwick. 1977. Sampling distribution of nitrates in irrigated fields. Agron. J. 69:588–592.

Sells, F., R.E. Karamanos, and R.G. Kachanoski. 1986. The spatial variability of nitrogen-15 and its relation to the variability of other soil properties. Soil Sci. Soc. Am. J. 50:105–110.

Shapiro S.S., and M.B. Wilk. 1965. An analysis of variance test for normality (complete samples), Biometrika. 52:591–611.

Sichel, H.S. 1952. New Methods in the statistical evaluation of mine sampling. London. Inst. Mining Met. Trans. 61:261–288

Sichel, H.S. 1966. The estimation of means and associated confidence limits for small samples from lognormal populations. In Symposium on Mathematical Statistics and Computer Appl., Johannesburg, S. Africa. S. Afr. Inst. Min. Metall., Johannesburg.

Sichel, H.S. 1987. Some advances in lognormal theory. APCOM 87. Proceedings of the Twentieth International Symposium on the Applications of Computers and Mathematics in the Mineral Industries. Vol. 3: GEOSTATISTICS. pp. 3–8. SAIMM, Johannesburg.

Snedecor, G.W., and W.G. Cochran. 1967. Statistical Methods. 6th edition, Iowa St. Univ., Ames, IA.

Sokal, R.R., and F.J. Rohlf. 1969. Biometry. W.H. Freeman, New York.

Standard Methods for the Examination of Water and Wastewater, 16th Ed. 1985. A.E. Greenberg et al. (eds.). Am. Publ. Health Assoc et al. Washington, D.C.

Steel, R.G.D., and J.H. Torrie. 1980. Principles and Procedures of Statistics; A Biometric Approach. McGraw-Hill, New York.

Tabor, J.A., A.W. Warrick, D.E. Myers, and D.A. Pennington. 1985. Spatial variability of nitrate in irrigated cotton: II. Soil nitrate and correlated variables. Soil. Sci. Soc. Am. J. 49:390–394.

Ugland, K.I., and J.S. Gray. 1982. Lognormal distributions and the concept of community equilibrium. Oikos. 39:171–178.

Warrick, A.W., and D.R. Nielsen. 1980. Spatial variability of soil physical properties in the field. In D. Hillel (ed.). Applications of soil physics. p. 385. Academic Press, New York.

Warrick, A.W., G.J. Mullen, and D.R. Nielsen. 1977. Scaling field measured soil hydraulic properties using a similar media concept. Water Resour. Res. 13:355–362.

White, R.E., R.A. Haigh, and J.H. Macduff. 1987. Frequency distributions and spatially dependent variability of ammonium and nitrate concentrations in soil under grazed and ungrazed grassland. Fert. Res. 11:193–208.

Appendix

Statistical Tables

Table A1. Coefficients (a_i) for the Shapiro-Wilk test of normality. From Shapiro and Wilk (1965)

$i \backslash n$	2	3	4	5	6	7	8	9	10	11
1	0.7071	0.7071	0.6872	0.6646	0.6431	0.6233	0.6052	0.5888	0.5739	0.5601
2	—	—	0.1677	0.2413	0.2806	0.3031	0.3164	0.3244	0.3291	0.3315
3	—	—	—	—	0.0875	0.1401	0.1743	0.1976	0.2141	0.2260
4	—	—	—	—	—	—	0.0561	0.0947	0.1224	0.1429
5	—	—	—	—	—	—	—	—	0.0399	0.0695

$i \backslash n$	12	13	14	15	16	17	18	19	20	21
1	0.5475	0.5359	0.5251	0.5150	0.5056	0.4968	0.4886	0.4808	0.4734	0.4643
2	0.3325	0.3325	0.3318	0.3306	0.3290	0.3273	0.3253	0.3232	0.3211	0.3185
3	0.2347	0.2412	0.2460	0.2495	0.2521	0.2540	0.2553	0.2561	0.2565	0.2578
4	0.1586	0.1707	0.1802	0.1878	0.1939	0.1988	0.2027	0.2059	0.2085	0.2119
5	0.0922	0.1099	0.1240	0.1353	0.1447	0.1524	0.1587	0.1641	0.1686	0.1736
6	0.0303	0.0539	0.0727	0.0880	0.1005	0.1109	0.1197	0.1271	0.1334	0.1399
7	—	—	0.0240	0.0433	0.0593	0.0725	0.0837	0.0932	0.1013	0.1092
8	—	—	—	—	0.0193	0.0359	0.0496	0.0612	0.0711	0.0804
9	—	—	—	—	—	—	0.0163	0.0303	0.0422	0.0530
10	—	—	—	—	—	—	—	—	0.0140	0.0263

$i \backslash n$	22	23	24	25	26	27	28	29	30	31
1	0.4590	0.4542	0.4493	0.4450	0.4407	0.4366	0.4328	0.4291	0.4254	0.4220
2	0.3156	0.3126	0.3098	0.3069	0.3043	0.3018	0.2992	0.2968	0.2944	0.2921
3	0.2571	0.2563	0.2554	0.2543	0.2533	0.2522	0.2510	0.2499	0.2487	0.2475
4	0.2131	0.2139	0.2145	0.2148	0.2151	0.2152	0.2151	0.2150	0.2148	0.2145
5	0.1764	0.1787	0.1807	0.1822	0.1836	0.1848	0.1857	0.1864	0.1870	0.1874
6	0.1443	0.1480	0.1512	0.1539	0.1563	0.1584	0.1601	0.1616	0.1630	0.1641
7	0.1150	0.1201	0.1245	0.1283	0.1316	0.1346	0.1372	0.1395	0.1415	0.1433
8	0.0878	0.0941	0.0997	0.1046	0.1089	0.1128	0.1162	0.1192	0.1219	0.1243
9	0.0618	0.0696	0.0764	0.0823	0.0876	0.0923	0.0965	0.1002	0.1036	0.1066
10	0.0368	0.0459	0.0539	0.0610	0.0672	0.0728	0.0778	0.0822	0.0862	0.0899
11	0.0122	0.0228	0.0321	0.0403	0.0476	0.0540	0.0598	0.0650	0.0697	0.0739
12	—	—	0.0107	0.0200	0.0284	0.0358	0.0424	0.0483	0.0537	0.0585
13	—	—	—	—	0.0094	0.0178	0.0253	0.0320	0.0381	0.0435
14	—	—	—	—	—	—	0.0084	0.0159	0.0227	0.0289
15	—	—	—	—	—	—	—	—	0.0076	0.0144

$i \backslash n$	32	33	34	35	36	37	38	39	40	41
1	0.4188	0.4156	0.4127	0.4096	0.4068	0.4040	0.4015	0.3989	0.3964	0.3940
2	0.2898	0.2876	0.2854	0.2834	0.2813	0.2794	0.2774	0.2755	0.2737	0.2719
3	0.2462	0.2451	0.2439	0.2427	0.2415	0.2403	0.2391	0.2380	0.2368	0.2357

Table A1. Continued

$i\backslash n$	32	33	34	35	36	37	38	39	40	41
4	0.2141	0.2137	0.2132	0.2127	0.2121	0.2116	0.2110	0.2104	0.2098	0.2091
5	0.1878	0.1880	0.1882	0.1883	0.1883	0.1883	0.1881	0.1880	0.1878	0.1876
6	0.1651	0.1660	0.1667	0.1673	0.1678	0.1683	0.1686	0.1689	0.1691	0.1693
7	0.1449	0.1463	0.1475	0.1487	0.1496	0.1505	0.1513	0.1520	0.1526	0.1531
8	0.1265	0.1284	0.1301	0.1317	0.1331	0.1344	0.1356	0.1366	0.1376	0.1384
9	0.1093	0.1118	0.1140	0.1160	0.1179	0.1196	0.1211	0.1225	0.1237	0.1249
10	0.0931	0.0961	0.0988	0.1013	0.1036	0.1056	0.1075	0.1092	0.1108	0.1123
11	0.0777	0.0812	0.0844	0.0873	0.0900	0.0924	0.0947	0.0967	0.0986	0.1004
12	0.0629	0.0669	0.0706	0.0739	0.0770	0.0798	0.0824	0.0848	0.0870	0.0891
13	0.0485	0.0530	0.0572	0.0610	0.0645	0.0677	0.0706	0.0733	0.0759	0.0782
14	0.0344	0.0395	0.0441	0.0484	0.0523	0.0559	0.0592	0.0622	0.0651	0.0677
15	0.0206	0.0262	0.0314	0.0361	0.0404	0.0444	0.0481	0.0515	0.0546	0.0575
16	0.0068	0.0131	0.0187	0.0239	0.0287	0.0331	0.0372	0.0409	0.0444	0.0476
17	—	—	0.0062	0.0119	0.0172	0.0220	0.0264	0.0305	0.0343	0.0379
18	—	—	—	—	0.0057	0.0110	0.0158	0.0203	0.0244	0.0283
19	—	—	—	—	—	—	0.0053	0.0101	0.0146	0.0188
20	—	—	—	—	—	—	—	—	0.0049	0.0094

$i\backslash n$	42	43	44	45	46	47	48	49	50
1	0.3917	0.3894	0.3872	0.3850	0.3830	0.3808	0.3789	0.3770	0.3751
2	0.2701	0.2684	0.2667	0.2651	0.2635	0.2620	0.2604	0.2589	0.2574
3	0.2345	0.2334	0.2323	0.2313	0.2302	0.2291	0.2281	0.2271	0.2260
4	0.2085	0.2078	0.2072	0.2065	0.2058	0.2052	0.2045	0.2038	0.2032
5	0.1874	0.1871	0.1868	0.1865	0.1862	0.1859	0.1855	0.1851	0.1847
6	0.1694	0.1695	0.1695	0.1695	0.1695	0.1695	0.1693	0.1692	0.1691
7	0.1535	0.1539	0.1542	0.1545	0.1548	0.1550	0.1551	0.1553	0.1554
8	0.1392	0.1398	0.1405	0.1410	0.1415	0.1420	0.1423	0.1427	0.1430
9	0.1259	0.1269	0.1278	0.1286	0.1293	0.1300	0.1306	0.1312	0.1317
10	0.1136	0.1149	0.1160	0.1170	0.1180	0.1189	0.1197	0.1205	0.1212
11	0.1020	0.1035	0.1049	0.1062	0.1073	0.1085	0.1095	0.1105	0.1113
12	0.0909	0.0927	0.0943	0.0959	0.0972	0.0986	0.0998	0.1010	0.1020
13	0.0804	0.0824	0.0842	0.0860	0.0876	0.0892	0.0906	0.0919	0.0932
14	0.0701	0.0724	0.0745	0.0765	0.0783	0.0801	0.0817	0.0832	0.0846
15	0.0602	0.0628	0.0651	0.0673	0.0694	0.0713	0.0731	0.0478	0.0764
16	0.0506	0.0534	0.0560	0.0584	0.0607	0.0628	0.0648	0.0667	0.0685
17	0.0411	0.0442	0.0471	0.0497	0.0522	0.0546	0.0568	0.0588	0.0608
18	0.0318	0.0352	0.0383	0.0412	0.0439	0.0465	0.0489	0.0511	0.0532
19	0.0227	0.0263	0.0296	0.0328	0.0357	0.0385	0.0411	0.0436	0.0459
20	0.0136	0.0175	0.0211	0.0245	0.0277	0.0307	0.0335	0.0361	0.0386
21	0.0045	0.0087	0.0126	0.0163	0.0197	0.0229	0.0259	0.0288	0.0314
22	—	—	0.0042	0.0081	0.0118	0.0153	0.0185	0.0215	0.0244
23	—	—	—	—	0.0039	0.0076	0.0111	0.0143	0.0174
24	—	—	—	—	—	—	0.0037	0.0071	0.0104
25	—	—	—	—	—	—	—	—	0.0035

Table A2. Quantiles of the Shapiro-Wilk test of normality. From Shapiro and Wilk (1965)

	Probability level					
n	0.010	0.020	0.050	0.100	0.500	0.900
3	0.753	0.756	0.767	0.789	0.959	0.998
4	0.687	0.707	0.748	0.792	0.935	0.987
5	0.686	0.715	0.762	0.806	0.927	0.979
6	0.713	0.743	0.788	0.826	0.927	0.974
7	0.730	0.760	0.803	0.838	0.928	0.972
8	0.749	0.778	0.818	0.851	0.932	0.972
9	0.764	0.791	0.829	0.859	0.935	0.972
10	0.781	0.806	0.842	0.869	0.938	0.972
11	0.792	0.817	0.850	0.876	0.940	0.973
12	0.805	0.828	0.859	0.883	0.943	0.973
13	0.814	0.837	0.866	0.889	0.945	0.794
14	0.825	0.846	0.874	0.895	0.947	0.795
15	0.835	0.855	0.881	0.901	0.950	0.975
16	0.844	0.863	0.887	0.906	0.952	0.976
17	0.851	0.869	0.892	0.910	0.954	0.977
18	0.858	0.874	0.897	0.914	0.956	0.978
19	0.863	0.879	0.901	0.917	0.957	0.978
20	0.868	0.884	0.905	0.920	0.959	0.979
21	0.873	0.888	0.908	0.923	0.960	0.980
22	0.878	0.892	0.911	0.926	0.961	0.980
23	0.881	0.895	0.914	0.928	0.962	0.981
24	0.884	0.898	0.916	0.930	0.963	0.981
25	0.888	0.901	0.918	0.931	0.964	0.981
26	0.891	0.904	0.920	0.933	0.965	0.982
27	0.894	0.906	0.923	0.935	0.965	0.982
28	0.896	0.908	0.924	0.936	0.966	0.982
29	0.898	0.910	0.926	0.937	0.966	0.982
30	0.900	0.912	0.927	0.939	0.967	0.983
31	0.902	0.914	0.929	0.940	0.967	0.983
32	0.904	0.915	0.930	0.941	0.968	0.983
33	0.906	0.917	0.931	0.942	0.968	0.983
34	0.908	0.919	0.933	0.943	0.969	0.983
35	0.910	0.920	0.934	0.944	0.969	0.984
36	0.912	0.922	0.935	0.945	0.970	0.984
37	0.914	0.924	0.936	0.946	0.970	0.984
38	0.916	0.925	0.938	0.947	0.971	0.984
39	0.917	0.927	0.939	0.948	0.971	0.984
40	0.919	0.928	0.940	0.949	0.972	0.985
41	0.920	0.929	0.941	0.950	0.972	0.985
42	0.922	0.930	0.942	0.951	0.972	0.985
43	0.923	0.932	0.943	0.951	0.973	0.985
44	0.924	9.933	0.944	0.952	0.973	0.985
45	0.926	0.934	0.945	0.953	0.973	0.985
46	0.927	0.935	0.945	0.953	0.974	0.985
47	0.928	0.936	0.946	0.954	0.974	0.985
48	0.929	0.937	0.947	0.954	0.974	0.985
49	0.929	0.937	0.947	0.955	0.974	9.985
50	0.930	0.938	0.947	0.955	0.974	0.985

Table A3. Quantiles of the D'Agostino test of normality. From D'Agostino (1971)

	Probability level					
n	0.005	0.010	0.025	0.050	0.100	0.900
50	−3.949	−3.442	−2.757	−2.220	−1.661	0.759
60	−3.846	−3.360	−2.699	−2.179	−1.634	0.807
70	−3.762	−3.293	−2.652	−2.146	−1.612	0.844
80	−3.693	−3.237	−2.613	−2.118	−1.594	0.874
90	−3.635	−3.100	−2.580	−2.095	−1.579	0.899
100	−3.584	−3.150	−2.552	−2.075	−1.566	0.90
150	−3.409	−3.009	−2.452	−2.004	−1.520	0.990
200	−3.302	−2.922	−2.391	−1.960	−1.491	1.032
250	−3.227	−2.861	−2.348	−1.926	−1.471	1.060
300	−3.172	−2.816	−2.316	−1.906	−1.456	1.080
350	−3.129	−2.781	−2.291	−1.888	−1.444	1.096
400	−3.094	−2.753	−2.270	−1.873	−1.434	1.108
450	−3.064	−2.729	−2.253	−1.861	−1.426	1.119
500	−3.040	−2.709	−2.239	−1.850	−1.419	1.127
550	−3.019	−2.691	−2.226	−1.841	−1.413	1.135
600	−3.000	−2.676	−2.215	−1.833	−1.408	1.141
650	−2.984	−2.663	−2.206	−1.826	−1.403	1.147
700	−2.969	−2.651	−2.197	−1.820	−1.399	1.152
750	−2.956	−2.640	−2.189	−1.814	−1.395	1.157
800	−2.944	−2.630	−2.182	−1.809	−1.392	1.161
850	−2.933	−2.621	−2.176	−1.804	−1.389	1.165
900	−2.923	−2.613	−2.170	−1.800	−1.386	1.168
950	−2.914	−2.605	−2.164	−1.796	−1.383	1.171
1000	−2.906	−2.599	−2.159	−1.792	−1.381	1.174

Table A4. Factors for calculating lower 0.95 confidence limit of lognormally distributed data. Interpolated from Land (1973)

	Sample number					
s	4	6	12	20	40	100
0.10	−2.006	−1.759	−1.679	−1.657	−1.648	−1.643
0.20	−1.865	−1.697	−1.646	−1.637	−1.636	−1.641
0.50	−1.626	−1.592	−1.608	−1.628	−1.651	−1.682
1.00	−1.522	−1.588	−1.677	−1.737	−1.796	−1.873
2.00	−1.686	−1.864	−2.075	−2.211	−2.347	−2.523
5.00	−2.877	−3.357	−3.919	−4.290	−4.659	−5.140
10.00	−5.328	−6.280	−7.416	−8.170	−8.921	−9.904

Table A5. Factors for calculating upper 0.95 confidence limit of lognormally distributed data. Interpolated from Land (1973)

	Sample number					
s	4	6	12	20	40	100
0.10	2.481	1.942	1.780	1.729	1.698	1.671
0.20	2.886	2.069	1.852	1.780	1.737	1.697
0.50	5.209	2.638	2.157	2.009	1.917	1.831
1.00	10.076	4.128	2.954	2.607	2.401	2.207
2.00	19.994	7.701	5.054	4.243	3.756	3.301
5.00	49.887	18.880	12.010	9.849	8.534	7.291
10.00	99.753	37.660	23.835	19.475	16.815	14.287

Computer Programs Available in USDA Technology Transfer Documents

1. NSTL91-1 PROGEN.BAS contains routines for reading, printing, and writing ASCII files to disk.

2. NSTL91-2 PROBIT.BAS performs probit analysis.

   MORMTEST.BAS performs test of normality using Shapiro-Wilk procedure and D'Agostino's procedure.

3. NSTL91-3 PSIEVAL.BAS evaluates Finney's PSI function.

   LSTATS.BAS calculates descriptive sample statistics for lognormally distributed data. UMVU, ML, and MM estimators are calculated along with Land's exact confidence interval method. Algorithms exist for computing and plotting sample histogram.

4. NSTL91-4 LPDFPLOT.BAS plots 2 parameter lognormal probability density function with user defined mean and standard deviation.

   LMCARLO.BAS performs Monte Carlo sampling from lognormal distribution with user-specified mean and standard deviation.

Nutrient Transformations in Soils Amended with Green Manures

Yadvinder-Singh, Bijay-Singh, and C.S. Khind

Advances in Soil Science, Volume 20

I. Introduction

The adoption of modern agricultural technologies has revolutionized food grain production bringing it up to levels to feed the quickly growing population of the world. Worldwide there is more grain per capita available today than ever before (Stewart, 1987). The application of plant nutrients to different crops has contributed substantially towards this spectacular increase. Herdt and Capule (1983) estimated that from 1965 to 1980 increase in fertilizer use contributed around 25% of the total increase in rice production in Asia.

In the past three decades, fertilizers have been the most commonly used sources for supplying nutrients to crop plants. In developing countries like India, fertilizer prices have been subsidized, thereby encouraging farmers to apply fertilizers in production-maximizing doses. The low cost, and readily available chemical fertilizers have pushed the use of organic manures, including green manures, the traditionally important sources of nutrients, into a decline the world over. However, in the face of a continuing energy crisis, increasing fertilizer prices, and growing concern for environmental quality, there has been, in recent years, a tremendous renewal of interest, on the part of both researchers and farmers, in the old practice of green manuring. Another issue of great concern associated with this practice is the sustainability of soil productivity as land is called upon to produce higher yields from a single crop and higher total yields under intensive cropping systems. Also, leguminous green manure crops are being seriously examined in countries where inorganic fertilizers are not likely to be available in sufficient quantities in the foreseeable future, and where the fertilizer crop price ratios are relatively high.

The value of leguminous green manure crops for improving soil fertility has been recognized since very early times. The benefits credited to them include increases in organic matter content and available plant nutrients and improvement in the microbiological and physical properties of the soil. Of these, the role of green manures in supplying plant nutrients, particularly N is most prominent. It has been estimated that global N flux from symbiotic fixation is about twice the input from fertilizers (Paul, 1988). The nitrogen fertilizer equivalence, an index used to compare green manure efficiency with chemical fertilizer, has been found to range from 40 to 200 kg N ha^{-1} in different crops but more typically is between 75 and 100 kg N ha^{-1} (Smith et al., 1987; Meelu et al., in press). Green manure must undergo decomposition and mineralization before its N becomes available to the succeeding crop. To utilize green manure N to the fullest extent, a more comprehensive understanding of N transformation processes, and the factors that govern them is essential for synchronizing the release of green manure N with uptake of N by crops.

The addition of organic matter in the form of green manures greatly influences the transformation and availability of N and several other essential plant nutrients through its impact on the chemical and biological properties of soils. On mineralization the green manures liberate their P in the soil in plant-available forms. Organic acids formed during the course of decomposition of green manures interact with soil components, thereby considerably enriching the pool of available P (Bin, 1983; Watanabe, 1984). In green-manure-treated soils, active biodegradation and the resultant extensive production of CO_2 can lead to the dissolution of Ca and Mg carbonates. In waterlogged soils, green manuring can markedly influence the chemical equilibria of S (Pan, 1985). The triggering of oxidation-reduction reactions, particularly in flooded soils, and increased chelation capacity brought about by the addition of green manures dictate the transformations and availability of micronutrients, particularly Fe and Mn.

The literature on green manuring is voluminous, but the investigations involving nutrient transformations have been carried out more often in the past decade. Recent attempts to reevaluate the usefulness of green manures in the context of sustainable intensive farming and environmental quality have many times been hindered by a lack of readily available and current information on nutrient transformations in soils amended with green manures. This paper attempts to compile and analyze the information available at present. After briefly reviewing the nutrient contributions of green manures, changes in different chemical and electrochemical properties of soils brought about by green manuring will be examined. These will be followed by a review of the transformations of N, P, K, S, and micronutrients in soils treated with green manures. Keeping in view the role of green manure as the basis of integrated nutrient management systems in sustaining food production and improving environmental quality, the influence of green manuring on the transformations of nutrients already present in soil or those applied through fertilizers is also discussed.

II. Nutrient Contributions by Green Manures

A. Biomass and N Accumulation

The benefits accruing from green manures are directly related to the amount of biomass and nutrients added to soil. Biomass production of green manure crops varies widely according to the species of the legume, environmental conditions, soil fertility, and crop management practices. Table 1 shows that biomass production in different symbiotic N_2 fixation systems varied from 1.3 to 24.0 t dry matter ha^{-1}, with largest values in the food legumes. In tropical legumes, age is a key factor controlling

Table 1. Estimates of biomass production and N accumulation in different symbiotic N fixation systems

N fixation system	Age (days)	Dry matter ($t\,ha^{-1}$)	N accumulation ($kg\,ha^{-1}$)
Tropical legumes	40–49	1.3–4.9	25–116
	50–60	2.9–8.9	60–225
Winter legumes			
Milk vetch	November to May	19.0–75.0[a]	65–338
Other legumes	November to May	1.7–6.7	56–209
Food legumes	80–120	4.0–24.0	40–240
Stem nodulating legumes	45–60	16.0–19.0	98–532
Legume trees	60–365	3.6–20.6	134–295
Azolla	40–365	29–390[a]	53–913

[a] Fresh matter

biomass accumulation (Yadvinder-Singh et al., 1991). Among the tree legumes, *Leucaena leucocephala* yields generally more biomass than other shrubby tropical legumes (Brewbaker, 1987). Stem-nodulating legumes possess large potential for biomass production in short periods. Biomass production in azolla shows a wide range, depending on a host of factors.

The contribution of N from leguminous green manure crops is their most commonly observed benefit and almost certainly their primary advantage. Nitrogen accumulation ranges from 25 to 199 $kg\,ha^{-1}$ for tropical legumes, 56 to 338 $kg\,ha^{-1}$ for winter legumes, 134 to 295 $kg\,ha^{-1}$ for legume trees, 40 to 240 $kg\,ha^{-1}$ for grain legumes, and 98 to 532 $kg\,ha^{-1}$ for stem-nodulating legumes (Table 1). Ladha et al. (1989) estimated that 55-day-old sesbania fixed as much as 303 $kg\,N\,ha^{-1}$ (5.5 $kg\,N\,day^{-1}$) when uninoculated and 383 $kg\,N\,ha^{-1}$ (6.96 $kg\,N\,day^{-1}$) when inoculated with *Azorhizobium*. In azolla, N accumulation ranged from 53 to 913 $kg\,N\,ha^{-1}$ depending upon multiplication methods.

B. Nutrient Composition

Table 2 shows the concentration of N and P, and the C:N ratios in the aboveground portions of several common green manure crops including azolla. The data were averaged for different studies. N concentration in different cultivated annual legumes ranged from 2.50 to 3.91%. The straws of food legumes have a comparatively low N content, ranging from 1.70% in cowpea/pea residues to 2.21% in mungbean, obviously because food legumes are harvested near maturity when straw N is translocated into grain N. The N content in the foliage of tree legumes used for green leaf manuring is usually high. The N content in azolla varies from 2.81 to 4.90% depending upon the species and environmental and experimental conditions.

Table 2. Total N and P contents and C:N ratios in above-ground portions of some important green manure crops

Green manure crop	Total concentration[a] (% dry weight) N	P	C:N ratio[a]
Sesbania (*Sesbania aculeata*)	2.62 (1.79–4.16) *n* = 21	0.32 (0.21–0.40) *n* = 17	16.4 (8.6–28.8) *n* = 21
Sunnhemp (*Crotolaria juncea*)	2.86 (2.00–3.12) *n* = 7	0.34 (0.18–0.40) *n* = 9	16.1 (13.5–24.0) *n* = 7
Cowpea (*Vigna unguiculata*)	2.69 (1.60–3.89) *n* = 6	0.28 (0.22–0.43) *n* = 6	17.1 (10.3–3.20) *n* = 4
Clusterbean (*Cyamopsis tetragonoloba*)	2.80 (2.30–3.39) *n* = 2	—	17.3 (12.7–22.0) *n* = 2
Milk vetch (*Astragalus sinicus*)	3.20	—	13.1
Crimson clover (*Trifolium incarnatum*)	2.50 (2.20–2.94) *n* = 7	—	15.0 (13.5–15.9) *n* = 3
Hairy vetch (*Vicia villosa*)	3.91 (3.60–4.22) *n* = 5	—	10.0 (9.5–10.5) *n* = 2
Mungbean residue (*Vigna radiata*)	2.21	—	16.1
Azolla	4.46 (1.95–6.57) *n* = 237	0.38 (0.063–1.74) *n* = 236	10.3 (9.0–12.9) *n* = 3
Leucaena (*Leucaena leucocephala*)	3.15 (2.42–3.50) *n* = 7	0.20 (0.13–0.27) *n* = 6	12.7[b]
Gliricidia (*Gliricidia maculata*)	3.49 (2.50–4.30) *n* = 6	0.22 (0.13–0.29) *n* = 4	10.42[b]
Sesbania sesban	3.98 (3.76–4.18) *n* = 6	0.24 (0.18–0.27) *n* = 3	10
Calopo (*Calopogonium mucunoides*)	4.20	—	8.5

[a] Figures in parenthesis indicate range

[b] Calculated by assuming 40% C in the above-ground portion

As shown in Table 2, C:N ratios vary widely among the legume species. Hairy vetch has been observed to have a narrow C:N ratio compared with other annual legumes. In sesbania (*Sesbania aculeata*), C:N ratios ranged from 8.6 to 28.8. The food legume residues have generally high C:N ratios. Legume trees and azolla, however, possess narrow C:N ratios, ranging from 8.0 in centrosema to 12.7 in leucaena. As a green manure crop grows, its protein content decreases and soluble carbohydrates, cellulose, and C:N ratio increase resulting in less decomposable material (Ishikawa, 1988). Different parts of a green manure plant also vary in N and lignin content, as well as in C:N ratios. Foliage contains higher N but lower lignin content and C:N ratio than stems or roots (Watanabe, 1984; Frankenberger and Abdelmagid, 1985; Palm et al., 1988).

The phosphorus content of green manure crops ranges from 0.20 to 0.34% (Table 2). Green leaf manures have lower P content (0.20 to 0.24%) than leguminous green manures and azolla in which P content ranges from 0.28 to 0.38%. The concentration of K, S, Ca, and Mg vary considerably depending upon the legume species (Table 3). The concentration of micronutrients in leguminous green manures is expected to be similar to those of nonleguminous crop plants.

III. Effect of Green Manuring on Chemical and Electrochemical Properties of Soils

The properties most likely to be affected by organic matter additions in the form of green manures are pH, redox potential, partial pressure of CO_2, and surface properties of soils. Several nutrient transformations in soils brought about by green manuring are in fact a direct consequence of changes in these soil properties.

A. Soil pH

Green manuring effects change in soil pH in two ways. Organic acids and CO_2 produced during decomposition of green manure can furnish protons to the soil, inducing a decrease in pH. However, organic reducing substances formed during decomposition of green manure may reduce Fe and Mn oxides causing soil pH to rise because protons are consumed in the course of the reduction of oxides. Increased soil pH may also result through mineralization of organic anions to CO_2 and H_2O, thereby removing H^+. The average ash alkalinity of legume crops (110 cmol kg^{-1}) is substantially greater than that of forage crops (69 cmol kg^{-1}) (Pierre and Banwart, 1973). The alkaline nature of the plant material arises from the dissociation of the organic acids (metabolized within the plant) in response to cation/anion imbalance caused by biological N fixation or NH_4^+ uptake. Plants reduce this imbalance by excreting H^+ ions into the rhizosphere. During green manuring, anions become separated from H^+ ions by being returned to the surface soil layers rather than to the subsoil, where the majority of roots originally released the H^+ ions. The

Table 3. K, S, Ca, and Mg contents in above-ground portions of some green manure crops

Green manure crop	Nutrient content (% dry weight basis) K	S	Ca	Mg
Sesbania	1.48 (0.62–3.50) $n = 7$	0.19	1.40	1.62
Cowpea	2.26 (1.53–2.35) $n = 3$	0.28 (0.23–0.33) $n = 2$	1.50	1.73
Leucaena leucocephala	1.73 (1.30–2.49) $n = 3$	—	1.88 (0.80–2.70) $n = 5$	0.41 (0.22–0.60) $n = 2$
Gliricidia sepium	2.44 (1.60–3.00) $n = 3$	—	1.89	0.43 (0.32–0.54) $n = 2$
S. sesban	1.98 (1.40–2.55) $n = 2$	0.58	—	0.27
Azolla	1.65 (1.21–2.09) $n = 4$	— 0.38–0.84 $n = 10$	1.04	1.24

direction and magnitude of change in soil pH after incorporation of green manure are determined by the relative number of protons consumed and released.

As shown in Figure 1, application of sesbania green manure under waterlogged conditions to acid, noncalcareous, and gypsum-amended sodic soils helped in rapidly achieving near-neutral soil pH—the commonly observed effect of submerging a soil. Similar results have been reported by Katyal (1977) by green manuring black clay (pH 7.8), lateritic loam (pH 6.4), and red sandy loam (pH 7.5) soils with *Gliricidia sepium*. More conspicuously in alkaline soils, pH dropped by nearly 1 to 2 units by green manuring during the first week of flooding. After the initial sudden drop, the pH increased and attained more or less steady values after 4 weeks. Ponnamperuma (1965) related the fall in pH of flooded soils to accumulation of CO_2 and soil reduction, and Motomura (1962) related it to production of organic acids. Cang et al. (1985) observed that the fall in pH was related to the amount of easily decomposable organic matter added through green manure. The subsequent increase in pH has been attributed to the reduction of Fe (Ponnamperuma, 1965; Islam and Islam, 1973). Temperature affected the change in soil pH by influencing the decomposition rate of green manure. The work of Cang et al. (1985) revealed that change in pH was indistinct at 15°C and was remarkable when the soil was incubated at 28°C.

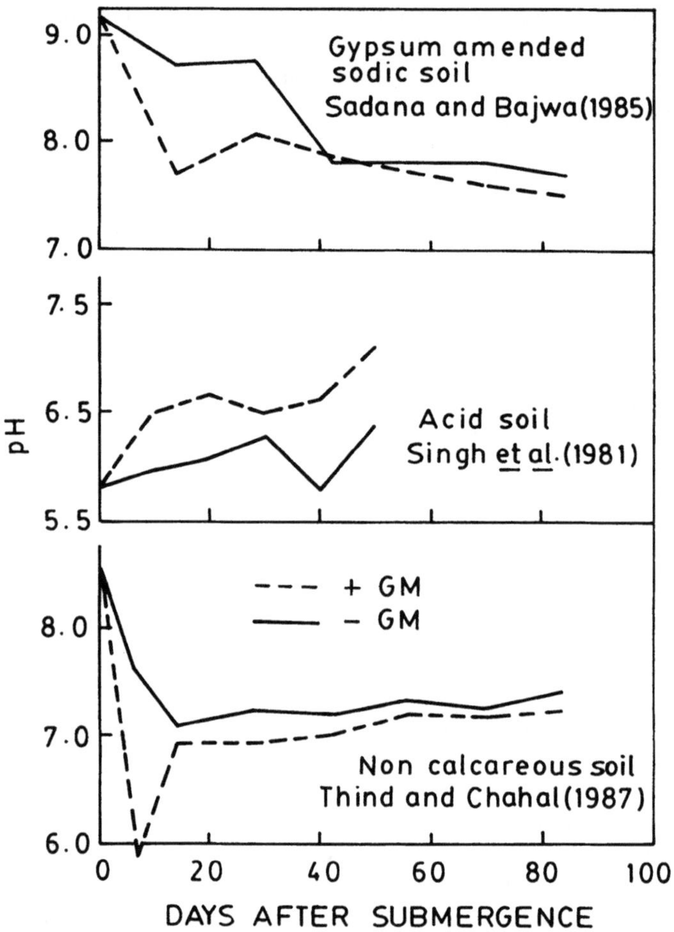

Figure 1. Effect of green manuring with sesbania on changes in pH of soils under waterlogged conditions.

Bajpai et al. (1980) observed that under aerobic soil conditions green manuring with sesbania reduced pH effectively during the initial 2-week incubation period in a loam soil (pH 7.2). However, in a saline sodic clay loam soil (pH 9.9) the effect of green manuring was observed only after 2 weeks. In an acid soil, Debnath and Hajra (1972) observed a significant increase in soil pH only during the first 30 days of incubation under aerobic conditions. In comparison with waterlogged situations, changes in soil pH with green manuring under upland conditions are less marked.

Green manuring may have a strong influence on short-term changes in soil pH that do not necessarily reflect the overall long-term changes. Morachan et al. (1972) reported that green manuring corn for 3 years with alfalfa residues at 16 t ha^{-1} reduced the soil pH from 5.3 to 5.1. Yadav and Singh (1986) reported that soil pH after 12 years of continuous sugarcane cropping was 8.0 and 8.5 under green manure and control treatments, respectively. Hargrove (1986) found that legume cover crops caused a significant reduction in soil pH in a depth of 300 mm. Swarup (1987) observed a decrease in soil pH of sodic soil by 0.2 units after harvesting a rice crop green manured with sesbania.

B. Redox Potential

Redox potential (Eh) is very closely linked to the O_2 status of the soils. The CO_2 evolved during microbial decomposition of green manure will diffuse relatively quickly when the soil air-filled porosity is large. Green manuring, therefore, is not likely to significantly affect the Eh of aerobic soils. The addition of green manure to waterlogged soils will enhance the rate of O_2 consumption (in microbiological respiration and chemical oxidation of organic reducing substances) resulting in CO_2 accumulation. This leads to a lower redox potential. The rate of reduction should depend on the relative rate of production of organic reducing substances and the supply of O_2. In a laboratory study, Yu (1985) observed that the diffusion current (a voltametric measure of reducing substances) for the red soil without the addition of green manures was only 0.39 $\mu A/cm^2$ at 700 mV, but it increased to 2.94 $\mu A/cm^2$ when the soil was amended with green manure.

Several studies have shown that green manures caused dramatic reduction in Eh of waterlogged soils differing in pH, organic matter, and easily reducible Fe and Mn contents (Katyal, 1977; Thind and Chahal, 1983; Sadana and Bajwa, 1985; Wen and Yu, 1988; Khind et al. 1987). The magnitude of depression was more pronounced during the early stages (1 to 4 weeks) of submergence. More reduction has been reported in noncalcareous than in calcareous or sodic soils. The rate of reduction is determined by the amounts of easily reducible Fe and Mn oxides, NO_3^-, and SO_4^{2-} on the one hand and by the amount of easily decomposable organic substances on the other. Katyal (1977) observed greater depression in Eh of black clay soil which contained smaller amounts of reducible Fe and Mn than of laterite or red soils. If the amount of organic reducing substances is smaller than the amounts of easily reducible Fe and Mn, the redox potential of the soil can be maintained at 100 to 200 mV. If it is large, the Eh can be negative, and under extreme reducing substances, it can even be lower than—100 mV. Green manuring can lower the Eh not only in the surface layer but also in the subsoil (Wen and Yu, 1988).

Green manuring affects the rate of formation and the thickness of the oxidized layer of submerged soils. If the amount of easily decomposable organic matter is less and the soil is very permeable, there will be no differentiation between oxidized and reduced layers, obviously due to the weak development of reducing conditions during submergence. Yu (1985) reported that application of green manure to a sandy soil reduced the thickness of the oxidized layer from 6 mm to 3 mm. Green manuring reduced Eh of the oxidized and reduced layers from 320 to 225 mV and from 68 to −79 mV, respectively. Addition of sesbania green manure (0.2% w/w) to a sandy loam soil containing 0.32% organic carbon reduced the thickness of the oxidized layer from 5 mm to 2 mm, but in a Crowley silt loam containing 0.87% organic carbon the thickness of the oxidized layer was unaffected and it remained 2 mm (C.S. Khind, unpublished data).

C. Partial Pressure of Carbon Dioxide

Katyal (1977) observed higher and sharper peaks of P_{CO_2} in waterlogged soils amended with green manure. The initial rise in P_{CO_2} after flooding is brought by aerobic respiration followed by the anaerobic decomposition of organic matter. The steep decline in P_{CO_2} after a peak may be due to the formation of sparingly soluble carbonates of Fe and Mn. The decrease in P_{CO_2} will, therefore, be slow in a soil with low contents of Fe and Mn. In a laboratory incubation study with a salt-affected soil, Sadana and Bajwa (1985) found marked increase in P_{CO_2} with the addition of green manure. The differences in P_{CO_2} persisted over a 12-week period. Organic reducing substances released during decomposition of green manures can also react with O_2. Yu (1985) observed that in subsoil of a red soil with very low organic matter, there was no detectable consumption of O_2 during submergence but O_2 content dropped from 11.8 to 4.2 mg l^{-1} within 1 to 2 min after the addition of an incubation solution of milk vetch green manure.

The enhanced CO_2 in the soil may increase the rate of photosynthesis, and it may regulate the activity of Fe and Mn. CO_2 has an important buffering function and in some soils with low CEC it may reduce phytotoxicity due to ammonia.

D. Electrical Conductivity

Generally it has been conjectured that little or no change in electrical conductivity (EC) occurs by applying organic manures because even with fertilizers which are soluble salts, the increase in EC has been short-lived. Under waterlogged conditions, EC of a soil solution increases with time, reaches a peak, and then decreases. Addition of readily decomposable organic matter in the form of a green manure would accelerate the decrease in Eh and accumulation of CO_2, and as a consequence result in the release of a large amount of ions into the soil solution and increased EC. Katyal (1977) reported that application of green manure to red, black, and alluvial soils caused a sharp increase in EC within 14 days of

Table 4. Effect of milk vetch green manuring on the activity of iron oxides of soils incubated under submerged conditions for 8 months (after He and Xu, 1981)

Soil	Milk vetch applied (%)	Fe_d (%)	$Fe_o:Fe_d$		
			Submerged	pF^2	Air-dried
Yellowish brown earth	0	2.41	0.35	0.14	0.20
	5	2.23	0.56	0.40	0.38
Red earth	0	4.77	0.13	0.06	0.06
	5	4.77	0.38	0.11	0.13
Laterite	0	13.0	0.05	0.02	0.04
	5	13.1	0.12	0.04	0.06

Fe_d, free iron oxide; Fe_o, amorphous iron oxide

flooding. Then there was a quick decline for up to 28 days followed by an asymptotic decrease. Sadana and Bajwa (1986) observed that addition of green manure along with gypsum caused a greater increase in EC than gypsum alone. Thind and Chahal (1987) observed that the increase in EC by green manuring under flooded conditions was more marked in a noncalcareous soil than in a calcareous one. The decrease in EC after an initial rise may be due to reprecipitation of Fe^{2+} and Mn^{2+} and consequent adsorption of other cations on the exchange sites, decrease in P_{CO_2}, and decomposition of organic acids (Ponnamperuma, 1972).

E. Activity of Iron Oxides

Green manures containing a large amount of active substances support reductive eluviation in soil which results in a gradual decrease in free iron oxide content and, therefore, positive charges carried by a soil particle. Using the ratio of amorphous iron oxide to free iron oxide ($Fe_o:Fe_d$) as an index of the activity of iron oxides in the soil, He and Xu (1981) observed a spectacular effect of adding milk vetch green manure to the soil before incubation in increasing this activity (Table 4). The change in the activity of iron oxide can have an important bearing on the surface properties of soils. For example, Wu and Chen (1981) observed a significant correlation between the extent of specific adsorption of Cu ions and the activity of iron oxides. Similar results have been reported for Cd and Zn (Wen and Yu, 1988). Since ferric oxides possess a high affinity for phosphate ions, a decrease in their content due to green manuring would mean a reduced fixation of phosphates by soils.

F. Organic Metabolites

During decomposition of a green manure several organic compounds accumulate in the soil. The decomposition proceeds in two different stages, the first involving the formation of organic acids and the second their conversion into gaseous products. In the aerobic soils the second stage of decomposition, in which organic compounds decompose further into CO_2, is normally faster than the first stage, and little accumulation of organic acids occurs under optimum soil temperature and moisture conditions. Ishikawa (1988) showed that organic acids can accumulate in significant amounts during decomposition of milk vetch green manure under waterlogged conditions. The amount of organic acids produced increased with temperature. The formation of organic acids was less when flooding began 5 days after the application of milk vetch green manure. This was due to the fast and simultaneous production and decomposition of organic acids under aerobic conditions.

Watanabe (1984) listed a large number of organic compounds which can accumulate in waterlogged soils, but observed that among volatile fatty acids acetic and butyric acids accumulated in large quantities with green manure incorporation. Ceccenti et al. (1990) reported that acetic acid decreased and butyric acid increased with time. Wang et al. (1967a) found that *Crotolaria juncea* green manure produced volatile fatty acids (mainly acetic), nonvolatile aliphatic acids (mainly tartaric), and phenolic acids (mainly p-hydroxybenzoic, vanillic, and ferulic). The relative abundance of butyric acid among the volatile fatty acids was higher from green manure than from graminaceous plant residues (Fujii et al., 1972; Takeda and Furusaka, 1975). Tsutsuki (1984) reported that alcohols were formed in significant amounts in submerged soil only when 1% green manure (*Gliricidia sepium*) was applied. The peak concentrations of alcohols (methanol and ethanol) were attained between 1 and 3 days after submergence. The kind and amount of alcohols formed varied with the type of soil and the kind of organic matter added. Wang et al. (1967b) observed accumulation of methanol, ethanol, n-propanol, and n-butanol in soils amended with *Crotolaria juncea.* An accumulation of aldehydes (mainly acetaldehyde) in the soils amended with 1% green manure was also detected under submerged conditions, but the amount formed was very low; 0.1 mmol at 35°C or 0.05 mmol at 20°C. An accumulation of amines (putresine, ethylamine, isobutrylamine) has been detected during anaerobic decomposition of clover plant residue (Fujii et al., 1972).

During decomposition of green manures large amounts of water-soluble products are formed with their relatively low (500 to 1000) molecular weights (Ceccenti and Ding, 1985). Some of these products are charged negatively and some positively (Yu, 1981). The amount of these products which have the ability to form a complex with some metal ions varies with the kind of green manure and is time dependent (Bao and Yu, 1986).

Application of green manure to soil can stimulate formation of ethylene (C_2H_4), which acts as a plant hormone for regulating root growth (Smith, 1976; Primrose, 1979). Small concentrations of volatile and nonvolatile fatty acids, C_2H_4, and H_2S can modify plant growth.

Organic acids can retard root elongation, restrict nutrient uptake, and reduce shoot weight (Watanabe, 1987). A concentration of acetic acid greater than 1 mmol restricted the uptake of P and K (Rao and Mikkelson, 1977).

IV. Nitrogen

Microbial decomposition constitutes the main process by which N held in the structure of plant tissues is released into the soil for reuse by plants. During decomposition of green manures, carbon is returned to the atmosphere as CO_2, while organic N is converted into NH_4^+– and NO_3^-–N through the process of N mineralization. Microbes involved in the decomposition of green manure also assimilate a part of the N leading to a process called N immobilization. Mineralization and immobilization of green manure N, therefore, constitute the two major biological processes controlling availability of N to plants. These two processes proceed continuously and simultaneously. Depending on their relative dominance, there is either a net increase or a net decrease in the inorganic N concentration, which is known as mineralization-immobilization turnover.

The composition of green manure affects the overall mineralization rate. The net N release from green manure is the balance of all the N transformation processes occurring in the different components of the green manure. The NH_4^+–N released through ammonification of green manure N is adsorbed rapidly into the cation exchange complex of the soil; a portion may be fixed in nonexchangeable forms, particularly in soils containing significant amounts of vermiculite and illite clay minerals. Under aerobic conditions, it is converted to NO_3^-–N through nitrification. The relative proportions of exchangeable and soil solution NH_4^+ are largely governed by the CEC of the soil and the nature of the cation exchange complex. Similar to N applied through chemical fertilizers, NH_4^+–N released from green manure may be partly lost through leaching as NO_3^-, NH_3 volatilization, and nitrification-denitrification. The efficiency of green manure should, therefore, depend on the rates of N release into the soil and the extent of N losses from the soil-plant system.

A. Kinetics of N Mineralization

Both conceptual and mathematical models have been used for studying N mineralization from organic substrates applied to soil. The simplest model is the mathematical formulation for the first-order kinetics based on the premise that only a part of the N present in the organic substrate is susceptible to decomposition, and that N mineralization proceeds at a rate proportional to its concentration. The equation used to estimate N mineralization potential and the first-order rate constant is:

$$N_m = N_o(1 - e^{-k.t})$$

where N_m is cumulative amount of N mineralized during a specific time interval t, N_o is N mineralization potential, and k is the first-order rate constant. Frankenberger and Abdelmagid (1985) used a nonlinear least-square method for computing N_o and k for N mineralization from different legume residues following first-order kinetics. The N_o values among different legume residues varied from 5 to 445 $\mu g\,N\,g^{-1}$ soil, reflecting the dissimilarity in the composition and amount of mineralizable N. The mineralization rate constant k, ranged from 0.061 to 0.325 week^{-1}, indicating that the mineralizable N fraction in various organs of leguminous plants (foliage, stems, and roots) was released at different rates. The required time of aerobic incubation at 23°C to mineralize one-half of N_o ($t_{1/2}$) for each of the specific organs of the four legumes ranged from 2.1 to 11.4 weeks. The $t_{1/2}$ for the foliage, stem, and root combinations ranged from 2.3 to 5.4 weeks.

The simple first-order reactions used to calculate N mineralization rates from organic residues have serious drawbacks. Data obtained from several studies indicate that plant residues (green manures) cannot have a constant decomposition rate. Initially, there is a rapid breakdown of sugars, proteins, and celluloses, the remaining material being more resistant. Also, the easily decomposable materials are partly synthesized into microbial products which decay at an intermediate rate. Bouldin (1988) proposed a simple two-component model to describe the N mineralization pattern of green manures. The organic material is treated as if there are two distinct components—one decomposing rapidly and the other slowly. The first fraction (fast-N) determines the potential supply of N to the first crop, and the second fraction (slow-N) determines the residual effects on N supply and soil organic matter. He proposed that 65% of the added N mineralizes during the first crop (4 to 6 months), 14% of the residual N mineralizes during the second crop, and 3.3% mineralizes during each succeeding year. Yadvinder-Singh et al. (1988) described the N mineralization kinetics of green manures incubated under aerobic conditions by using two simultaneous first-order reactions as:

$$N_m = N_{oq}(1 - e^{-kq.t}) + N_{os}(1 - e^{-ks.t})$$

where q and s correspond to quick and slow reactions. The quick reaction represents N mineralization rate for the labile green manure fraction, and the second term for the slowly decomposing nonlabile fraction. The model predicted the release of N from green manure well (Figure 2). The values of k for quick and slow reactions for sesbania green manure were calculated as 2.12 and 0.069 week^{-1}, respectively.

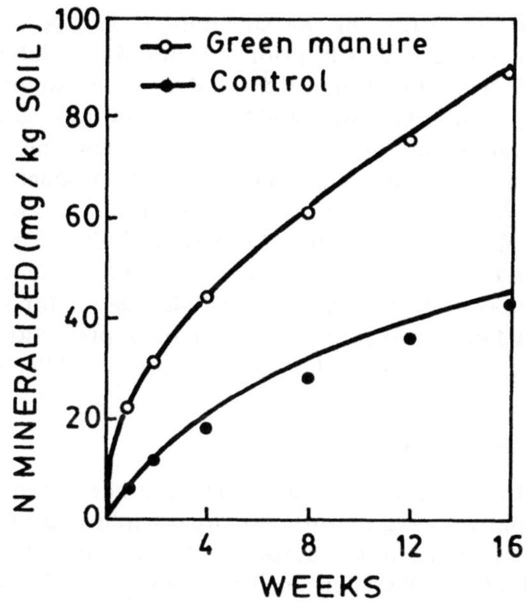

Figure 2. Kinetics of N Mineralization from sesbania green manure in a soil at field capacity moisture regime (after Yadvinder-Singh et al., 1988).

Wilson and Hargrove (1986) and Wagger (1989) used an exponential equation to describe N release patterns from different legume residues under field conditions:

$$Y = \mathrm{a} + \mathrm{b}e^{-\mathrm{c}.t}$$

where Y is legume N remaining (%) at time t (in weeks) and a, b, and c are constants. The exponential pattern supports the idea that residues contain constituents of varying degrees of resistance to decomposition with the readily decomposable ones disappearing rapidly and the more resistant ones slowly.

Gilmour et al. (1985) and Gale and Gilmour (1988) used a computer simulation model to estimate the net N and net C mineralization from alfalfa and clover residues under both aerobic and anaerobic conditions. The model partitioned the decomposing substrate into indigenous biomass and one created upon addition of substrate. The major inputs of the model were the experimentally determined C mineralization rate constants and the C:N ratio of the substrate. The introduction of the very rapid fraction, recalculation of the C:N ratio after that fraction had been mineralized, and reintroduction of the modified C:N ratio were the key features of the method. The plant tissues were represented by rapid, intermediate, and slow fractions. To account for the initial flush of

N mineralization, a fraction of the substrate C and N was subtracted from the total substrate C and N prior to computer simulation that was initiated at day 2. The model then treated decomposition as a two-phase process using the intermediate- and slow-phase rate constants. The decomposition rate under anaerobic conditions was controlled by the water-soluble organic C pool and CO_2 plus CH_4 production. Gale and Gilmour (1988) calculated first-order rate constants for aerobic decomposition of alfalfa as 0.123, 0.059, and 0.0095 day^{-1} for the rapid, intermediate, and slow phases, respectively. The rate constants for anaerobic decomposition were 0.118 and 0.024 day^{-1} for the rapid and intermediate fractions, respectively. The model predictions for net N mineralization agreed well with those observed ($r^2 = 0.946$).

B. Nitrogen Release Patterns

The decomposition and release of N from green manures generally proceed very rapidly during the first few weeks followed by much slower phases thereafter, both under aerobic (Wilson and Hargrove, 1986; Yadvinder-Singh et al., 1988; Jensen, 1989; Wagger, 1989) and anaerobic conditions (Shi et al., 1981; Singh et al., 1981; Bhardwaj and Dev, 1985; Ito and Watanabe, 1985; Khind et al., 1985; Beri et al., 1989a,b; Lin and Wen, 1990). The pattern of N release for azolla differs somewhat in that the release rate is relatively slower just after incorporation and becomes faster at a later period, probably because of the high lignin content (Shi et al., 1978). The data indicated that the initial rate of N mineralization is faster under anaerobic than under aerobic conditions.

1. Nitrogen Release Patterns in Upland Soils

The amount of N released from green manure after 2 weeks of incubation ranged from 12 to 38% under aerobic conditions (Table 5). In a laboratory incubation study, Yadvinder-Singh et al. (1988) observed that N was released rapidly from sesbania green manure and the amount released 4 and 8 weeks after incubation was 59 and 74%, respectively. Figure 3 illustrates the N mineralization pattern from different green manures. Wilson and Hargrove (1986) and Wagger (1989) observed an exponential N mineralization pattern of legume residues under field conditions. Nitrogen disappearance was very rapid during the first 4 weeks after placement, and decreased with time reaching a very slow rate by week 16. The net N release 14 and 112 days after incorporation was 52 and 69%, respectively (Wilson and Hargrove, 1986). Gale and Gilmour (1988) observed relatively slow rates of N mineralization from alfalfa residue and apparent net N release was only 12 and 28% in 14 and 30 days after incubation, respectively.

Table 5. Nitrogen mineralization from green manures under aerobic and anaerobic conditions

Green manure	Temperature (°C)	Time (days)	Net N mineralized (%)	Reference
		Aerobic conditions		
S. aculeata	30	14	42	Yadvinder-Singh et al. (1988)
		28	59	
Alfalfa	25	14	12	Gale and Gilmour (1988)
		30	28	
S. aculeata	30	16	20	Debnath and Hajra (1972)
		60	50	
M. littolatris	—	28	27	Ladd and Amato (1986)
Alfalfa	23	38	50	Frankenberger and Abdelmagid (1985)
Cowpea		38	50	
Soybean		16	50	
Vetches (field study using litter bags		14	52	Wilson and Hargrove (1986)
		28	60	
Crimson clover (field study using litter bags)	16.9–18.7	14	51	Wagger (1989)
		28	63	
Subterrarean clover	20	75	75	Muller (1987)
		Anaerobic conditions		
S. aculeata	30	14	31	Yadvinder-Singh et al. (1988)
Alfalfa	25	14	43	Gale and Gilmour (1988)
S. cannabina	32	7	45	Bhardwaj and Dev (1985)
		35	69	
S. sesban (field study using litter bags)		4	64	Palm et al. (1988)
		14	81	
S. rostrata (pot study, four soils)		83	44–81	Nagarajah et al. (1989)
S. aculeata	30	20	37	Singh et al. (1981)
		30	49	
Azolla (^{15}N-labeled azolla N in pots transplanted with rice)		60	45	Mian and Stewart (1985)
Azolla (pot study, four soils)		83	27–52	Nagarajah et al. (1989)

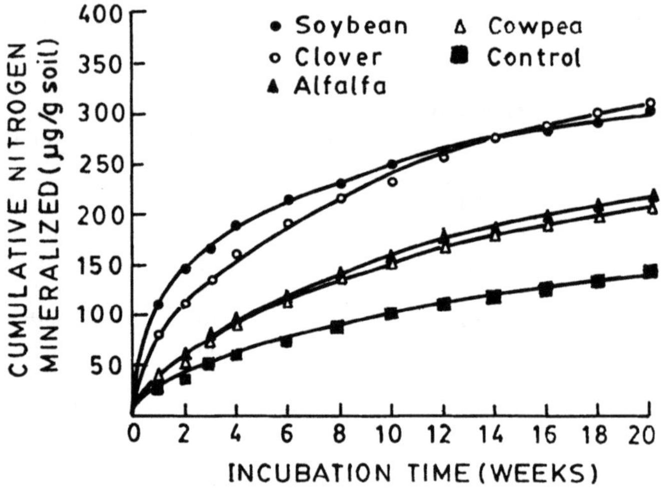

Figure 3. Nitrogen mineralization rates of different legumes incorporated into soil (after Frankenberger and Abdelmagid, 1985).

2. Nitrogen Release Patterns in Waterlogged Soils

In laboratory incubation experiments it has been observed by several workers (Joachim and Kandiah, 1929; Aspiras, 1966; Nagarajah et al., 1989) that release of NH_4^+–N from different green manures increased rapidly for up to 2 to 4 weeks, and thereafter it remained nearly constant for 8 to 12 weeks. Several other studies (Table 5) confirm that green manure N was rapidly mineralized. The peak of NH_4^+–N was observed after 7 to 15 days of incubation, and NH_4^+–N then decreased quickly due to N loss (Figure 4). Singh et al. (1981) reported that sesbania released 31% N in 20 days and 49% N in 30 days after flooding. The release of N from azolla species was faster than from sesbania, and it was 39 to 59% N in 20 days and 55 to 88% in 40 days after flooding. Ito and Watanabe (1985) observed patterns of N release from azolla species similar to those reported for legumes. Results from an incubation study by Singh et al. (1981) showed rapid loss of N from ammonium sulfate treatment under waterlogged conditions, indicating that net N mineralization from organic sources was a resultant of mineralization minus some unknown loss via denitrification and NH_3 volatilization. Results from several other studies revealed large losses of N from green manure during incubation (Singh et al., 1981; Bhardwaj and Dev, 1985; Yadvinder-Singh et al., 1988; Beri et al., 1989b). The actual release of N from green manure would thus be in excess of the values reported in Table 5.

According to Tusneem and Patrick (1971), NH_4^+–N released during mineralization was nitrified in the upper oxidized zone and after trans-

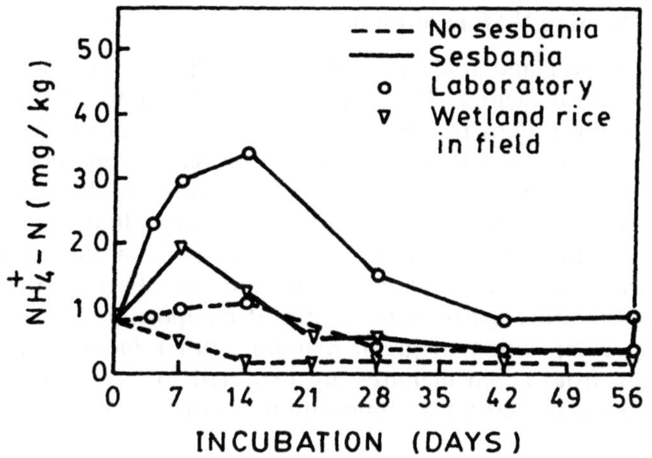

Figure 4. Effect of sesbania green manure incorporation on NH_4^+–N release patterns in flooded soils under laboratory and field situations (after Khind et al., 1985).

portation into the lower anaerobic zone it was lost via denitrification. Floodwater depth and the size of the mouth of the incubation vessel will affect the thickness of the oxidized layer at the soil-water interface. Tusneem and Patrick (1971) found nitrapyrin, a nitrification inhibitor applied at the rate of $20\,\mu g\,g^{-1}$, to be effective in preventing nitrification of NH_4^+ up to 50 days of incubation. Beri et al. (1989b) reported that the decrease in NH_4^+–N under saturated moisture conditions could be arrested by acetylene gas (10% by volume) to be replenished on alternate days.

In a field study without rice plants, Nagarajah (1988) observed that soil NH_4^+–N increased rapidly up to 10 to 20 days after flooding and then stabilized. In the presence of a rice crop, soil NH_4^+–N increased up to 15 days after transplanting (DT) and decreased to a very low level by about 40 DT largely because of plant uptake; the low level was maintained even after crop harvest. Palm et al. (1988) found that under field conditions N-rich leaves of *Sesbania sesban* decomposed and released about 73 and 88% N in 4 and 14 days after flooding, respectively. Beri et al. (1989b), also working in a field situation, observed that sesbania green manure released N immediately and a peak was observed at 7 DT, but a very low level was reached by 28 DT because of plant uptake and losses of N (Figure 4). In a field study in China, Lin and Wen (1990) observed that decomposition of green manure proceeded very rapidly in the initial 3 months, particularly in the first month, then it gradually became slower. Mian and Stewart (1985) found that *Azolla caroliniana* released 45% of N contained in it in 60 days.

C. Factors Affecting N Mineralization

The decomposition and release of mineral N from green manure are affected by several factors. These include green manure characteristics (substrate quality), environmental factors, soil characteristics, and management factors. Various chemical and physical properties of a soil which affect the soil microbial and faunal populations are likely to have a marked effect on the rate of N mineralization in soil. Although only a few studies on decomposition and N mineralization from the leguminous green manure crops incorporated into the soil have been carried out, a great deal of literature exists on plant litter decomposition. In addition to the chemical composition of green manures, other factors affecting N mineralization from green manures are expected to have effects almost similar to those reported for decomposition of crop residues.

1. Characteristics of Green Manure

Green manure characteristics determining decomposition and N mineralization kinetics include N content, C:N ratio, lignin content, polyphenol content, and concentration of various other organic compounds. The proportion of each organic constituent varies according to legume species and the plant part. When added into soils, different organic components of plant biomass decompose at different rates. The water-soluble materials are the first to be decomposed, followed by celluloses and hemicelluloses. The ether- and alcohol-soluble fractions are decomposed slowly but at comparable rates, and lignins are the most resistant organic compounds in plant material (Alexander, 1977). Obviously, decomposition rates of organic residues tend to vary according to the proportion in which different organic compounds are present in them. For example, young succulent, high N, and low lignin residues will decompose more readily than more mature, low N, and high lignin residues.

a. Total N Content

The nitrogen content of plant material is an important factor controlling the rate of decomposition and N mineralization. A high N content of plant material favors N mineralization, encouraging a high rate of decomposition and ensuring that N mineralization exceeds immobilization by a considerable extent. Frankenberger and Abdelmagid (1985) found a highly significant positive correlation between total N content in legume residues and the cumulative amount of N mineralized in soil (Figure 5). A critical N value of 1.73%, a dividing line between the positive and negative contribution of the residues in releasing N, was obtained. Iritani and Arnold (1960) reported a mean critical N value of 1.74% with three different incubation methods. Jensen (1989) observed a net mineralization of N after 50 days of incubation of pea residues containing 2.35% N (C:N

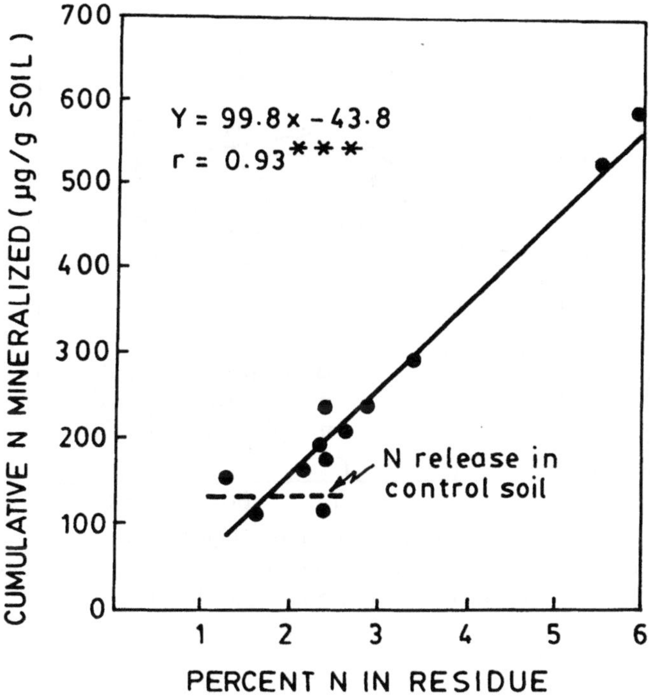

Figure 5. Relationship between cumulative amounts of N mineralized in soil and %N in legume residues (after Frankenberger and Abdelmagid, 1985).

ratio 17), but no net mineralization of N occurred even after 220 days of incubation with pea straw containing 1% N (C:N ratio 39).

In a laboratory incubation study under flooded conditions, Beri et al. (1989a) observed no significant difference in the amount of NH_4^+ released from 60-day-old sunn hemp (2% N), sesbania (2.2% N), cowpea (1.6% N), and clusterbean (2.3% N). Nagarajah (1988) reported that several leguminous green manures with almost similar N content and C:N ratio released nearly similar amounts of NH_4^+–N under waterlogged conditions. The initial rate of N release was, however, faster in sesbania (2.72% N), than sunn hemp (1.68% N), cowpea (2.43% N), mungbean (2.42% N), and pigeonpea (2.64% N). Soybean, which had the lowest N content (1.10% N), released the lowest amount of N (Figure 6). Ito and Watanabe (1985) noted that rates of N release from three species of fresh azolla were related to their N content. *Azolla pinnata*, with a high N content (4.88%), released more NH_4^+–N and at a faster rate than *Azolla mexicana*, which had the lowest N content (3.46%). Singh et al. (1981), however, observed large differences in net N mineralization (55 to 88%) from three species of azolla in spite of the comparable N content

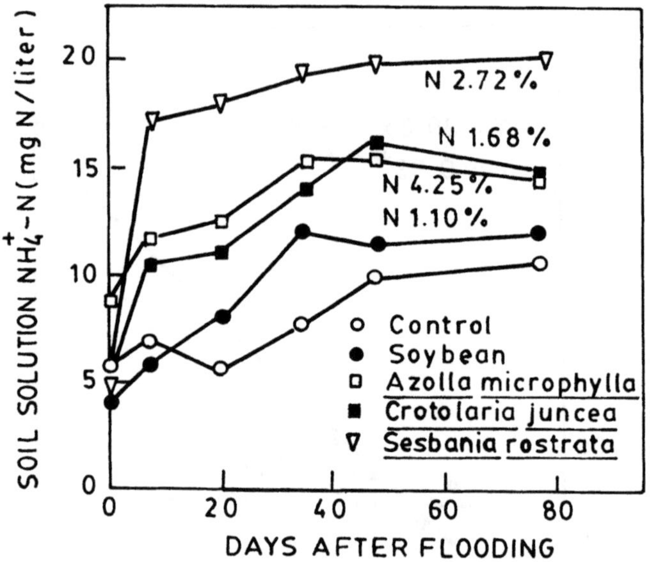

Figure 6. Effect of incorporation of different green manures on kinetics of soil-solution NH_4^+–N in a flooded soil (after Nagarajah, 1988).

(~5%). De Haan (1977) also found no correlation between %N of the added plant tissue and the rate of decomposition. It suggests that in addition to N content, other substrate quality parameters may regulate the mineralization-immobilization balance.

b. C:N Ratio

Because crop residues including green manures contain an almost similar amount of C (approximately 40% on a dry weight basis), the C:N ratio should have a similar relationship with decomposition and N mineralization-immobilization turnover as discussed for N content. The organic residues with wide C:N ratios often have a slow rate of decomposition. Frankenberger and Abdelmagid (1985) found that the relationship between percentage of N mineralized in 20 weeks (Y) and the C:N ratio of the legume residues incorporated into the soil (X) best fit the inverse curvilinear relationship:

$$Y = \mathrm{b}X^{\mathrm{m}}, \qquad r = 0.88^{**}$$

This study revealed that, compared with N content, the C:N ratio of legume residues was a slightly inferior criterion for predicting N releases. The critical C:N ratio for N mineralization was calculated as <19. Black (1968) reported that in many cases, the critical C:N ratio falls between 15 and 33. These values are obviously not constant and depend on factors such as temperature, time allowed for N mineralization, and supply of mineral N in soil. Haynes (1986) reported critical C:N ratios ranging from 27 to 167. Under field conditions, it appears that the concept of a fixed critical C:N ratio that controls N mineralization-immobilization balance is not of general validity.

Weeraratna (1979) found that, among the five green leaf manures, *Centrosema pubsescens* leaves with high N content and low C:N ratio and crude fibre content released the maximum amount of mineral N under anaerobic conditions. Sunflower leaves, which contained the lowest N and the highest C:N ratio and crude fiber levels, released the lowest amount of N. Wagger (1989) observed that under field conditions the rate of N release was greater with hairy vetch (3.22 to 4.85% N, C:N ratio 8 to 12) than with crimson clover (2.48 to 3.12% N, C:N ratio 13 to 16).

c. Lignin Content

For a given C:N ratio, an increase in lignin content of the plant material is know to retard its rate of decomposition and release of N (Alexander, 1977). According to Melillo et al: (1982), the lignin content of residues exerts more control over the rate of decomposition than does N content. De Haan (1977) and Frankenberger and Abdelmagid (1985) reported a significant linear relationship between the lignin content of crop residues including legumes and the cumulative amount of N mineralized.

In a study on five rice soils in the Philippines, it was found that at the same N equivalent, the rate of NH_4^+–N released from *Azolla microphylla* was lower than that from *Sesbania rostrata* green manure despite its lower C:N ratio (Nagarajah et al., 1989). This was attributed to the much higher lignin content of azolla (20% vs 9%). Higher lignin content had been considered as the reason for lower N availability in azolla than in milk vetch green manure (Shi et al., 1981). In a pot experiment, Zhu et al. (1984) observed that N recovery by rice was 42.4% (n = 10) for leguminous green manures and 27.3% (n = 6) for azolla.

Herman et al. (1977) observed an interaction between lignin content and C:N ratio in controlling the decomposition rate of plant material. It was found that the decomposability index, [(C:N) * (% lignin)] * (% carbohydrate$^{-0.5}$), was correlated inversely with total CO_2 evolution from decomposing roots of three grass species during a 47-week incubation. The index was also found to accurately predict decomposition of other plant materials, including legume residues. Bruulsema and Christie (1987) found that the decomposability index of medium red clover was higher than that of alfalfa, predicting a lower decomposing rate due to a higher lignin content and a lower C:N ratio. Their data suggested that the lignin content and the C:N ratio of legumes influenced the proportion of plough down N made available to the succeeding maize crop.

Table 6. Chemical composition of winter cover crop residue as affected by age (mean of 2 years) (after Wagger, 1989)

Legume crop	Date of incorporation	Total N ($g\,kg^{-1}$)	C:N ratio	Cellulose ($g\,kg^{-1}$)	Hemicellulose ($g\,kg^{-1}$)	Lignin ($g\,kg^{-1}$)
Crimson clover	18–19 April	29.4	13.5	194	125	67
Crimson clover	1–7 May	25.6	15.5	251	140	94
Hairy vetch	18–19 April	42.2	9.5	175	76	84
Hairy vetch	1–7 May	37.5	10.5	232	89	107

d. Polyphenol Content

In general, as the polyphenol content of litter increases, the rate of decomposition and N release decreases (Williams and Gray, 1974; Swift et al., 1979). Polyphenols may also decrease the rate of decomposition by direct inhibition of fungal and/or faunal activity (King and Heath, 1967; Harrison, 1917). Vallis and Jones (1973) found that although leaves and litter of the legumes *Desmodium intortum* and *Phaseolus atropurpureus* had similar N and lignin content, net N mineralization from the former during 32 weeks was less (3%) than that from *P. atropurpureus* (46%), partly due to a much higher polyphenol content in the leaves of *D. intortum*.

e. Effect of Plant Age

The chemical composition of most crop plants changes dramatically during their growing period. As the plant matures, its content of N, protein, and water-soluble constituents steadily decreases, while the amounts of fiber, hemicelluloses, celluloses, and lignin, and the C:N ratio increase. Therefore, tissues of immature plants (as in case of green manures) generally decompose more rapidly than those of mature plant materials. Singh and Rai (1975) observed that in soils amended with 60- and 75-day-old sesbania plants, NH_4^+–N reached its maxima after 15 and 30 days of aerobic incubation, respectively. Fulkerson (1983) observed a greater response of succeeding maize to plough down of less mature legume plant material rather than mature and lignified material. Wagger (1989) found that about 2 weeks delay in the desiccation of crimson clover and hairy vetch reduced their N contents but increased the C:N ratio and cellulose, hemicellulose, and lignin contents (Table 6). The faster rate of N release was observed with the early desiccation dates (Figure 7). For example, after 2 and 4 weeks in the field, crimson clover killed early contained only 49 and 37%, respectively, of the initial N compared with 79 and 64% for the corresponding intervals associated with the late desiccation. John et al. (1989b) reported that N

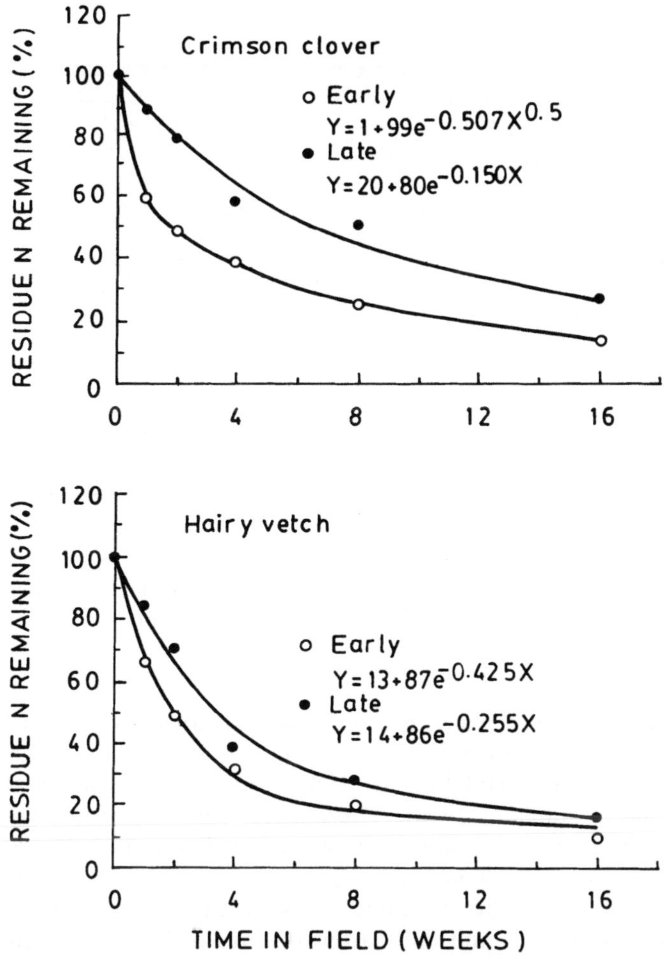

Figure 7. Percentage of initial N remaining in crimson clover and hairy vetch residue at various times during the 1984 growing season for the early and late cover crop desiccation dates (after Wagger, 1989).

mineralization from cowpea green manure (C:N ratio 15) was higher at 15 days after transplanting of rice than from cowpea residue incorporated after maturity (C:N ratio 21). Bhardwaj and Dev (1985), however, observed no significant effect of varying age of sesbania from 45 to 65 days on percentage of N released under waterlogged conditions.

f. Effect of Plant Parts

Different parts of a plant differ widely in their chemical composition and thus in their rate of decomposition and N mineralization. The stems and

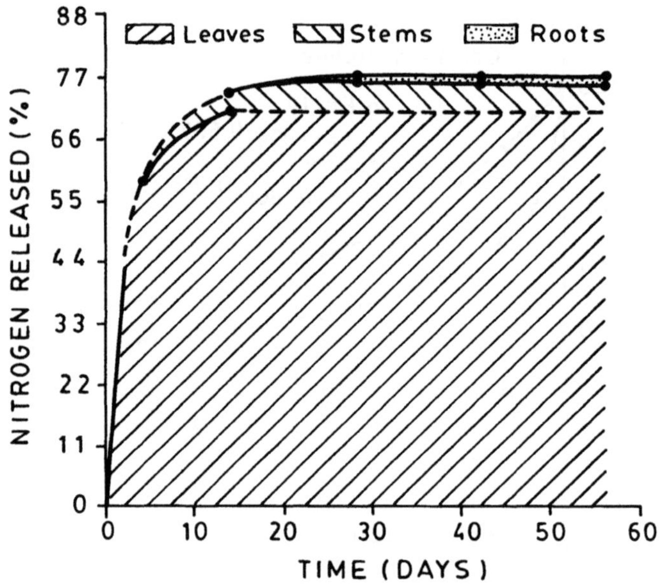

Figure 8. Cumulative N release from leaves, stems, and roots of *Sesbania sesban* incorporated in a flooded rice field (after Palm et al., 1988).

roots of green manure plants generally possess lower N contents, higher lignin contents and wider C:N ratios than those of the foliage portion (Frankenberger and Abdelmagid, 1985; Liu, 1988; Palm et al., 1988; Ventura and Watanabe, 1990). These parts will, therefore, decompose slowly and may result in immobilization of N. Palm et al. (1988) further found that leaves of *Sesbania sesban* green manure, which contained 88% of the total N in the above-ground parts, released about 73% of the N during the first 4 days and 88% after 14 days of incubation under submerged soil conditions. In contrast, release of N from stems and roots was marginal compared with that from leaves (Figure 8). Stems and roots, besides having lower N contents and wider C:N ratios, were more lignified than leaves. Frankenberger and Abdelmagid (1985) showed greater N release from foliage of four leguminous residues than that from stems and roots. There was a greater N release from roots of cowpea than from its foliage because of lower C:N ratio of the former (14 vs 15.9). Amato et al. (1987) found that after 1 year, the residual ^{15}N accounted for 56 and 63%, respectively, of input ^{15}N from legume tops and roots. Waid and Chulan (1981) reported that the percentage of N mineralization after 16 weeks of incubation was 19, 32, and 37 from laterals, radicals, and nodules of *Centrosema pubsescens*, respectively. Jensen (1989) noted that green tops and fresh roots of pea plants harvested around flowering with about 3.9% N (C:N ratio 10) decomposed quickly, while about 35%

of organic N from the residue was mineralized in the soil after 50 days of incubation at 20°C. Top growth of legumes is among the most rapidly degradable plant materials. It is high in protein, low in lignin, generally low in other inhibitory polyphenolic compounds, and has relatively fine structure.

2. Environmental Factors

Among the environmental or climatic factors, temperature and moisture are the most significant and are discussed below.

a. Temperature

The rate of decomposition of organic residues increases with temperature, as it exerts a profound influence on the growth and activity of microorganisms. Although each species of the soil population has a temperature optimum, the overall optimum range in soils is generally 30° to 40°C (Stott and Martin, 1989). The rate of decomposition is very slow near the freezing point and above 40°C it abruptly declines (Alexander, 1977). In a laboratory experiment with grass litter, Floate (1970) found that NH_4^+–N production during a 12-week period was 61.0, 16.5, and 0.10 mg N $(100\,g)^{-1}$ of added material at 30°, 10°, and 5°C, respectively. Jenkinson and Ayanaba (1977) noted that the decomposition process for ryegrass proceeded four times faster in Nigeria than that in the cooler climates of England. Westcott and Mikkelson (1985) reported that microbial decomposition of purple vetch green manure was stimulated by increasing the temperature from 20° to 30°C. The average Q_{10} across treatments was 1.83. Relatively cool conditions during crop growth may retard N availability from green manure. Ishikawa (1963) found that decomposition and N release from milk vetch green manure increased as the temperature increased from 18° to 30°C. About 70% of the N contained in green manure was mineralized after 30 days of incubation at 30°C, although the mineralization rate was substantially high even at 18°C (Figure 9). Aspiras (1966) reported that NH_4^+–N release from *S. sesban* increased when the temperature increased from 20° to 40°C. In one soil, however, it declined at temperatures higher than 30°C.

The combined effect of high temperature and moisture is more prominent than that of temperature alone. At low temperatures, the effect of soil moisture is less significant than at more optimum temperatures. Conditions of high temperature and optimum moisture generally favor microbial growth and thus decomposition. Thus, the rate of green manure decomposition in tropical and subtropical climates should be substantially higher than that observed in cool climates. No inflormation is available in the literature on the effect of fluctuating temperatures, and freezing and thawing cycles on the decomposition and N release from green manures in soils.

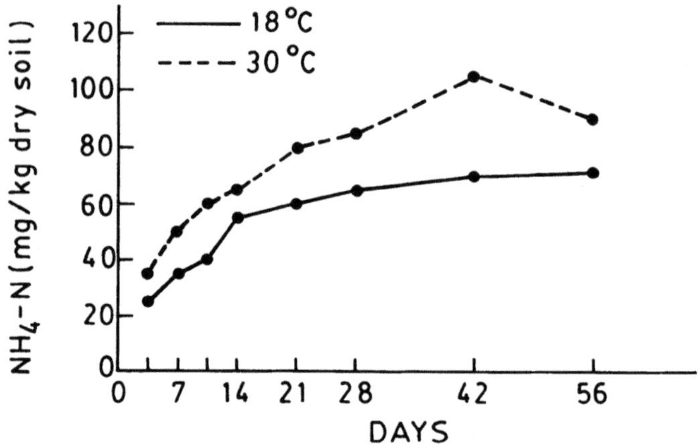

Figure 9. Effect of temperature on release of NH_4^+–N in a flooded soil treated with milk vetch green manure (after Ishikawa, 1988).

b. Soil Moisture

Soil moisture status exerts a marked influence on the rate and course of organic residue decomposition. Favorable moisture conditions for optimum organic residue decomposition in soils range from about 50 to 90% of the water-holding capacity (−50 to −15 kPa). Above 90%, biological activity and decomposition rate are adversely affected due to lack of oxygen (Stott and Martin, 1989). However, a characteristic feature of anaerobic decomposition is the low metabolic efficiencies of anaerobic microbial populations, which leads to more rapid release of NH_4^+–N than would ordinarily be expected on the basis of the C:N ratio of the decomposing material and the slow rate of decomposition. Tusneem and Patrick (1971) observed faster rates of N mineralization during the first 2 months of incubation under waterlogged than under optimum moisture conditions (Figure 10). A decline in NH_4^+–N accumulation under waterlogged conditions was observed at later stages, possibly due to N loss through nitrification-denitrification. Williams et al. (1968) concluded that the N content of rice straw required for its decomposition in submerged soils was one-third (0.5 vs 1.5% N) of the average N content required for aerobic decomposition. Gale and Gilmour (1988) observed that net N mineralized from alfalfa legume residues (C:N ratio 13.1) at 7 days of incubation under anaerobic conditions was equal to N mineralized at 30 days under aerobic conditions (Table 7). Net N mineralization was observed for the first 2 weeks of anaerobic decomposition and appeared to cease thereafter, which coincided with the cessation of decomposition. Lin and Wen (1990) suggested that shortage of water, especially in summer, was more effective in retarding decom-

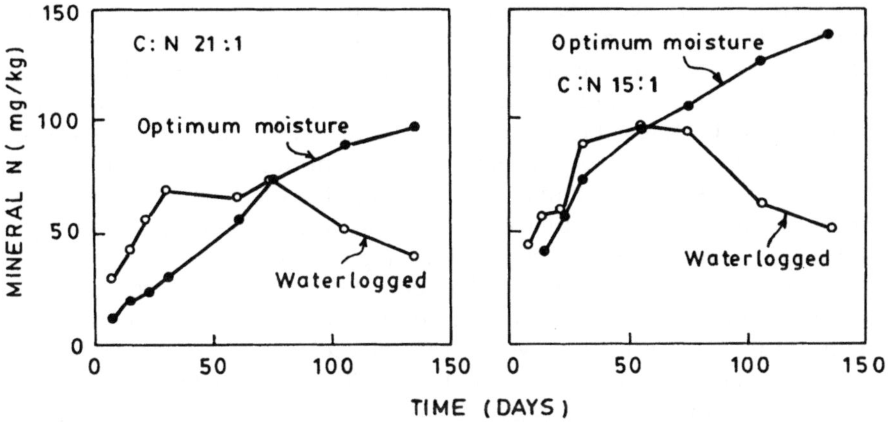

Figure 10. Mineralization of organic N at two C:N ratios of added plant material under waterlogged and optimum moisture conditions (after Tusneem and Patrick, 1971).

Table 7. Nitrogen mineralization from alfalfa (3.24% N) under aerobic and anaerobic conditions (after Gale and Gilmour, 1988)

	Net N mineralized at different days (%)				
Treatment	2	7	14	22	30
Aerobic	8.0	10.5	12.3	23.6	27.8
Anaerobic	11.7	26.5	43.2	40.7	38.9

position of green manure in upland fields than submergence in paddy fields. Aulakh et al. (1983) observed that the rate of N mineralization from clover residue was higher in experimental fields which received more rainfall. The effect of drying and rewetting on the decomposition of plant residues is unclear (Haynes, 1986).

3. Management Factors

The way green manures are managed exerts a profound influence on the decomposition and N mineralization of green manures in the soil. The management factors include the method of incorporation of green manure, quantity of green biomass added, size of the green biomass pieces, and cropping pattern.

a. Method of Incorporation of Green Manures

Several workers (McCalla and Russel, 1948; Parker, 1962; Brown and Dickey, 1970) have shown that decomposition and mineralization of N are slower from the surface-applied residues. This effect is commonly attributed to lower moisture and nutrient availability to the microorganisms at the surface. Nutrient limitations for surface residues are, however, generally not applicable to nutrient-rich green manures. McCalla and Duley (1943) observed that when moisture loss was eliminated from the surface residues, placement had no effect on decomposition of the high-N alfalfa. Thus, killing of legume cover crops using herbicide in conservation tillage systems and desiccation of green biomass left on the surface will lead to slow decomposition.

Using a nylon mesh bag technique, Wilson and Hargrove (1986) measured rates of C and N release from crimson clover residues under field conditions. Litter bags were either placed on the surface of the no-tillage plots or buried in the conventional-tillage plots. About one-half of N was lost from the buried bags in 2 weeks, yet 8 weeks were required for comparable loss from the litter bags placed on the surface (Figure 11). Groffman et al. (1987) found that clover residues incorporated into conventionally tilled soil decayed nearly twice as fast as the clover remaining on the surface of no-till soil. Nitrogen release from the residues reflected the pattern of decomposition. Similar observations have been recorded for hairy vetch (*Vicia villosa*) residues by Varco et al. (1987). The half-life for residue decomposition was less than 15 days, and 45 to 75 days with conventional tillage and no-tillage, respectively. About 47% of vetch-N was recovered as mineral N at 15 days with conventional tillage. In contrast, no more than 10% of the added vetch N was recovered as mineral N with no-tillage. Lathwell et al. (1989) observed that after 49 days of incubation there was twice as much mineral N in the pots where mucuna leaves were mixed rather than applied to the surface. Triplett et al. (1979), however, concluded that there was no effect of tillage on the availability of N to maize planted into alfalfa meadow.

In studies using ^{15}N-labeled azolla, N release was largest when azolla was incorporated into the soil (Ito and Watanabe, 1985). Only a small amount of exchangeable NH_4^+–N originated from azolla grown in water. N release was intermediate when azolla was placed on the soil surface. The effect of tillage practices on decomposition and mineralization of N from organic residues will obviously be affected by climate and soil.

b. Particle Size of Green Manures

Large pieces of crop residues generally decompose more slowly than smaller ones (Sims and Frederick, 1970; Allison, 1973). In small pieces more surface area is exposed for microbial colonization and action of enzymes so that the decomposition process is accelerated. For materials with a wide C:N ratio, this results in the rapid immobilization of inorganic soil N. For example, Sims and Frederick (1970) observed that the fine pieces of maize stalk pith immobilized six times as much

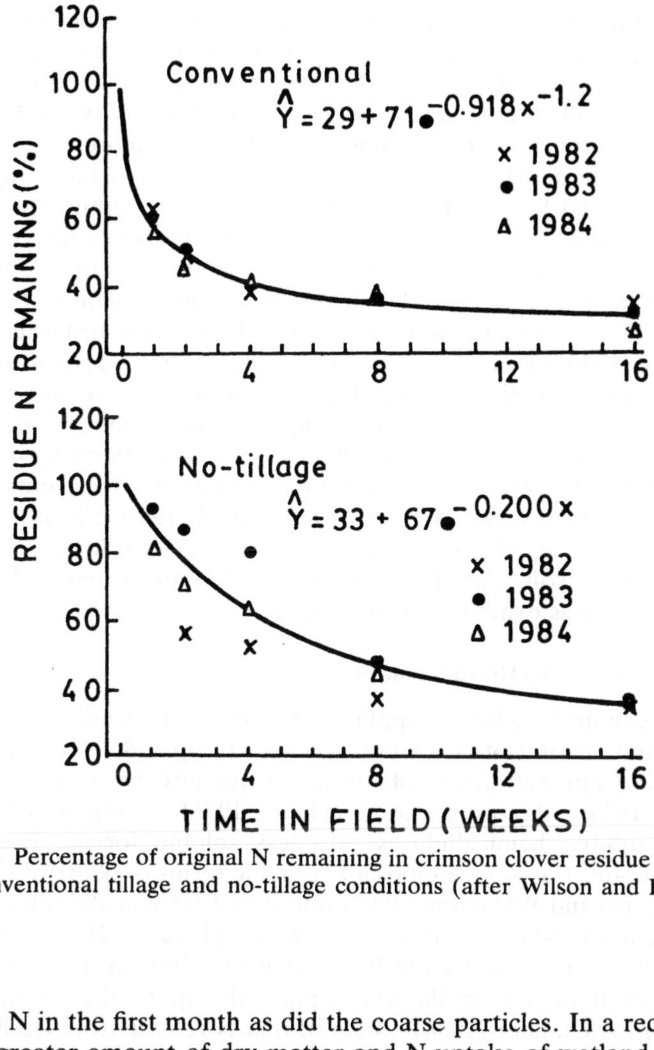

Figure 11. Percentage of original N remaining in crimson clover residue with time under conventional tillage and no-tillage conditions (after Wilson and Hargrove, 1986).

inorganic N in the first month as did the coarse particles. In a recent field study, a greater amount of dry matter and N uptake of wetland rice was observed in the plots amended with relatively smaller chopped pieces (7.5 cm) of sesbania green manure (Yadvinder-Singh and Bijay-Singh, unpublished data).

c. Amount of Green Manure

Pinck and Allison (1961) considered that percentage decomposition of plant C in soils is nearly independent of input levels, provided that C additions do not exceed an amount equivalent to 1.5% of dry soil. In incubation studies carried out for 3 to 6 months, Sauerbeck (1966) and Jenkinson (1981) found that the percentage of added C evolved as

CO_2 and stabilized in humus was not significantly influenced by residue additions up to 2%. A wide range of C additions were, however, not used in these studies. Ladd et al. (1983b) investigated decomposition of ^{14}C-and ^{15}N-labeled legume (*Medicago littoralis*) at 0.8 to 3.2 t ha^{-1}. The greater the amount of plant material added, the more extensive was its decomposition. The maximum concentration of legume input corresponded to <0.12% of plant C in soil to 7.5 cm depth. The residual organic ^{15}N and ^{14}C data showed that mineralization rates were proportional to the amount of legume residue added. Since the C:N ratio of the added legume material was 10.9, metabolism of plant C even when added in the highest amounts would not have been restricted due to inadequate supplies of N for microbial metabolism. Jenkinson (1977) suggested that differences in the decomposition rates for different input levels may result from the soil having a limited number of sites for protection and stabilization of organic residues. In a laboratory incubation study under both aerobic and anaerobic conditions, Weeraratna (1979) observed relatively very low rates of apparent N release from green leaf manures applied at 100 t ha^{-1}, equivalent to 1.53 to 1.80% of organic C additions in dry soil. The inhibitory effect of green manure additions in large quantity was exhibited in these studies.

d. Effect of Drying the Green Manures

Drying green manure before application is generally considered to reduce the rate and amount of N release into soil. In upland crops, field drying reduced the mineralization rate of green manure by as much as 20% (Lohnis, 1926). According to Joachim (1931) drying green manure converts soluble hemicelluloses into less soluble forms. This retards decomposition. Easily decomposable organic forms of N are also reduced by drying. Ito and Watanabe (1985) found that fresh azolla released up to 2.5 times more NH_4^+–N than dried azolla (Figure 12). The NH_4^+–N released from fresh azolla reached a plateau after 16 days, whereas N mineralization in dry azolla was rapid only up to the fourth day of incubation. Alexander (1988) observed that N release from fresh and dry azolla (10 t fresh weight ha^{-1}) after 4 weeks of flooding was 43.6 and 35.6%, respectively. In a laboratory incubation study, drying of sesbania green manure at 60°C for 24 h, however, showed no marked influence on the N release pattern in a soil under a field capacity moisture regime at 35°C (Yadvinder-Singh and Bijay-Singh, unpublished data). Gu and Wen (1981) have also reported that incorporation of fresh or dried milk vetch into the soil produced a similar yield response in rice.

e. Effect of Cropping

Several researchers have shown that the presence of plants slows the rate of decomposition of added ^{14}C-labeled residues (Fuher and Sauerbeck,

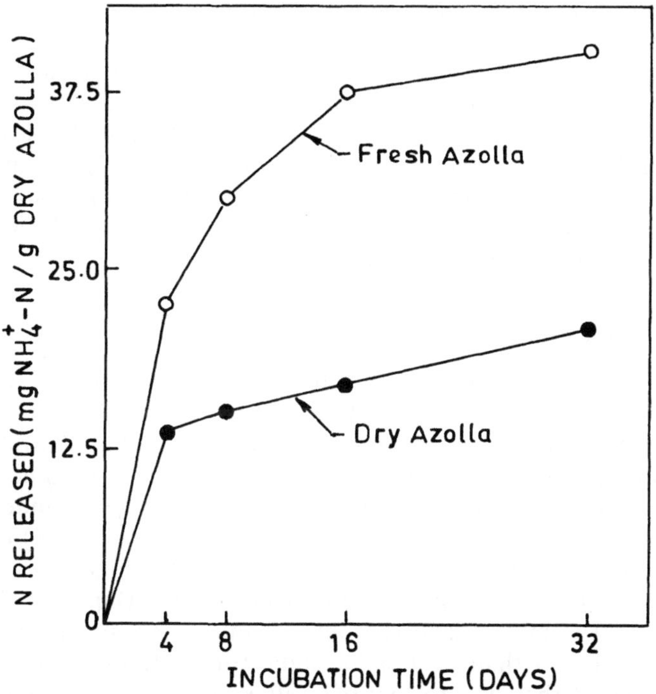

Figure 12. Ammonium-N released from fresh and dry *Azolla pinnata* incorporated into flooded soil (after Ito and Watanabe, (1985), with revision suggested by Watanabe).

1968; Jenkinson, 1977). Such an effect has been attributed to the desiccating effect of vegetation on soils during the summer (Jenkinson, 1977). The production and supply of readily available C in the rhizosphere can also reduce apparent N mineralization by accelerating N immobilization. Janzen and Radder (1989) observed that N mineralization from green manure (*Lathyrus tingitanus*) was higher in the fallow treatment (39 mg N kg^{-1} soil) than in plots cropped with wheat or canola (*Brassica napus*) (20 mg N kg^{-1} soil). Haider et al. (1977), however, observed higher mineralization of organic N in planted than in unplanted soils. Possibly crop competition for available N suppressed microbial activity, thereby inhibiting N mineralization (Merckx et al., 1987; Van Veen et al., 1989). This effect may not be predominant due to the high N content of green manures. The suppressed net mineralization of green manure N in cropped soils was explained by enhanced immobilization of N into microbial biomass.

The mineralization of green manure N was significantly affected by cropping history. The highest values for green manure N mineralization were observed in plots with cereal grain crops receiving fertilizers, and the lowest values were observed in the treatments which included perennial crops of alfalfa and forage grass (Janzen and Radder, 1989). The low rate of N mineralization from the 6-year rotation with alfalfa and crested wheat grass may be due to high immobilization of N following the addition of large amounts of carbon substrate from forage ploughed under. The potential contribution of green manure to the short-term N fertility will, therefore, be greater if soils are left uncropped rather than if they are cropped immediately after application.

4. Effect of Soil Properties

The decomposition and release of N from organic materials is subject to the influence of various soil properties. Soil pH is most important in this regard. Decomposition typically proceeds more readily in neutral than in acid or saline-sodic soils. Consequently, liming of acid soils accelerates the decay of plant residues and soil organic matter (Alexander, 1977; Yadvinder-Singh and Beauchamp, 1986). Singh and Rai (1975) observed that the salinity and alkalinity of soils depressed the mineralization of N during decomposition of sesbania and *Melilotus alba*. Bajpai et al. (1980) also found that N mineralization from sesbania green manure was faster in normal (pH 7.2) than in a saline-sodic (pH 9.9) soil. Lin and Wen (1990) suggested that due to low pH values, the decomposition rate in Chinese soils from the subtropics and tropics may not be so high as is generally thought.

Many investigators have observed that organic residues decompose more slowly in soils with higher clay contents, especially with clays having higher exchange capacities (Lynch and Cotnoir, 1956; Allison, 1973; Sorensen, 1975). In a 10-year study by Jenkinson (1977), soils with higher clay contents retained greater amounts of C of added ^{14}C-labeled residues. The presence of clay reduced total C loss as CO_2 through increased efficiency of C conversion to biomass and through forming complexes with decomposition products and new humus colloids (Burns, 1978). Zhu et al. (1984) found that N release from vetch green manure plus rice straw after 32 days of flooding was 15.6 and 26.2% in heavy clayey and clay loam soil, respectively (Table 8).

Nagarajah et al. (1989) studied N mineralization from green manures in four flooded Philippines rice soils. After 83 days of flooding, the apparent net N release from sesbania and azolla green manure ranged from 44 to 81% and 27 to 52%, respectively. There seemed to be no single factor controlling the rate of N release from green manure in different soils. From the results of 12 field experiments in a semiarid climate in Australia, Amato et al. (1987) found that the influence of soil physical and chemical properties on the amount of ^{15}N released from legume residues was small. Van Faassen and Dijk (1987) showed that N mineralization from

Table 8. Effect of soil texture on N mineralization from vetch plus rice straw after submergence in a pot experiment (after Zhu et al., 1984)

	Net N mineralized at different days (%)			
Soil texture	32	42	62	82
Clay loam	26.2	27.9	33.6	41.1
Heavy clay	15.6	16.9	22.4	28.7

organic manures was faster in an acid sandy soil than in a calcareous sandy loam soil in which bacteria with a C:N ratio of about 4 predominated. In the acid sandy soil fungi with a C:N ratio of about 10 predominated.

Deficiencies of nutrient elements in soils may not generally limit mineralization of immature leguminous green manures, except in highly deficient soils. Singh and Rai (1973, 1975) reported that N mineralization of green manure was enhanced after the addition of P to saline-alkali soils. In a pot experiment, Subramani and Kannaiyan (1987) found that a foliar spray of urea over the azolla mat increased the decomposition of azolla, and the rate of decomposition increased with increasing urea concentration from 50 to 2000 $\mu g\ ml^{-1}$.

D. Relationship Between N and C Mineralization

The releases of C and N from organic residues differ in that C generally gets volatilized as CO_2 or CH_4 while N tends to be conserved, partly by decomposing microorganisms. The relative rates of decomposition of labeled C and N from plant residues reflect such conservation of N (Paul, 1976; Amato and Ladd, 1980; Ladd et al., 1981a). In four south Australian soils Ladd et al. (1981b) found that 60 to 65% of ^{15}N labeled legume (*Medicago littoralis*) residues did not decompose even after 32 weeks. In contrast, more than 50% of legume ^{14}C was lost in 4 weeks. The decomposition pattern of residues as a function of time was, however, similar with respect to ^{14}C and ^{15}N (Figure 13).

Westcott and Mikkelson (1987) observed a highly significant correlation between net N mineralization and CO_2 production from milk vetch green manure. They observed that the slope of the regression line declined from 0.45 at 10 days to 0.21 at 30 days of incubation. The actual amount of N mineralized per unit of CO_2 evolved was underestimated because of the loss of N by different mechanisms, immobilization, and fixation by clays. Gilmour et al. (1985) showed that CO_2 evolution from leguminous residues can be used as a mean for the estimation of both the total amount and the rate of N mineralization for a given period. The relationship between net N and net C mineralization from clover residue under aerobic conditions turned out to be:

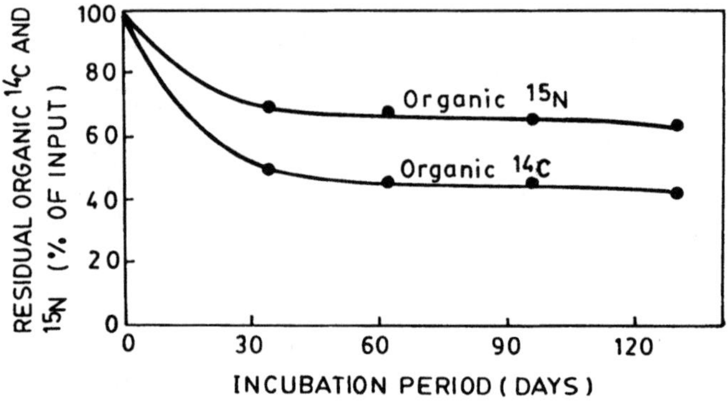

Figure 13. Decomposition of ^{14}C- and ^{15}N-labeled *Medicago littoralis* leaf material in a laboratory study (after Amato and Ladd (1980), reprinted with permission from Soil Biol. Biochem, 12:405–411, studies of nitrogen immobilization and mineralization in calcareous soils, V, Copyright 1980, Pergamon Press PLC.

$$N\min\ (\mathrm{g\,kg^{-1}}\,N) = 0.98C + 79 \quad (r = 0.962)$$

where C represents the carbon mineralized, $\mathrm{g\,kg^{-1}}$ C. Alfalfa and clover residues did not differ in terms of C and N mineralization. Recently, Gale and Gilmour (1988) studied the relationship between net N and C mineralized during the net mineralization phase for alfalfa incubated under aerobic and anaerobic conditions. Under anaerobic conditions there was one slope whereas, in the aerobic system, two slopes were observed implying that different fractions of the substrate were under decomposition at different times. For the first 14 days, the slope of the net N vs net C mineralized was 0.17 ($r^2 = 0.963$), suggesting a substrate with a high C:N ratio. With the onset of net N mineralization the slope of this relationship was 1.3 ($r^2 = 0.863$), which was similar to the slope found for the anaerobic system.

E. Effect of Green Manures on Urease and Nitrifier Activities

Urease enzyme originates in the soil from both soil microorganisms and plant roots (Bremner and Mulvaney, 1978) and is widely spread among plant species (Thompson, 1980). It has been found in leaves, roots, and bark of plants with actively growing tissues possessing greater activity than the senescing ones (Shim et al., 1973). Many researchers have, thus, observed a marked increase in the urease activity of soils when amended with green manures (Table 9).

Since the hydrolysis of urea is accelerated by an increase in urease activity in soils, Bajpai et al. (1984) observed that the rate of urea hydrolysis was enhanced by the addition of sesbania green manure in

Table 9. Effect of green manuring (GM) on urease activity of soils

Reference	Urease activity		Units
	−GM	GM	
Thomas and Shantaram (1984)	2.9	3.9	μg NH_4^+–N g^{-1} h^{-1}
Bolten et al. (1985)	1.7	2.5	μmol NH_4^+–N g^{-1} min^{-1}
Saini (1990)	8.8	18.1	μg urea hydrolyzed g^{-1} h^{-1}

Table 10. Effect of green manuring (GM) on first-order rate constant of urea hydrolysis, time required for complete hydrolysis, and half-life of urea in two soils at field capacity moisture regime (FC) and under waterlogged conditions (WL) (after Bajpai et al., 1984)

Treatment	Normal soil (pH 7.3)		Saline-alkali soil (pH 9.8)	
	FC	WL	FC	WL
	Rate constant, k (h^{-1})			
−GM	0.0665	0.0345	0.0165	0.0147
+GM	0.0909	0.0508	0.0250	0.0173
	Time for complete urea hydrolysis (days)			
−GM	3.4	6.0	13.4	15.0
+GM	2.4	4.3	8.8	12.7
	Half-life, $t_{1/2}(h)$			
−GM	10.0	20.0	42.0	47.0
+GM	7.6	10.5	30.0	36.0

both normal (pH > 7.5) and saline-alkali soil (pH 9.8). Nevertheless, urea hydrolysis followed first-order kinetics in both control and green manure amended soils incubated at field capacity moisture regime or under waterlogged conditions. Green manuring reduced the time required for completion of the hydrolysis of urea, half-life of urea, and first-order rate constant (Table 10). In the saline alkali soil, an enhanced rate of urea hydrolysis was partly due to the reduction in soil pH, an important consequence of green manuring. Similarly, Saini (1990) reported that while hydrolysis of urea was complete in 72 h in the control treatment, no urea was detected in sesbania-treated soil after 24 h. The first-order rate constant (k) for hydrolysis of urea was computed as 0.06 h^{-1} in control and 0.125 h^{-1} in green manure amended soil. Application of green manure also reduced the lag phase of urea hydrolysis from 1.18 h in the control soil to 0.40 h. Carmona et al. (1990) reported that in soybean straw amended soil 90% of the urea was hydrolysed in 2 days as com-

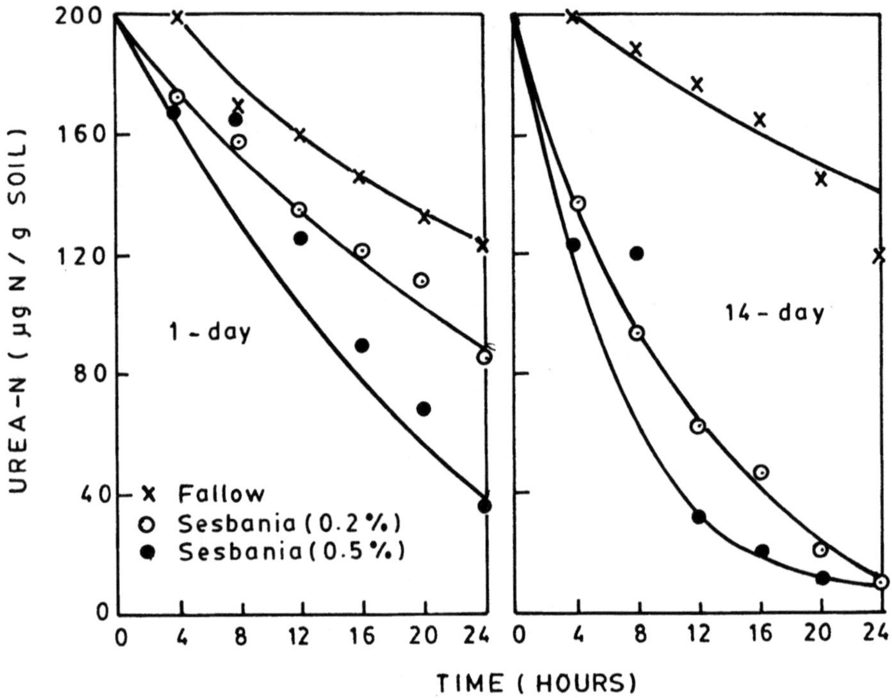

Figure 14. Effect of decomposition period of sesbania green manure applied at two rates on urea hydrolysis in a flooded soil (after Khind et al., 1991).

pared with 48% in unamended soil. Khind et al. (1991) reported that urea hydrolysis was faster in soil preincubated with sesbania green manure for 14 days rather than 1 day. The rate of hydrolysis was increased further when the application rate of green manure was increased from 0.2 to 0.5% (dry weight basis) (Figure 14).

The addition of organic matter in the form of green manure influenced both the population and activity of nitrifying microorganisms in soils. In a long-term field experiment, Bolten et al. (1985) observed a marked reduction in the population of *Nitrosomonas* in the soil amended with green manure. The *Nitrosomonas* population of 2.3×10^4 and 15.5×10^4 was recorded in soils from green manure amended and control plots, respectively. Similarly, Ramaswami and Raj (1973) observed that green manuring resulted in a significant reduction in the nitrifying power of soil, possibly due to changes in soil microbial ecology.

F. Losses of N from Soils Amended with Green Manure

The biologically fixed N_2 might well be as vulnerable as industrially fixed N to various mechanisms of N loss from the soil-plant system. The rapid

Table 11. Balance sheet of labeled azolla-N and urea-N in wetland rice (after Watanabe et al., 1989)

Treatment	^{15}N recovery (%)		^{15}N unaccounted for (%)
	Crop	Soil	
^{15}N azolla-N[a] + ^{14}N azolla-N[b]	39.3	49.7	11.1
^{15}N azolla-N + ^{15}N azolla-N	63.0	36.6	0.4
^{15}N urea-N + ^{14}N urea-N	27.3	40.5	32.2
^{14}N urea-N + ^{15}N urea-N	48.0	22.0	30.0

[a] 50% N applied at transplanting of rice.

[b] 50% N applied at 42 days after transplanting.

release of N from green manures can, therefore, result in losses of N from soil either as NH_3 through volatilization, or as N_2 and N_2O through denitrification of NO_3^- or through leaching of NO_3^-. Ito and Watanabe (1985) observed that N losses of more than 60% in 6 weeks occurred when ^{15}N-labeled azolla was placed on the soil surface. Although mechanisms were not identified, losses were only about 30% when azolla was incorporated into the soil. Watanabe et al. (1989) observed that ^{15}N losses in flooded rice were about 30% from urea and 0.4 to 11% from azolla (Table 11). In a pot study with flooded rice, Mian and Stewart (1985) calculated 11% losses of N in 60 days from ^{15}N-labeled azolla as compared with 14% from ammonium sulfate. In field experiments, Ladd et al. (1981a, 1983a) and Ladd and Amato (1986) reported that unrecovered ^{15}N from legume residues ranged between 9 and 23%. Huang et al. (1981) reported that in flooded soils losses of applied N from ammonium sulfate, green manure, and their combined application (50% + 50%) averaged 23.2, 16.2, and 20.4%, respectively, suggesting that losses of fertilizer N could be reduced when applied along with green manure. Possibly mineralization of green manure N and immobilization of fertilizer N proceed simultaneously immediately after fertilizer application. Azam et al. (1985) reported that sesbania green manure (C:N ratio 20) increased immobilization of fertilizer N under upland conditions and resulted in decreased plant uptake of fertilizer N. Losses of fertilizer N mainly through denitrification, were reduced by 30% (Table 12). The immobilization of fertilizer N was evidenced by a relatively high amount of ^{15}N in the microbial biomass and humus fractions. Recently, John et al. (1989a), however, observed that unrecovered ^{15}N in the soil-plant system was not affected by cowpea green manure. The application of green manure had no influence on pNH_3 and floodwater $(NO_2^- + NO_3^-)$-N. In this study, cowpea green manure was incorporated 15 days before transplanting rice and 29 kg urea-N ha^{-1} was applied 15 days after transplanting rice. In a 2-year field experiment on three sites in western Canada, Janzen et al. (1990) reported that mean

Table 12. Effect of sesbania green manure (GM) and fertilizer N on ^{15}N balance in soil (after Azam et al., 1985)

Treatment	^{15}N in 5-week-old maize (shoots + roots %)	^{15}N in soil (%) Total	Microbial biomass	Humus fraction	^{15}N unaccounted for (%)
$^{15}N-(NH_4)_2SO_4$	19.9	40.5	2.4	15.0	39.6
^{15}N–GM	5.0	89.2	11.0	34.1	5.8
$^{15}N-(NH_4)_2SO_4$ + GM	14.0	58.5	9.0	23.3	27.5
^{15}N–GM + $(NH_4)_2SO_4$	4.7	86.3	19.4	35.7	9.0

unrecovered N from ^{15}N-labeled green manure (flatpea) and $(NH_4)_2SO_4$ was 32 and 45%, respectively.

1. Denitrification

The increased availability of organic C due to green manuring can affect both the rate and extent of denitrification in soils. High levels of readily decomposable organic matter can also indirectly enhance the potential for denitrification through a general stimulation of microbial respiration, causing rapid O_2 consumption and an acceleration of the onset of anaerobiosis. Bolten et al. (1985) reported significant increase in the population of the denitrifiers in the soil amended with green manure. In a laboratory incubation study with a flooded soil, Mian (1985) observed that 32% of the azolla N was released as NH_4^+–N in 60 days, and 96% of this was lost through denitrification (Figure 15). In the presence of rice plants, denitrification loss was 25 to 28% of the N released in 60 days. In the $(NH_4)_2SO_4$ treatment, denitrification started after 30 days, whereas in the $NaNO_3$ treatment after 3 days of flooding. The 30-day delay was attributed to the time required for the development of a surface oxidized layer where nitrification could occur.

Working in wetland rice fields, John et al. (1989a) observed that prior cropping of cowpea to flowering and subsequent incorporation as green manure ($2.5\,t\,ha^{-1}$ dry matter) at 15 days of rice transplanting had no apparent effect on the evolution of $(N_2 + N_2O)-^{15}N$ from labeled urea applied to rice at 15 days after transplanting. The cumulative recovery of $(N_2 + N_2O)-^{15}N$ over 19 days was 0.15 and $0.13\,kg\,ha^{-1}$ for fallow and green manure treatments, respectively. The low floodwater $(NO_2^- + NO_3^-)$–N following urea application ($<0.14\,g\,m^{-3}$ or 0.3% of the applied N) suggested that denitrification loss from urea was controlled by the supply of NO_3^- rather than by the availability of organic C. The interval between the incorporation of green manure and the application of fertilizer together with water management practices during crop growth

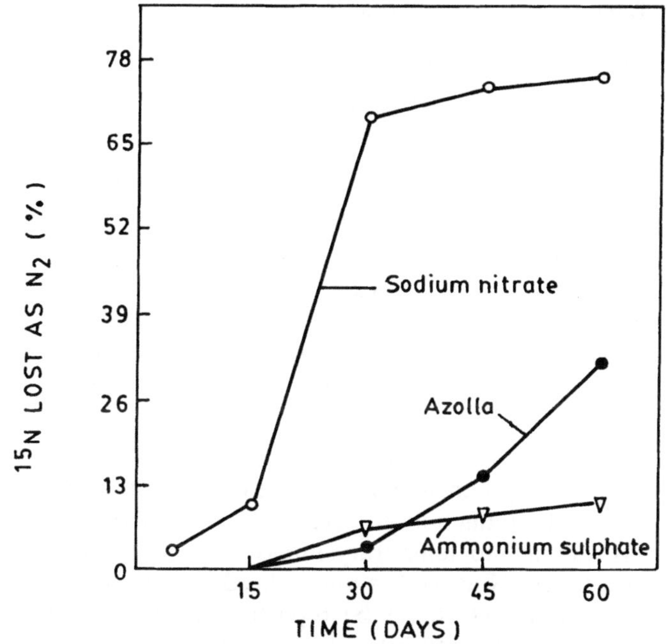

Figure 15. Denitrification loss of nitrogen released from *Azolla caroliniana*, ammonium sulfate, and sodium nitrate in a flooded soil (after Mian, 1985).

would markedly influence N loss through denitrification. Alternate flooding and drying conditions may favor nitrification-denitrification in soils under wetland rice. Under dryland conditions of Canada, Aulakh et al. (1983) concluded that incorporation of clover green manure followed by a partial fallow resulted in substantially less gaseous loss of N than with the standard summer-fallowing practices, and at the same time significantly increased mineral N accumulation in soils. Direct measurements of denitrification losses of green manure–N are not available.

2. Leaching

Green manuring can influence leaching losses of NO_3^-–N by affecting the physical properties of soil. Since the hydraulic conductivity and permeability of soils are favorably affected by green manuring (Meelu et al., in press), NO_3^- losses may be greater from the green manured fields. The leaching losses of green manure–N are normally expected to be smaller than inorganic fertilizers, because it must be mineralized before it can be lost via leaching. In a laboratory incubation experiment, Muller (1987) found that 17% of the subterranean clover–N, 51% of the $(NH_4)_2SO_4$–N and 69% of the KNO_3–N was lost via leaching (Table 13).

Table 13. Leaching of N (in leachate as % of applied) derived from subterranean clover, $(NH_4)_2SO_4$, and KNO_3 through 50 cm long columns in a laboratory experiment at 20°C (after Muller, 1987)

	Irrigation periods (days)			
	I	II	III	Total
Treatment	11–30	31–40	66–75	I–III
Subterranean clover	3.6	6.9	6.6	17.1
$(NH_4)_2SO_4$	8.2	25.0	17.5	50.7
KNO_3	25.5	36.5	7.4	69.4
Total leachate (mm)	88	72	72	232

Velu et al. (1988) reported that leaching losses of N from a clay loam soil were lower from green manure plus urea (50% N + 50% N) than from urea alone. The reduction in leaching loss under the green manure treatment was ascribed to temporary immobilization of applied N, and adsorption of fertilizer N as well as mineralized N from soil by the added organic matter. Bhagat et al. (1988) reported that leaching losses of N from urea alone (split application) and green manure plus urea (50% N + 50% N) were similar. The leaching loss of N from basally applied urea was, however, about three times the N loss under green manure treatment. In a column study, the addition of sesbania green manure (2% by weight in the 0 to 10 cm soil depth and precincubated at 25°C for 10 days) caused urea–N to move to greater depth as compared with control except in the case of the highest (20 cm) depth of water when the urea peaks were observed at the same depth (15 cm) in both control and sesbania-treated columns (Saini, 1990).

Winter cover crops will reduce NO_3^- leaching by depleting both soil NO_3^- and water. This occurs most actively in late winter and spring when the potential for leaching is greatest in the cool climates of North America. Similarly, off-season growing of green manure crops can reduce leaching loss of N from the soil. Varco et al. (1987) reported that N accumulated in legumes is less subjected to leaching than fertilizer N during the summer growing season (Table 14). Only top 20 cm soil was sampled in this study conducted on a well-drained soil in a fairly wet year. Losses of N were smaller in the semiarid climate of Australia (Ladd and Amato, 1986), but still slightly more fertilizer N than alfalfa–N was lost via leaching.

3. Ammonia Volatilization

Under flooded conditions, mineralization of green manure results in accumulation of NH_4^+–N which is susceptible to NH_3 volatilization

Table 14. ^{15}N balance for labelled fertilizer and hairy vetch added to small soil cylinders without plants and incubated in the field (after Varco, 1986)

		Percentage of ^{15}N recovered after					
		30 days			120 days		
N source	Tillage	Inorganic N	Organic + residue N	Lost	Inorganic N	Organic + residue N	Lost
NH_4NO_3	NT	32	8	60	12	9	79
	CT	35	16	49	5	11	84
Hairy vetch	NT	8	55	37	4	44	52
	CT	24	62	14	2	42	56

NT, no tillage; CT, conventional tillage

losses, especially in high pH soils. Green manuring may, however, reduce NH_3 volatilization losses through the temporary immobilization of N and the production of acidic decomposition products. The addition of green manure particularly in flooded soils, affects NH_3 volatilization also through its influence on soil pH and P_{CO_2}. The enhanced CO_2 production in the green manure amended soils buffers the soil against pH changes.

In an incubation study, Venkatakrishnan (1980) observed that sesbania green manure additions at $5.4\,t\,ha^{-1}$ (1.79% N on dry weight basis; C:N ratio 28.8) initially caused a reduction in the amount of NH_3 volatilized from both sodic (pH 10.7) and reclaimed (pH 8.3) sandy loam soils under flooded conditions. After 10 days of flooding, marked NH_3 volatilization occurred from the sodic soil amended with green manure, and 23% of the green manure N was lost in 63 days. In the reclaimed green manure amended soil, losses were significantly less than from the control. In a laboratory experiment, Cornforth and Davis (1968) observed substantial NH_3 volatilization losses from gliricidia green leaf manure on four acid soils. The losses were proportional to the CEC of the soil. In a fine sand (pH 5.3, CEC $2.6\,cmol\,kg^{-1}$) N losses over 14 days were as high as 34% (Figure 16). In a laboratory study using a forced draft chamber technique, Khind et al. (1989) reported a very small loss of N (1.0% of applied N) in a sodic soil (pH 10.0) amended with sesbania green manure ($6\,t\,ha^{-1}$ dry biomass; 2.5% N) and 100 kg urea-N ha^{-1}. The loss of N from urea treated soil was 4.5%. In a waterlogged sodic soil (pH 10.6), using a closed incubation system, Rao and Batra (1983) observed a 4.5% N loss through NH_3 volatilization from sesbania green manure (applied to supply $60\,mg\,N\,kg^{-1}$ soil) as compared with 28.9% from urea. Carmona et al. (1990), however, observed 12.3 and 28% ammonia volatilization losses of applied urea-N after 4 days from untreated and soybean straw treated soil, respectively. The differences in cumulative N loss were insignificant after more than 8 days. Lathwell et al. (1989) observed much greater N

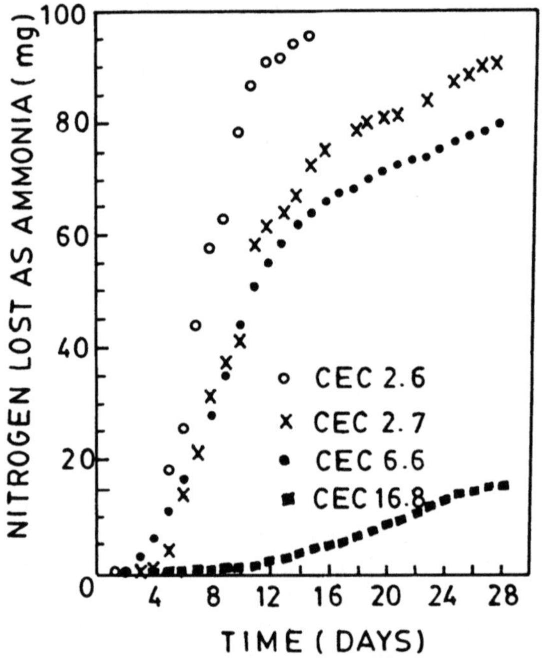

Figure 16. Cumulative NH_3 volatilization losses from different soils incubated with Gliricidia green leaf manure (10% dry weight basis) (after Cornforth and Davies (1968), by permission of the publishers, Butterworth-Heinemann Ltd ©.)

Table 15. Effect of green manure (GM) on loss of applied N (at 90 kg ha^{-1}) through ammonia volatilization measured over 15 days in flooded rice fields

Experimental conditions	Total N loss (kg ha^{-1}) Urea basal	Urea split	GM + urea split	Reference
Alfisol, Typic Hapluldalf (acidic)	2.0	1.4	1.2	Bhagat et al. (1988)
Silty clay loam, pH 8.4	11.9	7.6	5.1[a]	Santra et al. (1988)
Clay loam, pH 7.7	11.0	11.8	7.7	Saravanan et al. (1988)
Silty clay loam, pH 7.6	16.5	20.3	11.7	Mishra et al. (1990)

[a] All basal

losses from pots where mucuna leaves were surface-placed as compared with mixed treatment.

Results from several field studies showed that the N loss was generally lower when N was applied as urea in combination with sesbania green manure on 1:1 basis than that with urea alone (Table 15). Volatilization

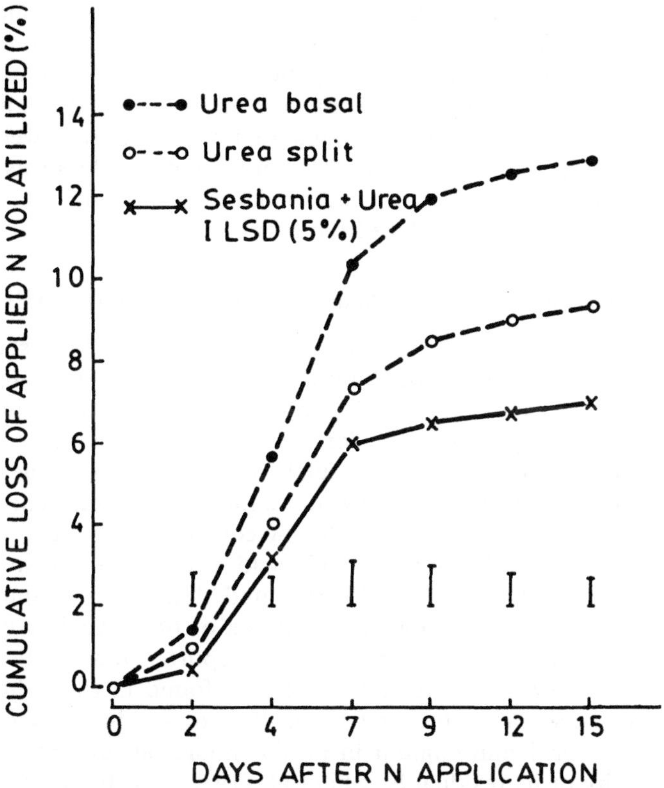

Figure 17. Effect of green manuring on cumulative loss of applied urea-N through ammonia volatilization in a flooded rice field (after Santra et al., 1988).

of NH_3 started within 2 days after application of fertilizer (Figure 17). The NH_4^+–N content and pH of floodwater were correlated significantly with loss of N. At IRRI, John et al. (1989b) found that incorporation of cowpea green manure or residue 15 days prior to transplanting had no effect on N loss from urea either incorporated prior to transplanting or broadcast at 15 days after transplanting, as determined by pNH_3 in floodwater and using the ^{15}N technique. Biswas (1988), however, observed that NH_3 loss from the combined application of sesbania green manure and urea was slightly less than from either sole sesbania or sole urea. Unlike John et al. (1989b), Biswas (1988) incorporated green manure immediately prior to rice transplanting. The volatilization losses of NH_3 from green manures and crop residues under upland conditions have rarely been investigated.

The increase in floodwater pH through photosynthetic activity of the aquatic photosynthetic and blue-green algae and bacteria of flooded rice soils will stimulate NH_3 volatilization from the surface applied urea. However, under azolla, floodwater pH may not increase due to the interruption of light. Furthermore, applied N is absorbed by azolla and on decomposition, used by rice. Watanabe and Liu (1990) reported that 50% of urea-N was lost when applied to floodwater at 2 weeks after transplanting rice, and when applied to azolla mat, the loss was reduced to 42%. Allowing azolla to grow for 2 weeks after applying urea and then incorporating into the soil reduced the loss to 25%.

G. Effect of Green Manures of Chemical Characteristics of Soil N

A few studies (Srivastva et al., 1984; Azam et al., 1985) have shown that green manuring increased the humic and fulvic acid fractions of soil organic matter. Azam et al. (1985) reported that the fulvic acid fraction contained a major part of ^{15}N derived from sesbania green manure. Also, the contribution of soil N to fulvic acid N decreased due to green manuring. It was suggested that green manure N was preferentially used by soil microbes for the synthesis of humic compounds or green manuring resulted in the mineralization of fulvic acid already present in soil. On the contrary, Srivastava et al. (1984) found that green manuring with sesbania contributed more towards synthesis of fulvic acid as compared with humic acid fraction. Shi et al. (1981) reported that while only 2% of residual ^{15}N derived from milk vetch was found in the light fraction (sp. gravity <1.80) after the first crop, 48% of residual ^{15}N originated from azolla. High lignin content in azolla should be responsible for the low availability of its residual N. Shi et al. (1981) also found that 35% of ^{15}N derived from milk vetch was present in the humic acid and fulvic acid fractions compared with about 20% from azolla.

Green manuring favorably affects soil microbial biomass. Azam et al. (1985) noted that soil biomass N increased by about 90% by green manuring. When green manure was applied alone, up to 10% of its N was found in microbial biomass. Bolten et al. (1985) observed a 26.6% increase in microbial biomass with the continuous use of leguminous green manures over the application of chemical fertilizers. Ladd et al. (1981a) found that large amounts of legume N were incorporated into microbial biomass during transformations in soil. These studies suggest that microbial biomass is an important sink of N applied as green manure.

V. Phosphorus

A. Mineralization of P from Green Manures

When incorporated into the soil, green manures decompose and through mineralization P is released in available forms (Bin, 1983; Singh, 1984). Although the rate of decomposition of green manure is influenced by a number of soil and environmental factors, P mineralization is closely related to the analogous transformations of N (Thompson et al., 1954). Phosphorus release would be most rapid under soil and climatic conditions favoring ammonification (Alexander, 1977).

Phosphorus content of the added organic matter is perhaps the most important factor in regulating the mineralization of P from green manures. Phosphorus in residues high in total P was more readily available than P contained in residues low in total P (Fuller et al., 1956). Immobilization of P occurred if the added plant residues contained less than 0.2% (Fuller et al., 1956) or 0.3% P (Kaila, 1954; Singh and Jones, 1976). Some workers have reported that mineralization of green manure P is dependent on the C:P ratio of the green manure. The critical C:P ratios for P mineralization to occur have, however, been found to vary from as low as 55 to as high as 300 (Barrow, 1960; Hannapel et al., 1964; Blair and Boland, 1978). The reasons for such a wide range of C:P ratios are not clear. Enwezor (1976) observed immobilization of P in pea residues (C:P ratio 112) incubated under aerobic conditions for 12 weeks. Phosphorus content and C:P ratio of green manures would depend on age of the green manure crops and their species, and available P status of the soil in which these are grown. The total P content of 6- to 8-week-old green manures is generally more than 0.3%. The C:P ratio is around 100. This suggests that during decomposition of green manures, immobilization of soil P should not be expected. The data recorded by Blair and Boland (1978) suggested that P mineralization would be fast from young (19-day-old) plants. Fuller et al. (1956) found that P contained in shoot material was more readily mineralized than P in roots.

B. Availability of P in the Soil

Several studies have shown that the availability of P can be increased through the incorporation of green manures into the soil both under laboratory (Blair and Boland 1978; Singh et al., 1981; Hundal et al., 1987; Yadvinder-Singh et al., 1988) and field (Chatterjee et al., 1979; Yadav and Singh, 1986) conditions. The effects of green manure on P transformations in soils are expected to be markedly different under upland and waterlogged conditions. In waterlogged soils, green manuring increases the availability of P through the mechanisms of reduction, chelation, and favorable changes in soil pH. Soil reduction has been shown to cause marked changes in the amount of plant available P (Ponnamperuma, 1972) and sometimes the P-sorption properties of the soil (Patrick and Khalid, 1974; Khalid et al., 1977). Changes in soil pH due to green manuring can influence solubility with P becoming more soluble between pH 5.5 and 7.5. The decomposing green manures release organic acids and CO_2, both tending to lower the soil pH in neutral and alkaline soils and raise solubility of calcium phosphate. Figure 18

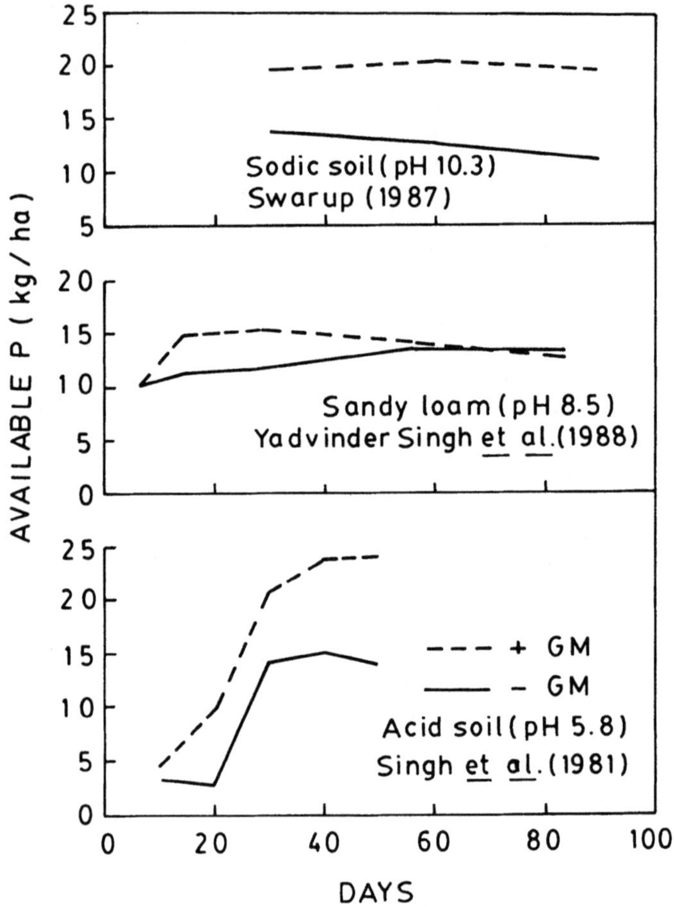

Figure 18. Effect of green manuring on changes in available P in three soils varying in pH under flooded conditions.

illustrates the role of green manuring in enhancing the availability of P in different soils. As expected, the effect of green manuring in increasing the availability of P was more pronounced in strongly acidic and alkaline soils than in normal soils. Singh et al. (1981) reported that pH changes in an acid soil amended with organic residues were significantly correlated with available P.

The decomposition products of green manures have significant chelation capacity which lowers the activity of polyvalent cations (Ca, Fe, Al) that form insoluble salts with P and thus liberate P from basic phosphates of these elements (Singh, 1962; Agboola, 1974; Bhattacharya and Das, 1975). Saha and Mandal (1979) have reported that increased P

Table 16. Langmuir parameters for P sorption by soil incubated with three green manures[a]

Green manure	20 days incubation		40 days incubation	
	k	b	k	b
Cowpea	0.037	0.375	0.036	0.333
Sesbania	0.044	0.400	0.043	0.353
Sunn hemp	0.058	0.462	0.053	0.387
No manure	0.120	0.600	0.070	0.522

k = bonding energy, $ml\,mg^{-1}\,P$; b = P adsorption maxima, $mg\,P\,g^{-1}$

[a] After Hundal et al. (1988). By permission of the publishers, Butterworth-Heinemann Ltd ©

availability in blue-green algae amended soil was due to the solubilizing effects of organic acids released during decomposition of algae cells. Prabhakar et al. (1972) reported that the addition of *Gliricidia maculata* green manure increased the availability of native soil P by 26 to 37% in four soils. Silva et al. (1985) found that the application of green manures in an acid soil increased the availability of P in rock phosphate and yield of beans and oats. Increased availability of P by green manuring in waterlogged soils may also be caused by the increase in labile P or decrease in buffering capacity, but both of these effects are mainly mediated through changes in the iron chemistry of the soils. Hundal et al. (1988) observed that green manure incorporation significantly reduced the binding energy and P sorption capacity of flooded soils (Table 16). The phosphorus sorption characteristics of the soil were changed, possibly due to the release of P during the mineralization of green manure and accumulation of intermediate organic acid complexed metal cations, thereby inducing solubilization of soil native P or reduced fixation of added inorganic P. Srivastava and Jafri (1973) reported that the availability of fertilizer P to sugarcane was 50.5 and 38.5% with and without green manuring, respectively. Another significant effect of green manuring is the increase in labile organic P in soil solution (Singh and Rai, 1973; Hundal et al., 1987). This fraction of the soluble P may be very important as to the management of P supply to plants.

The effects of green manure on P transformations are expected to be smaller under upland than waterlogged conditions. In a laboratory incubation study, Yadvinder-Singh et al. (1988) observed that green manuring with sesbania did not appreciably increase the available P content of a sandy loam soil over no green manure control. Singh and Rai (1975) and Bajpai et al. (1980) have reported slightly greater build-up of inorganic P due to green manuring in normal than in saline and alkaline soils, possibly because of greater microbial activity in the former. However, under aerobic conditions even in normal soils the increase in available P was so small that green manure was not considered to be a good source of P to the crops (Debnath and Hajra, 1972). Fuller et al. (1956) and Blair and Boland (1978) concluded that the absence of plants in most of the organic P mineralization studies has led to the general belief that the turnover of P from plant residues is low. They indicated that as P uptake by plants is increased, there is less inorganic P in solution that is thus subject to chemical fixation or microbial immobilization. Using ^{32}P-labeled white clover plants, Blair and Boland (1978) reported that after 48 days 12.0 and 29.3% of total P was released in low and high P soils in the presence of plants, respectively. The greater P release in high P soil was perhaps due to high P content of the plant material added (0.33 against 0.16% P).

The availability to crops of the P in green manures is governed by temperature. At high temperature and adequate moisture levels, green manures can be just as effective as fertilizers to supply P to crops. Some research workers (White et al., 1947; Blair and Boland, 1978; Handal et al., 1987) have studied the utilization of P by crops from soils amended with ^{32}P-labeled green manures. White et al. (1947) reported that cowpea green manure was as effective a source of P as the KH_2PO_4 for sudan grass grown over a period of 5 weeks and that utilization of P in the sand culture was three to five times as great as in the silt loam soil. The differences between soil and sand was attributable largely to their P-fixing capacities. The utilization of green manure-^{32}P was around 3%. Blair and Boland (1978) observed that P derived from white clover by 36-day-old oat was 2.8 and 5.2% on low and high P soils, respectively. Hundal et al. (1987) observed that utilization by 8-week-old rice of P from three green manures ranged from 11.2 to 13.2%. The per cent utilization was higher for cowpea having more P and a narrow C:P ratio than sesbania and sunn hemp green manures.

Krishna Rao et al. (1961) and Subbiah and Mannikar (1964) have reported that green manure crops could tap subsoil P and make it available to the shallow-rooted crops. In the study carried out by Subbiah and Mannikar (1964), sesbania was found to derive about 77% (mean of three

Table 17. Characteristics of roots of some green manure crops and utilization of difficultly available nutrients (after Gu and Wen, 1981)

Crop	CEC (cmol (+) kg^{-1} dry root)	Respiratory intensity (mg CO_2 g^{-1} dry root h^{-1})	Increment of dry matter (%)		
			Apatite	Orthoclase	Serpentine
Milk vetch	47.0	5.01	312	40	30
Vetch	41.9	3.91	183	15	53
Wheat	20.5	1.86	102	12	21

soils) of the fertilizer P from the subsoil (20 to 30 cm). The soil having the lowest amount of available P had the highest fertilizer P recovery from the subsoil. Deep penetration of the legume rooting system also allows exudates to exist at greater soil depths. Upon decomposition microbial metabolites of these exudates may serve as solubilizing and/or chelating agents for plant nutrients fixed in the unavailable forms.

VI. Potassium, Calcium, and Magnesium

Leguminous green manure plants, because of their typical root characteristics, possess a strong ability to absorb the difficultly available K from the soil (Table 17). Many workers have reported increased availability of K in soils due to green manuring (Kute and Mann, 1969; Debnath and Hajra, 1972; Katyal, 1977; Tiwari et al., 1980; Nagarajah et al., 1986; Swarup, 1987). The release of CO_2 and organic acids from the green manures possibly act on insoluble soil minerals, thereby releasing nutrients into the soil solution (Agboola, 1974). Nagarajah et al. (1986) observed marked increases in soil-solution K with the addition of sesbania green manure in three soils. The increase in the soil-solution K concentration was due to the release of Fe^{2+} and Mn^{2+} under highly reduced conditions created by green manuring. Katyal (1977) observed that increase in soil-solution concentration of K of flooded soil was directly related to the concentration of water-soluble Fe^{2+} and Mn^{2+}. Potassium contained in the green manure is also eventually released on decomposition and becomes available to the plants. The influence of green manuring in increasing the availability of K in soil is illustrated by the data listed in Table 18. In a laboratory incubation study conducted for 120 days, Debnath and Hajra (1972) observed that increase in available K due to green manuring occurred in the first 8 days. The work of Hargrove (1986) indicated that increase in the available K due to green manuring in the surface layer was due to the recycling of K from the subsurface layer (Table 18).

Katyal (1977) and Khind et al. (1987) observed significant increases in water-soluble Ca and Mg with green manuring of flooded soils. The peaks of Ca and Mg concentration were observed after 8 days of application of green manure. Like K, the increased in soil solution concentration of Ca and Mg in the green manure amended waterlogged soils is due to the release of Fe^{2+} and Mn^{2+} from the insoluble Fe^{3+} and Mn^{4+} oxides and hydrates under highly reduced conditions. In the green manure amended soils, active biodegradation coupled to an extensive CO_2 production can lead to the dissolution of Ca and Mg carbonates. Ploughing of crimson clover (*Trifolium incarnatum* L.) tessuls resulted in an increase in the levels of P, K, Ca, and Mg in soil (Groffman et al., 1987).

Table 18. Effect of green manuring on available K content in soil

Experimental Conditions		Exchangeable K ($mg\,kg^{-1}$ soil) −GM	+GM	Reference
Laboratory incubation study (30°C, 60% water-holding capacity), *Sesbania aculeata* at 0.5 g (dry weight) $(100\,g)^{-1}$ soil	8 days	128	342	Debnath and Hajra (1972)
	16 days	165	347	
	120 days	155	347	
3-year field study on wetland rice (six croppings); green leaf manuring (*Sesbania cannabina* + *Ipomea carnea*) 10 t dry matter ha^{-1}	0–225 mm	97	111	Chatterjee et al. (1979)
3-year field study on grain sorghum; mean of 4 legume cover crops	0–75 mm	105	114	Hargrove (1986)
	75–150 mm	100	89	
	150–300 mm	109	83	
12 years of continuous cropping of sugarcane	0–225 mm	63	80	Yadav and Singh (1986)
2-year field study on wetland rice, sunn hemp and sesbania green manures	0–150 mm	121	143	Sharma and Mittra (1988)
1-year field study on a sodic soil under wetland rice (*S. aculeata*)	0–150 mm	195	270	Swarup (1987)
	150–300 mm	200	275	
3-year field study on potato (sesbania + sunn hemp)	0–225 mm	95	103	Sharma et al. (1988)

VII. Sulfur

Green manures decompose rapidly and may increase the supply of plant available S. The principal end product of microbial decomposition of organic S in soil is SO_4^{2-} under aerobic conditions and S^{2-} under anaerobic conditions. The volatile S is also evolved during the decomposition process, particularly under anaerobic conditions. It is presumed that N and S will be mineralized in approximately the same ratio as these occur in organic residues. However, several studies have shown that the amounts of S mineralized during incubation were not significantly correlated with the total or mineralized N (Biederbeck, 1978). It is generally

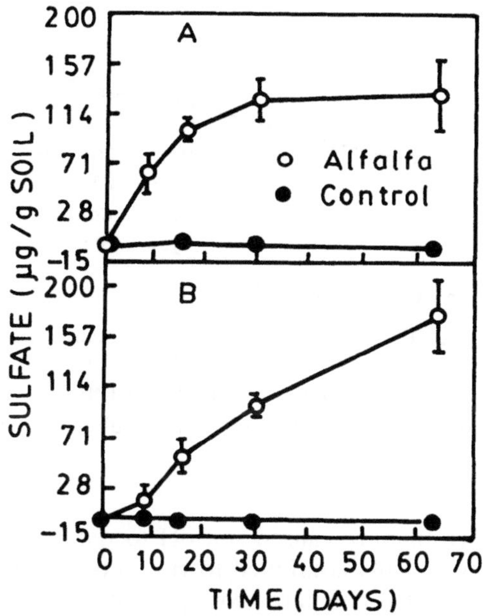

Figure 19. Mineralization of S in dark red (A) and red yellow (B) latosols amended with alfalfa, 5% dry weight basis (after Duxbury et al., 1989).

believed that organic residues with S content more than 0.15% will undergo S release during decomposition (Biederbeck, 1978; Zhu et al., 1984). Barrow (1960) observed S release from organic materials with a C:S ratio below 200 and immobilization of S above a C:S ratio of 420. Since the leguminous green manures usually contain >0.15% S and have a C:S ratio below 200, it is expected that S will be released during their decomposition.

There is a paucity of literature on S transformations in the green manure amended soils. Since S mineralization in soils is catalyzed by a wide variety of heterotrophic microorganisms, it can be expected that environmental and soil factors also affect the rate of S mineralization. Little is known about the effect of these factors on S mineralization from green manures. Such effects are expected to be somewhat similar to those discussed under N mineralization. In a laboratory incubation study on two oxisols (dark red and gray yellow latosols), Duxbury et al. (1989) observed fast mineralization of S from alfalfa green manure. The pattern of S mineralization differed markedly in the two soils (Figure 19).

Duxbury et al. (1989) studied the effect of green manuring on SO_4^{2-} adsorption in two latosols. The addition of alfalfa green manure immediately reduced adsorption of SO_4^{2-} relative to the unamended soil, and this effect was maintained even after adjustment of pH to that of

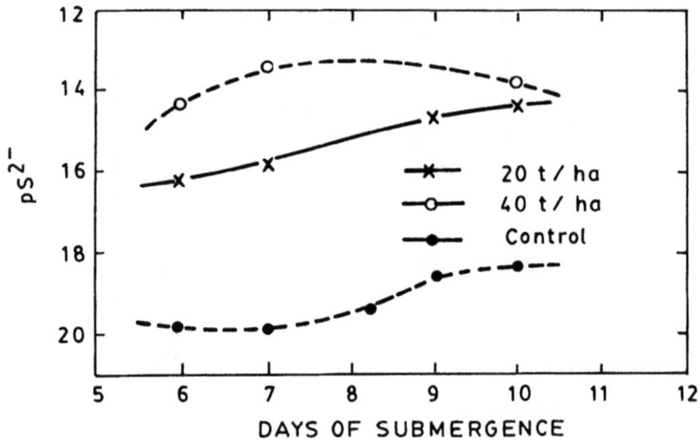

Figure 20. Effect of green manuring on S^{2-} in a submerged red soil (after Pan, 1985).

control soil. Green manure caused soil pH to rise immediately by 0.5 to 1.0 pH units. The soil pH continued to rise upon incubation with alfalfa to about pH 7.5 to 8.0 as compared with pH 4.4 to 5.3 in the control soils.

In waterlogged soils, a reduction of SO_4^{2-} occurs under highly reduced conditions. This reduction is a biological process but is controlled by physicochemical conditions. Obviously, it is related directly to reducing the strength of the medium and indirectly to the organic matter status of the soil. On a red soil containing very little organic matter, the S^{2-} content was very low even after it was submerged at a high temperature. The S^{2-} content increased several-fold however, after the addition of green manure (Figure 20). Green manure also increased the concentration of H_2S under submerged conditions (Pan, 1985).

Green manure would not only supply S to crops through the mineralization process, but it would also influence S availability by altering the physicochemical properties of soil. In a greenhouse study, Jordan and Baker (1959) found that a green manure crop of peas supplied as much S to a growing crop of beans as most of the inorganic S sources and much more than the equivalent addition of elemental S. Using ^{35}S-labeled green manure, Dhillon and Dhillon (1991) observed that cowpea and clusterbean supplied about 20% S to toria (*Brassica compestris* var. toria).

VIII. Micronutrients

Green manures do not constitute an important source of micronutrients as legumes do not spectacularly differ from nonleguminous crops in accumulating micronutrients. Transformations of micronutrients in soils are, however, conspicuously influenced by green manuring through changes in oxidation-reduction regimes, particularly in the submerged soils, and through increased chelation capacity brought about by the addition of organic matter as green manure. Among different micronutrients, transformations of Fe and Mn are closely related to the oxidation-reduction regimes in the soil and, thus, are more influenced by green manuring of soil.

A. Iron and Manganese

Submergence over a period of 10 to 12 weeks increased Fe^{2+} and Mn^{2+} in the soil solution of calcareous of noncalcareous soils (Thind and Chahal, 1983), black, red and lateritic soils (Katyal, 1977), sodic soils (Sadana and Bajwa, 1985), neutral soils (Khind et al., 1987), and acidic soils (Nagarajah et al., 1989), but a peak concentration of Fe^{2+} and Mn^{2+} was conspicuous when the soils were green manured. In all of the soils, peak Fe^{2+} concentration was recorded about 2 weeks after submergence. For Mn^{2+}, the peak concentration generally appeared at 1 or 2 weeks after submergence; the only exceptions were saline sodic soils exhibiting very low concentrations of soil-solution Mn^{2+}, in which peaks were observed between 4 to 6 weeks after submergence. A typical example of the dynamics of Fe^{2+} and Mn^{2+} in a green manured soil is shown in Figure 21.

The extent of the increase of total Fe^{2+} and Mn^{2+} due to reduction of oxides of these elements by organic reducing substances added to the soil as green manures under submerged conditions varied from very low (<2%) in sodic soils (Sadana and Bajwa, 1985) to as high as 30-fold in an acid lateritic soil (Katyal, 1977). Peak concentrations of Fe^{2+} and Mn^{2+} in the solution of green manure treated submerged soils were markedly influenced by soil pH. Significant relationships of pFe^{2+} and pMn^{2+} with soil pH were observed when data from different soils (Katyal, 1977; Thind and Chahal, 1983; Sadana and Bajwa, 1985; Nagarajah et al., 1989) were pooled together:

$$pFe^{2+} = -5.98 + 0.61\,pH, \quad r = 0.87, n = 11$$
$$pMn^{2+} = -6.20 + 0.69\,pH, \quad r = 0.85, n = 10$$

Although no significant trend could be observed, the calcium carbonate and organic matter content of soils seem to be the other important factors.

The water-soluble Fe^{2+} and Mn^{2+} in the soil solution can be distinguished into ionic and chelated forms. Organic matter added in the form of green manure, in addition to affecting the oxidation-reduction conditions of the soil, also functions as a chelating agent for water-soluble Fe^{2+} and Mn^{2+}. Depending upon the kind of green manure, the mean apparent stability constant (log K) of chelated Fe^{2+} ranged between 2.4

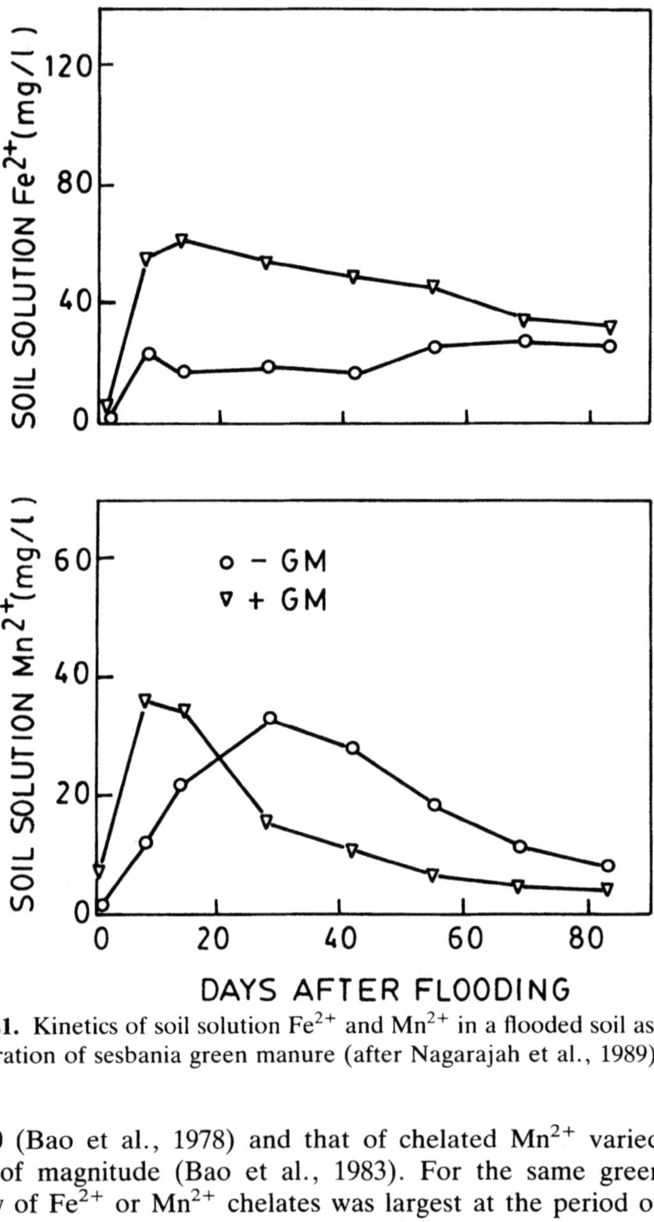

Figure 21. Kinetics of soil solution Fe^{2+} and Mn^{2+} in a flooded soil as affected by incorporation of sesbania green manure (after Nagarajah et al., 1989).

and 4.0 (Bao et al., 1978) and that of chelated Mn^{2+} varied by three orders of magnitude (Bao et al., 1983). For the same green manure, stability of Fe^{2+} or Mn^{2+} chelates was largest at the period of intensive decomposition and smallest at the final stage of decomposition. The magnitude of the stability constants was reduced by the presence of both negatively charged clay particules and iron oxides carrying positive charges (Bao et al., 1978, 1983).

The dynamics of ionic and chelated Fe^{2+} under submerged soil conditions exhibiting the role of green manuring both as reducing and

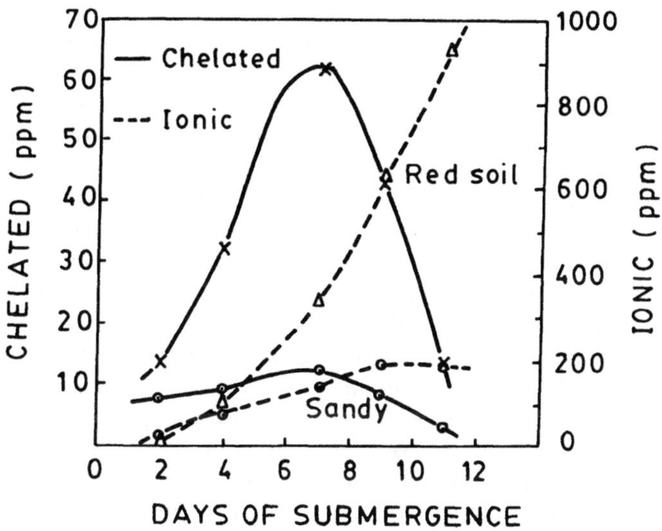

Figure 22. Effect of vetch green manure incorporation (5% dry weight basis) on dynamics of ionic and chelated Fe^{2+} in two flooded soils (after Bao et al., 1978).

chelating agents is shown in Figure 22. The work of Bao et al. (1978, 1983) and Bao and Yu (1986) revealed that at the beginning of submergence the rate of formation of Fe^{2+} as a result of reduction in the presence of green manure lags behind that of organic reducing substances. Later on, most of the Fe^{2+} in the soil solution was found to be in the chelated form. Water-soluble chelated Fe^{2+} reached its peak value after 4 to 7 days when because of intensive decomposition of green manure the amount of chelating agent is at its maximum. The peak period for Mn^{2+} complexes was earlier than that for Fe^{2+} complexes. Complexing of Fe^{2+} and Mn^{2+} ions may have a beneficial effect on rice growth. It has been suggested that this is the reason why rice plants can grow normally on some strongly reducing soils where concentration of water-soluble Fe^{2+} is quite high but with a considerable part of the ions in the complexed form (Yu, 1985). After the rice field is drained, water-soluble complexes are precipitated, forming orange- or red-colored coatings. If green manure has been applied, the coatings can be frequent and brilliantly colored.

The role of green manuring in increasing the availability of Fe and Mn has been studied in terms of different availability indices and increased supply to crop plants. Bijay-Singh et al. (1992) observed that in a calcareous sandy loam soil incubated with sesbania ($2\,g\,kg^{-1}$ soil), DTPA-extractable Fe and Mn increased by several-fold over the initial values during a 4- to 8-week period. The extent of the increase was more than twofold in the submerged soil over the soil maintained at the field capacity moisture regime. Studies carried out by Sharma and Katyal (1982) in field microplots revealed that a boost in low availability of soil iron (DTPA extractable) achieved through a 10-day soil submergence superimposed over sesbania green manure application was maintained at least for 6 weeks even if flooding was discontinued and soil was kept at around a field capacity moisture regime. A significant increase in DTPA-extractable Fe and Mn has also been reported following green manuring of wetland rice with mungbean (*Vigna radiata*) straw (Meelu and Rekhi, 1981). Swarup (1987) extracted soil Fe and Mn with ammonium acetate to confirm the rapid increase in the availability of these metals in the soil following green manuring with sesbania. Chahal and Khera (1988) studied the effect of green manuring on different forms of Fe under both submerged and 50% saturation moisture regimes. Green manure was obviously more effective in increasing water-soluble plus exchangeable Fe ($0.5\,M$ $CaCl_2$ extractable) under flooded conditions. Nevertheless, the beneficial effects of green manure in solubilizing Fe under 50% saturation moisture regime were almost equal to the effects produced by submerging the soil in the absence of green manure. From a field experiment on a coarse-textured soil, Takkar and Nayyar (1986) found that Fe deficiency in wetland rice was more effectively corrected by sesbania green manure than by soil-applied ferrous sulfate. Similarly, from a pot experiment with a red loam soil, Gopala Rao (1956) reported increased Fe and Mn availability (extractable with Morgan's reagent) and the shoot weight of rice green manured with *Gliricidia maculata*.

The management of iron chlorosis in rice nurseries is a serious problem. The success of green manure plus submergence in mobilization of soil iron as a result of intense reduction and subsequent retention in available form at a sufficiently high level during the growth of rice nurseries has been demonstrated by Sharma and Katyal (1982) and Maskina et al. (1985). The data in Table 19 demonstrate that green manuring with sesbania produced higher dry matter yield and Fe^{2+} content of 30-day-old seedlings than was recorded in those with soil application of ferrous sulfate and the control. Obviously, symptoms of iron chlorosis were absent in green manure treated rice seedlings.

B. Zinc and Copper

Studies reporting the effect of green manuring on the chemistry of Zn and Cu in soils maintained near field capacity moisture regimes are rarely available in the literature. Recently, Bijay-Singh et al. (1992) reported that when an initially dry calcareous sandy loam soil was incubated at field capacity with sesbania, it was observed that changes with time in DTPA-extractable Zn and Cu in the amended soil were not different than in the unamended soil.

Table 19. Effect of green manure and soil application of iron sulfate on dry matter yield, incidence of chlorosis, and Fe^{2+} content of rice seedlings (after Sharma and Katyal, 1982)

Treatment	Dry matter (g (25 plants)$^{-1}$)	Symptoms[a]	Fe^{2+} content (ppm)
Iron sulfate	4.1	4	60
Green manure	7.2	0	80
Control	3.9	4	52
L.S.D. ($p = 0.05$)	1.22	—	—

[a] 0, symptoms absent; 5, symptoms most pronounced

Under flooded soil conditions, soluble and exchangeable Fe and Mn increase at the expense of organic or oxide-held metals, and this process provides surfaces of high adsorptive capacity on which Zn and Cu may be bound. Thus, several workers (Katyal, 1977; Iu et al. 1981; Khind et al. 1987; Thind and Chahal, 1987; Bijay-Singh et al. 1992) have reported that water-soluble or DTPA-extractable Zn declined with the duration of flooding whether or not the soil was green manured. As liberation of water-soluble Fe^{2+} and Mn^{2+} was enhanced by green manuring, it helped in a further lowering of soil-solution Zn concentration in most of the investigations. Only in a sodic soil having pH 10.2 was green manure able to increase the availability of applied Zn (Swarup, 1987). Obviously, this was due to a significant reduction in soil pH brought about by green manuring under submerged conditions. The availability of Zn in soil may also increase by green manuring due to the mining of subsoil Zn. When vetch was grown for 3 years in a Zn-deficient soil in Yunnan Province in China, the Zn content of surface soil (0 to 10 cm) increased from 92.5 to 170 ppm and the available Zn increased from 2.9 to 4.9 ppm (Chen, 1983). It controlled the Zn deficiency of maize and increased its yield by 20 to 70%.

The behavior of Cu in a green manured soil differs from that of Zn because there exists some competition for Cu between the adsorption sites on the oxides and soluble organic matter, presumably with chelating ability. Copper may thus be present in the soil solution at a higher concentration in complexed form than is Zn (Hodgson et al. 1966). Bijay-Singh et al. (1992) reported that application of sesbania green manure to a flooded calcareous sandy loam soil could maintain, throughout the 12-week incubation period, at least a 1.5 times higher concentration of DTPA-extractable Cu than in the unamended soil. A significant increase in DTPA-extractable Cu by incorporating mungbean straw as green manure into the soil was also recorded by Meelu and Rekhi (1981).

IX. Conclusions

Leguminous green manure crops and azolla possess a remarkable potential for biological N_2 fixation. Legumes also possess a strong ability to absorb difficultly available nutrients in the soil and can help in concentrating plant nutrients in the surface layers of soil. Green manures, by virtue of their containing large amounts of readily decomposable organic matter, dictate transformations in the soil of several plant nutrients, including N, by altering the physicochemical properties of the soil. The redox potential of soils, particularly under submerged conditions, is conspicuously reduced by green manuring. After decomposition, the organic matter in green manures produces large quantities of CO_2 that can improve the availability of nutrients in the soil in a number of ways. Acetic and butyric acids can accumulate in large amounts in the green manure treated flooded soils. The changes in redox potential, accumulation of CO_2, and production of organic acids can cause both short- and long-term changes in soil pH, which in turn influence transformations of several plant nutrients in the soil. Green manuring can also cause positive changes in the activity of iron oxides and, therefore, of the surface properties of soils.

Nitrogen and other nutrients contained in green manures are converted into plant-available forms in the soil through the process of mineralization. The kinetics of N mineralization can differ under aerobic and anaerobic conditions. A few studies conducted so far show that the kinetics of N mineralization from green manures can best be described by using two or three simultaneous first-order or exponential functions. To model N fluxes in the green manure treated soils, precise information on potentially mineralizable N (N_o) is needed under different soil and climatic conditions. The simultaneous release and loss of N from green manures does not allow accurate prediction of the N release kinetics under flooded conditions.

Nitrogen from green manures is rapidly mineralized for absorption by plants. In flooded fields under tropical conditions as much as 80% of green manure-N can be mineralized within 2 weeks of its incorporation. The chemical composition of green manures (e.g., N content, C:N ratio, lignin content), as determined by the age and constituent plant parts profoundly affects N mineralization. No single chemical constituent has, however, been found to control the release of N from green manures. Among environmental factors, temperature and moisture are the most prominent. In the initial periods of decomposition of green manure, the rate of N release is faster under anaerobic than under aerobic conditions. No information is available on the interactive effects of moisture and temperature. Under dry and warm conditions the release of N may be limited. The influence of the environment is more pronounced on the surface-placed green manures. Rapid mineralization occurs with tillage and with incorporation of green manure. Drying and increasing particle size of green manure can slow down the process of N mineralization. Because the particle size of green manure, soil disturbance, and environmental conditions markedly affect the rate of N mineralization, measurements made in a few laboratory investigations without plants are unlikely to be quantitatively similar to those made in the field. Even the results from field studies will be applicable over a rather narrow range of climatic conditions.

Green manuring can cause a marked increase in urease activity and the rate of urea hydrolysis in soils. The rapid release of N from green manure can lead to substantial losses of N from the soil-plant system. A limited number of studies show that N losses from green manures are lower than with equivalent quantities of fertilizer N. Even the integrated management of green manures and fertilizers can result in reduced losses of N. Ammonia volatilization losses can be substantial from the surface applied green manures, particularly in the soils with high pH and/or low CEC. N derived from green manure is much less susceptible to leaching than fertilizer N. Keeping in view the large addition of readily available C substrate to denitrifiers, loss of green manure and fertilizer N through denitrification need to be precisely estimated in the soils amended with green manures. In order to use green manures more effectively, research on release and losses of N under field conditions needs to be carried out using ^{15}N-labeled green manures. The efficiency of green manure-N can be increased by synchronizing rate of N mineralization with that of N uptake by crop. One way to achieve it can be by breeding legumes which release N slowly, perhaps by selection for legumes with high lignin-, polyphenol-, or tannin-rich tissues resistant to microbial decay.

In addition to supplying N, green manure can increase the availability of P, K, and S in soils. The solubility of P in soils is markedly increased through the incorporation of green manures, particularly in flooded soils. Green manure can influence the sorption characteristics of both P and S in soils. The addition of green manure in flooded soils increases SO_4^{2-} reduction. Only a few studies have been carried out on the mineralization of plant nutrients other than N contained in green manures. Because of the intensification of reduction conditions and the release of chelating agents, the amount of exchangeable Fe^{2+} and Mn^{2+} in soil will increase following green manuring. The effects of green manure on Zn and Cu availability in aerobic soils are expected to be small. In flooded soils, green manuring can exaggerate Zn deficiency. More information, particularly on a long-term basis, should be obtained on the effects of green manuring on the changes in the soil-solution concentrations of different nutrient elements.

References

Agboola, A.A. 1974. Problems of improving soil fertility by the use of green manuring in tropical farming system. In: Organic Materials as Fertilizers. pp. 147–152. FAO Soils Bull. No. 17, Rome, Italy.

Alexander, D. 1988. Fertilizer requirement of rice-rice-green manure cropping system. Int. Rice Res. Newsl. 13(5):27.

Alexander, M. 1977. Symbiotic nitrogen fixation. In: Introduction to Soil Microbiology. pp. 305–330. Wiley, New York.

Allison, F.E. 1973. Soil Organic Matter and its Role in Crop Production. Elsevier Science Publishing Company, New York.

Amato, M., and J.N. Ladd. 1980. Studies of nitrogen immobilization and mineralization in calcareous soils: V. Formation and distribution of isotope labelled biomass during decomposition of ^{14}C and ^{15}N-labelled plant material. Soil Biol. Biochem. 12:405–411.

Amato, M., J.N. Ladd, A. Ellington, G. Ford, J.E. Mahoney, A.C. Taylor, and P. Walsgot. 1987. Decomposition of plant material in Australian soil. IV. Decomposition of in-situ of ^{13}C- and ^{15}N-labelled legume and wheat materials in a range of southern Australian soils. Aust. J. Soil Res. 25:95–105.

Aspiras, R.B. 1966. Some Factors Affecting Ammonification in Flooded Soils. M.S. thesis, University of Philippines, College, Laguna.

Aulakh, M.S., D.A. Rennie, and E.A. Paul. 1983. The effect of various clover management practices on gaseous N losses and mineral N accumulation. Can. J. Soil Sci. 63:593–605.

Azam, F., K.A. Malik, and M.I. Sajjad. 1985. Transformation in soil and availability to plants of ^{15}N applied as inorganic fertilizer and legume residues. Plant Soil 86:3–13.

Bajpai, P.D., R. Arya, and B.R. Gupta. 1980. Comparative studies on decomposition pattern of some plant materials in two different soil conditions during winter season. Indian J. Agric. Res. 14:91–102.

Bajpai, S., O.P. Srivastava, and A.N. Pathak. 1984. Effect of organic matter on urea hydrolysis in soil. In: Nitrogen in Soils, Crops and Fertilizers. pp. 155–159 Bull No. 13, Indian Society of Soil Science, New Delhi.

Bao, Xue-ming, and Yu, Tian-ren. 1986. Stability constants of Fe (II) complexes in soil. [in Chinese with English summary] Acta Pedol. Sin. 23:40–43.

Bao, Xue-ming, Liu, Zhi-guang, and Yu, Tian-ren. 1978. Studies on oxidation-reduction processes in paddy soils. IX. Forms of water soluble ferrrous iron. Acta Pedol Sin. 15:174–181.

Bao, Xue-ming, Ding Chang-pu, and Yu, Tian-ren. 1983. Stability constants of Mn (II)—Complexes in soils as determined by a voltametric method. Z. Pflanzenern. Bodenkd. 146:285–294.

Barrow, N.J. 1960. A comparison of the mineralization of nitrogen and of sulphur from decomposing organic materials. Aust. J. Agri. Res. 11:960–969.

Beri, V., O.P. Meelu, and C.S. Khind. 1989a. Biomass production, N accumulation, symbiotic effectiveness and mineralization of green manures in relation to yield of wetland rice. Trop. Agric. (Trinidad) 66:11–16.

Beri, V., O.P. Meelu, and C.S. Khind. 1989b. Studies on *Sesbania aculeata* Pers. as green manure for N accumulation and substitution of fertilizer N in wetland rice. Trop. Agric. (Trinidad) 66:209–212.

Bhagat, R.M., B.B. Kanwar, T.S. Verma, and R.S. Minhas. 1988. Nitrogen economy in lowland rice culture. Oryza 25:255–260.

Bhardwaj, K.K.R., and S.P. Dev. 1985. Production and decomposition of *Sesbania canabina* (Retz) Pers. in relation to its effect on the yield and wetland rice. Trop. Agric. (Trinidad) 62:233–236.

Bhattacharya, D., and B. Das. 1975. Mobilization of iron in soil and its influence on phosphorus availability through the chelating effects of aqueous extracts of some green manuring plant materials. J. Indian Chemical Soc. 52:726–729.

Biederbeck, V.O. 1978. Soil organic sulphur and fertility. In M. Schnitzer and S.U. Khan (eds.). Soil Organic Matter. pp. 273–310. Elsevier Scientific Publishing Company, New York.

Bijay-Singh, Yadvinder-Singh, U.S. Sadana, and O.P. Meelu. 1992. Effect of green manure, wheat straw and organic manures on DTPA extractable Fe, Mn, Zn and Cu in a calcareous sandy loam soil at field capacity and under waterlogged conditions. J. Indian Soc. Soil Sci. 40:000–000.

Bin, Jiao 1983. Utilization of green manure for raising soil fertility in China. Soil Sci. 135:65–69.

Biswas, T.K. 1988. Nitrogen Dynamics and Nitrogen-15 Balance in Lowland Rice as Affected by Green Manure and Urea Application. Ph.D. thesis. Indian Agricultural Research Institute, New Delhi.

Black, C.A. 1968. Soil-Plant Relationships (2nd ed.) Wiley, New York.

Blair, G.J., and O.W. Boland. 1978. The release of phosphorus from plant material added to soil. Aust. J. Soil Res. 16:101–111.

Bolten, Jr.H., L.F. Elliot, R.T. Papandick, and D.F. Bezdicek. 1985. Soil microbial biomass and selected enzyme activities. Effect of fertilizers and cropping practices. Soil Biol. Biochem. 17:297–302.

Bouldin, D.R. 1988. Effect of green manure on soil organic matter content and nitrogen availability. In: Green Manure in Rice Farming. pp. 151–163. International Rice Research Institute, Los Banos, Philippines.

Bremner, J.M., and R.L. Mulvaney. 1978. Urease activity in soils. In R.G. Burns (ed.). Soil Enzymes, pp. 149–196. Academic Press, London.

Brewbaker, J.L. 1987. Leucaena: a group of multipurpose treses for tropical agroforesty. In: Ten Years of Development in Agroforestry. International Council for Research in Agro-Forestry, Nairobi, Kenya.

Brown, P.L., and D.D. Dickey. 1970. Losses of wheat straw residue under simulated field conditions. Soil Sci. Soc. Am. Proc. 34:118–121.

Bruulsema, T.W., B.R. Christie. 1987. Nitrogen contribution to succeeding corn from alfalfa and red clover. Agron. J. 79:96–100.

Burns, R.G. 1978. Enzyme activity in soil. Some theoretical and practical consideration. In R.G. Burns (ed.). Soil Enzymes. pp. 295–340. Academic Press, New York.

Cang Dong-ging, Wang Jing-hua and Zhang Xiao-nian. 1985. Acidity In Tu Tian-ren (ed.). Physical Chemistry of Paddy Soils. pp. 131–156. Science Press, Beijing.

Carmona, G., C.B. Christianson, and B.H. Byrens. 1990. Temperature and low concentration effect of the urease inhibitors N-(n-butyl thiophosphoric triamide (NBTPT) on ammonia volatilization from urea. Soil Biol. Biochem. 22:933–937.

Ceccenti, B., and C. Ding. 1985. Anaerobic decomposition of green manures in flooded soils in the presence of Cr (VI). 2. Charge and size characterization of decaying products and their metal complexes. Agrochimica 29:50–65.

Ceccenti, B., C.P. Ding, M. De Nobili, M. Calcinoi, and R. Tersitano. 1990. Volatile fatty acids and N dynamics in soils amended with plant residues in soils amended with plant residues under flooding. Trans. 14th Int. Congr. Soil Sci. 4:766–767.

Chahal, D.S., and S.S. Khera. 1988. Effect of green manuring on iron transformations in soils under submerged and unsubmerged conditions. J. Indian Soc. Soil Sci. 36:433–438.

Chatterjee, B.N., K.I. Singh, A. Pal, and S. Maiti. 1979. Organic manures as substitutes for chemical fertilizers for high yield rice varieties. Indian J. Agric. Sci. 49:188–192.

Chen Shi-ping. 1983. Green manures in multiple cropping systems in China. In Fertilizer Use under Multiple Cropping Systems. pp. 143–150. FAO Fertilizer and Plant Nutrition Bulletin 5, FAO, Rome.

Cornforth, I.S., and J.B. Davis. 1968. Nitrogen transformations in tropical soil. I. The mineralization of nitrogen rich organic materials added to soil. Trop. Agric. (Trinidad) 45:211–221.

De Haan, S. 1977. Humus, its formation, its relation with the mineral part of the soil and its significance for soil productivity. In Soil Organic Matter Studies. pp. 21–30, Vol. 1. IAEA/FAO, Vienna.

Debnath, N.C., and J.N. Hajra. 1972. Transformation of organic matter in soil in relation to mineralization of carbon and nitrogen availability. J. Indian Soc. Soil Sci. 20:95–102.

Dhillon, K.S., and S.K. Dhillon. 1991. Relative contribution of green manures in sulphur nutrition of toria (*Brassica compestris*). J. Nuclear Agric. Biol. 20:128–133.

Duxbury, J.M., P.P. Motavalli, and W. Goedert. 1989. Ion movement in Cerrado Soils: The effects of inorganic and organic amendments on sulphur (S) availability. Trop. Soils Tech. Rep. 1986–1987. pp. 315–318. North Carolina State Univ., Raleigh, NC.

Enwezor, W.O. 1976. The mineralization of nitrogen and phosphorus in organic materials of varying C:N and C:P ratios. Plant Soil 44:237–240.

Floate, M.J.S. 1970. Decomposition of organic materials from hill soils and pastures. III. The effect of temperature on mineralization of carbon, nitrogen, and phosphorus from plant materials and sheep faeces. Soil Biol. Biochem. 2:187–196.

Frankenberger W.T., Jr., and H.M. Abdelmagid. 1985. Kinetic parameters of nitgoren mineralization rates of leguminous crops incorporated into soil. Plant Soil 87:257–271.

Fuher, F., and D. Sauerbeck. 1968. Decomposition of wheat straw in the field as influenced by cropping and rotation. In: Isotopes and Radiation in Soil Organic Matter Studies. pp. 241–250. IAEA, Vienna.

Fujii, K., M. Kobayashi, and E. Takahashi. 1972. Amines in the mixture of plant residue and sand incubated under aerobic and waterlogged conditions. J. Sci. Soil Manure, Japan. 43:160–164.

Fulkerson, R.S. 1983. Research review of forage production. Crop Science Department, University of Guelph, Guelph, Ontario.

Fuller, W.H., D.R. Nielsen, and R.W. Miller. 1956. Some factors influencing the utilization of phosphorus from crop residues. Soil Sci. Soc. Am. Proc. 20:218–224.

Gale, P.M., and J.T. Gilmour. 1988. Net mineralization of carbon and nitrogen under aerobic and anaerobic conditions. Soil Sci. Soc. Am. J. 52:1006–1010.

Gilmour, J.T., M.D. Clark, and G.C. Sigua. 1985. Estimating net nitrogen mineralization from CO_2 evolution. Soil Sci. Soc. Am. J. 49:1398–1402.

Gopala Rao, H.G. 1956. Effect of green manuring on red loamy soils freshly brought under swamp paddy cultivation with particularly reference to availability of iron and manganese. J. Indian Soc. Soil Sci. 4:225–231.

Groffman, P.M., D.F. Herdrix, C. Han, and D.A. Crossby, Jr. 1987. In J.F. Power (ed.). The Role of Legumes in Conservation Tillage Systems. pp. 7–8. Soil Conservation Society of America, Ankeny.

Gu Rong-Shen and Wen Qi-Xiao. 1981. Cultivation and application of green manure in paddy field of China. In: Proc. Symp. on Paddy Soil. pp. 207–219. Science Press, Beijing; Springer-Verlag, New York.

Haider, K., J.P. Martin, and E. Rietz. 1977. Decomposition in soil of ^{14}C-labelled coumaryl alcohols, free and linked into dehydropolymer and plant lignins and model humic acids. Soil Sci. Soc. Am. J. 41:556–562.

Hannapel, R.J., W.H. Fuller, and R.H. Fox. 1964. Phosphorus movement in a caclareous soil. 2. Microbial activity and organic P movement. Soil Sci. 97:421–426.

Hargrove, W.L. 1986. Winter legumes as a nitrogen source for no-till grain sorghum. Agron. J. 78:70–74.

Harrison, A.F. 1971. The inhibitory effect of oak leaf litter tannins on the growth of fungi in relation to their decomposition. Soil Biol. Biochem. 3: 167–172.

Haynes, R.J. 1986. Mineral Nitrogen in the Plant-Soil System. Academic Press, Inc. New York.

He, Qun, and Xu, Zuyi 1981. Influence of transformation of iron oxides on soil structure. In Proceedings of Symposium on Paddy Soil. pp. 699–703. Institute of Soil Science, academa, ed., Science Press, Beijing, Springer-Verlag, Berlin, Heidelberg, New York.

Herdt, R.W. and C. Capule. 1983. Adoption, spread, and production impact of modern rice varieties in Asia. Int. Rice. Res. Inst., Manila, Philippines.

Herman, W.A., W.B., McGill, and J.F. Dormaar. 1977. Effect of initial chemical composition on decomposition of roots of three grass species. Can. J. Soil Sci. 57:202–215.

Hodgson, J.F., W.L Lindsay, and J.F. Trierweiler. 1966. Micronutrient cation complexing in soil solution. II. Complexing of zinc and copper in displaced solution from calcareous soils. Soil Sci. Soc. Am. Proc. 30:723–726.

Huang, D.M., J.H. Gao, and P.L. Zhu. 1981. Transformation and distribution of organic and inorganic fertilizer nitrogen in rice-soil system (in Chinese, English summary). Acta Pedol. Sin. 18:107–121.

Hundal, H.S., C.R. Biswas, and A.C. Vig. 1987. The utilization by rice of P from different ^{32}P labelled green manures. Biol. Wastes 22:97–105.

Hundal, H.S., C.R. Biswas, and A.C. Vig. 1988. Phosphorus sorption characteristics of flooded soil amended with green manures. Trop. Agric. (Trinidad) 65:185–187.

Iritani, W.M., and C.Y. Arnold. 1960. Nitrogen release of vegetable crop residues during incubation as related to their chemical composition. Soil Sci. 89:74–82.

Ishikawa, M. 1963. Soil scientific and plant nutritional study on the milk vetch manuring of rice. Toyama Agric. Exp. Stn. Spec. Stud. Rep. 5.

Ishikawa, M. 1988. Green manure in rice. The Japanese experience. In: Green Manure in Rice Farming. pp. 45–61. International Rice Research Institute, Los Banos, Philippines.

Islam, A., and W. Islam. 1973. Chemistry of submerged soils and growth and yield of rice. I. Benefits from submergence. Plant Soil 39:555–565.

Ito, O., and I. Watanabe. 1985. Availability to rice plants of nitrogen fixed by Azolla. Soil Sci. Plant Nutr. 31(1):91–104.

Iu, K.L., I.D. Pulford, and H.J. Duncan. 1981. Influence of waterlogging and lime or organic matter additions on the distribution of trace metals in an acid soil. Plant Soil 59:327–333.

Janzen, H.H., and G.D. Radder. 1989. Nitrogen mineralization in a green manure-amended soil as influenced by cropping history and subsequent crop. Plant Soil 120:125–131.

Janzen, H.H., J.B. Bole, V.O. Biederbeck, and A.E. Slinkare. 1990. Fate of N applied as green manure or ammonium fertilizer to soil subsequently cropped with spring wheat at three sites in Western Canada. Can. J. Soil Sci. 70:313–323.

Jenkinson, D.S. 1977. Studies on decomposition of plant material in soil. V. The effects of plant cover and soil type on the loss of carbon from ^{14}C-labelled rye grass decomposing under field conditions. J. Soil Sci. 28:417–423.

Jenkinson, D.S. 1981. The fate of plant and animal residues in soil. In D.J. Greenland and M.H.B. Hayes (eds.). The Chemistry of Soil Processes. Wiley, New York.

Jenkinson, D.S., and A. Ayanaba. 1977. Decomposition of ^{14}C labelled plant material under tropical condition. Soil Sci. Soc. Am. J. 41:912–915.

Jensen, E.S. 1989. The role of pea cultivation in the N economy of soils and succeeding crops. In P. Plancquaert and R. Haggar (eds.). Legumes in Farming Systems. pp. 3–15. Martinus Nijhoff Publishers, Dordrecht.

Joachim, A.W.R. 1931. The principles of green manuring and their application in Ceylon. In: A Manual of Green Manuring. pp. 5–34. Department of Agriculture, Peradeniya, Sri Lanka.

Joachim, A.W.R., and S. Kandiah. 1929. Laboratory and field studies on green manuring under paddy land (anaerobic) conditions. Trop. Agric. (Sri Lanka) 72:253–271.

John, P.S., R.J. Buresh, R. Prasad, and R.K. Pandey. 1989a. Nitrogen gas (N_2 + N_2O) flux from urea applied to lowland rice as affected by green manure. Plant Soil 119:7–13.

John, P.S., R.K. Pandey, R.J. Buresh, and R. Prasad. 1989b. Lowland rice response to urea following three cowpea cropping systems. Agron. J. 81:853–857.

Jordan, J.V., and G.O. Baker. 1959. Sulphur studies in North Idaho soils using radiosulphur. Soil Sci. 88:1–6.

Kaila, A. 1954. Microbiological fixation and mineralization of phosphorus during the decomposition of organic matter. Z. Pflenaehr Dung. 64:27–35 (c.f. Soil Fertil. 17:946.).

Katyal, J.C. 1977. Influence of organic matter on the chemical and electrochemical properties of some flooded soils. Soil Biol. Biochem. 9:259–266.

Khalid, R.A., W.H. Patrick, Jr., and R.D. Delaune. 1977. Phosphorus sorption characteristics of flooded soils. Soil Sci. Soc. Am. Proc. 41:305–310.

Khind, C.S., A.S. Josan, and V. Beri. 1985. N release from sesbania green manure and effect of time of application of N fertilizer on lowland rice. Int. Rice Res. Newsl. 10(4):26–27.

Khind, C.S., A. Jugsujinda, C.W. Lindau, and W.H. Patrick, Jr. 1987. Effect of sesbania straw in a flooded soil on soil pH, redox potential, and water-soluble nutrients. Int. Rice Res. Newsl. 12(3):42–43.

Khind, C.S., A. Garg, and M.S. Bajwa. 1989. Effect of sesbania green manure and wheat straw on ammonia volatilization loss in wetland soil. Int. Rice Res. Newsl. 14(2):31–32.

Khind, C.S., A. Garg, and M.S. Bajwa. 1991. Effect of sesbania, rice straw and pre-incubation on urea hydrolysis in wetland soil. Int. Rice Res. Newsl. 16(2):18.

King, H.G.C., and G.W. Heath. 1967. The chemical analysis of small samples of leaf material and the relationship between the disappearance and composition of leaves. Pedobiologia 7:192–197.

Krishna Rao, D.V., T. Seshagiri Rao, and P, Krishnamurthy. 1961. Green manure as a source of nitrogen to rice grown in regurs of the Nizam Sagar Project area. J. Indian Soc. Soil Sci. 9:119–123.

Kute, S.B., and H.S. Mann. 1969. Effect of green manuring on the composition of soil and wheat crop and the balance of major plant nutrients in the soil after the crop. Indian J. Agric. Sci. 39:10–17.

Ladd, J.N., and M. Amato. 1986. The fate of nitrogen from legume and fertilizer sources of soils successively cropped with wheat under field conditions. Soil Biol Biochem. 18:417-425.

Ladd, J.N., J.M. Oades, and M. Amato. 1981a. Microbial biomass formed from ^{14}C, ^{15}N-labelled plant material decomposing in soils in the field. Soil Biol. Biochem. 13:119–126.

Ladd, J.N., J.M. Oades, and M. Amato. 1981b. Distribution and recovery of nitrogen from legume residues decomposing in soils sown to wheat in the field. Soil Biol. Biochem. 13:251–256.

Ladd, J.N., M. Amato, R.B. Jackson, and J.H.A. Butler. 1983a. Utilization by wheat crops of nitrogen from legume residues decomposing in soils in the field. Soil Biol. Biochem. 15:231–238.

Ladd, J.N., R.B. Jackson, M. Amato, and J.H.A. Butler. 1983b. Decomposition of plant material in Australian Soils. 1. The effect of quantity added on decomposition and on residual microbial biomass. Aust. J. Soil Res. 21:563–570.

Ladha, J.K., S. Miyan, and M. Garcia. 1989. Sesbania rostrata as a green manure for lowland rice: Growth, N_2 fixation, Azorhizobium sp. inoculation, and effects on succeeding crop yields and nitrogen balance. Biol. Fertil. Soils 7:191–197.

Lathwell, D.J., D.R. Bouldin, W.T. Bowen, and F. Costa. 1989. Soil- and crop management systems for acid Savanna soil using green manures and crop residues as a source. In Tropsoils Technical Report (1981–1987). pp. 305–309. North Carolina State Univ., Raleigh, NC.

Lin, X., and Q. Wen. 1990. Decomposition of plant material in calcareous soils in north China plain. Trans. 14th Int. Congr. Soil Sci. Commission III:351–352.

Liu Chung-Chu. 1988. Integrated use of green manure in rice fields in South China. In Green Manure in Rice Farming. pp. 319–331. International Rice Research Institute, Los Banos, Philippines.

Lohnis, F. 1926. Nitrogen availability of green manures. Soil Sci. 22:253–290.

Lynch, D.L., and L.J. Cotnoir, Jr. 1956. The influence of clay minerals on the breakdown of certain organic substrates. Soil Sci. Soc. Am. Proc. 20:367–370.

Maskina, M.S., P.S. Sandhu, and O.P. Meelu. 1985. Effect of integrated use of organic and inorganic nitrogen sources on growth and nutrient composition of rice seedlings. Oryza 22:11–16.

McCalla, T.M., and F.L. Duley. 1943. Disintegration of crop residues as influenced by subtillage and plowing. Agron. J. 35:306–315.

McCalla, T.M., and J.C. Russel. 1948. Nitrate production as affected by sweet clover residues left on the surface of the soil. Agron. J. 40:411–421.

Meelu, O.P., and R.S. Rekhi. 1981. Mung straw management and nitrogen economy in rice culture. Int. Rice Res. Newsl. 64(4):21.

Meelu, O.P., Yadvinder-Singh, and Bijay-Singh. In press. Green manuring for soil productivity improvement. FAO Soils Bulletin.

Melillo, J.M., J.D. Aber, and J.F. Muratore. 1982. Nitrogen and lignin control of hardwood leaf litter decomposition dynamics. Ecol. Monogr. 63:621–626.

Merckx, R., A. Dijkstra, A. den Hartog, and J.A. Van Veen. 1987. Production of root-derived material and associated microbial growth in soil at different nutrient levels. Biol. Fertil. Soils 5:126–132.

Mian, H.H. 1985. Detection of denitrification, by a ^{15}N tracer technique, of nitrogen released from azolla and blue green algae in a flooded soil. Aust. J. Soil Res. 23:245–252.

Mian, M.H., and W.D.P. Stewart. 1985. A ^{15}N tracer study to compare N supply by Azolla and ammonium sulfate to IR8 rice plants grown under flooded conditions. Plant Soil 83:371–379.

Mishra, B., R.D. Sharma, and J.R. Murthy. 1990. Effect on neem cake and shellac coating of urea and green manure on ammonia volatilization and nitrogen use efficiency for rice. J. Indian Soc. Soil Sci. 38:224–228.

Morachan, Y.B., W.C. Moldenhaner, and W.E. Larson. 1972. Effect of increasing amounts of organic residues on continuous corn. I. Yield and soil physical properties. Agron. J. 64:199–203.

Motomura, S. 1962. The effect of organic matter on the formation of ferrous iron in soil. Soil Sci. Pl. Nutr. 8:20–29.

Muller, M.M. 1987. Leaching of substerrarean clover derived N from a loam soil. Plant Soil 102:185–191.

Nagarajah, S. 1988. Transformation of green manure nitrogen in lowland rice soils. In Green Manure in Rice Farming. pp. 193–208. International Rice Research Institute, Los Banos, Philippines.

Nagarajah, S., M.C.G. Robielos, and H.U. Neune. 1986. Effect of incorporation of sesbania, azolla and rice straw on nitrogen release pattern in flooded rice soils. International Rice Research Institute, Saturday Seminar, Department of Soils, Aug. 2.

Nagarajah, S., H.U. Neue, and M.C.R. Alberto. 1989. Effect of sesbania, azolla and rice straw incorporation on the kinetics of NH_4, K, Fe, Mn, Zn and P in some flooded soils. Plant Soil 116:37–48.

Palm, O., W.L. Weerakoon, M.A.P. DeSilva, and R. Thomas. 1988. Nitrogen mineralization of Sesbania sesban used as green manure for lowland rice in Sri Lanka. Plant Soil 108:210–209.

Pan, Shu-Zhang. 1985. In Yu, Tian-ren (ed.). Physical Chemistry of Paddy Soils. pp. 92–110. Science Press, Beijing, China.

Parker, D.T. 1962. Decomposition in the field of buried and surface applied cornstalk residue. Soil Sci. Soc. Am. Proc. 28:559–562.

Patrick, W.H. Jr., and R.A. Khalid. 1974. Phosphate release and sorption by soils and sediments: Effect of aerobic and anaerobic conditions. Science 186:53–55.

Paul, E.A. 1976. Nitrogen in terrestrial ecosystem. In J.O. Nriagu (ed.). Environmental Biochemistry, Vol. 1: Carbon, Nitrogen, Phosphorus, Sulfur and Selenium Cycles. pp. 225–243. Ann Arbor Science, Ann Arbor, MI.

Paul, E.A. 1988. Towards the year 2000: directions for future nitrogen research. In J.R. Wilsen (ed.). Advances in Nitrogen Cycling in Agriculture Ecosystems. pp. 417–425. C.A.B. International, Wallingford.

Pierre, W.H., and W.L. Banwart. 1973. Excess base and excess base/nitrogen ratios of various crop species and plant parts. Agron. J. 65:91–96.

Pinck, L.A., and F.E. Allison. 1961. Adsorption and release of urease by and from clay minerals. Soil Sci. 91:183–188.

Ponnamperuma, F.N. 1965. Dynamics aspects of flooded soils and the nutrition of the rice plant. In Mineral Nutrition of the Rice Plant. pp. 295–328. Johns Hopkins Press, Baltimore, MD.

Ponnamperuma, F.N. 1972. The chemistry of submerged soils. Adv. Agron. 24:29–96.

Prabhakar, A.S., S.V. Patil, and K. Krishnamurthy. 1972. Influence of organic manures, ammonical and nitrate nitrogen on the availability of soil and applied phosphorus. J. Indian Soc. Soil Sci. 20:413–415.

Primrose, S.B. 1979. Ethylene and agriculture: The role of microbe. J. Appl. Bact. 46:1–25.

Ramaswami, P.A., and D. Raj. 1973. Effect of green manure and nutrient applications on the changes in soil microbial populations. Madras Agric. J. 60:995–1000.

Rao, D., and D.S. Mikkelsen. 1977. Effect of acetic, propionic and butyric acids on young rice seedlings growth. Agron. J. 69:923–928.

Rao, D.L.N., and L. Batra. 1983. Ammonia volatilization from applied nitrogen in alkali soils. Plant Soil 70:219–228.

Sadana, U.S., and M.S. Bajwa. 1985. Manganese equilibrium in submerged sodic soils as influenced by application of gypsum and green manuring. J. Agri. Sci. Camb. 104:257–261.

Sadana, U.S., and M.S. Bajwa. 1986. Effect of gypsum and green manuring on electrochemical and chemical changes in submerged sodic soils. Oryza 23:89–95.

Saha, K.C., and L.N. Mandal. 1979. Effect of algal growth on the availability of phosphorus, iron and manganese in rice soils. Plant Soil 52:139–149.

Saini, P. 1990. In Situ Leaching and Transformations of Urea Applied to Wetland Rice Soils. M.Sc. thesis, Punjab Agricultural University, Ludhiana, India.

Santra, G.H., D.K. Das, and L.N. Mandal. 1988. Loss of nitrogen through ammonia volatilization from flooded rice fields. J. Indian Soc. Soil Sci. 36:652–659.

Saravanan, A., V. Velu, and K.M. Ramanathan. 1988. Effect of sources and methods of nitrogen application on volatilization loss of ammonia and yield of rice under submerged soils of Cauvery delta, India. Oryza. 25:143–148.

Sauerbeck, D. 1966. A critical evaluation of incubation experiments on the priming effect of green manure. In The Use of Isotopes in Soil Organic Matter Studies. pp. 209–221. Report FAO/IAEA Technical Meeting, Pergamon, Oxford.

Sharma, A.R., and B.N. Mittra. 1988. Effect of green manuring and mineral fertilizer on growth and yield of crops in rice based cropping on lateritic soil. J. Agric. Sci. Camb. 110:605–608.

Sharma, B.D., and J.C. Katyal. 1982. Changes in DTPA-iron and management of iron chlorosis in rice nurseries. Plant Soil 69:123–126.

Sharma, R.C., P.M. Govindakrishnan, R.P. Singh, and H.C. Sharma. 1988. Effect of farmyard manure and green manures on crop yields and nitrogen needs of potato based cropping system in Punjab. J. Agric. Sci. Camb. 110:499–504.

Shi Shu-lian, L.L. Cheng, H.H. Lin, C.L. Shu, and C.H. Wen. 1978. Effect of azolla on the fertility of paddy soils (in Chinese, English summary). Acta Pedol. Sin. 15:54–60.

Shi Shu-lian, Lin Xin-Xiong and Wen Qi-Wiao. 1981. Decomposition of plant materials in relation to their chemical composition in paddy soil. In Proceedings Symposium on Paddy Soil. pp. 306–310. Institute of Soil Science, Academia, Sinica.

Shim, K., W.E. Splittstoesser, J.S., Titus. 1973. Changes in urease activity in apple trees as related to urea applications. Physiol. Plant 28:327–334.

Silva, E.M.R. De., D.L. De Almeida, A.A. Franco, and J. Dobereiner. 1985. Effect of green manuring on the availability of phosphate on an acid soil. Reviste Brasileira de Ciencia do Solo 9(1):85–88.

Sims, J.L., and L.R. Frederick. 1970. N immobilization and decomposition of corn residue in soil and sand as affected by residue particle size. Soil Sci. 109:355–361.

Singh, A. 1962. Studies on the modus operandi of green manures in tropical climates: A critical review of literature. Indian J. Agron. 7:69–75.

Singh, B.B., and J.P. Jones. 1976. Phosphorus sorption and desorption characteristics of soil as affected by organic residues. Soil Sci. Soc. Am. Proc. 40:389–394.

Singh, N.T. 1984. Green manures as source of nutrients in rice production. In Organic Matter and Rice. pp. 217–228. International Rice Research Institute, Los Banos, Philippines.

Singh, P.K., B.C. Panigrahi, and K.B. Satapathy. 1981. Comparative efficiency of Azolla, blue-green algae and other organic manures in relation to N and P availability in a flooded rice soil. Plant Soil 62:35–44.

Singh, S., and R.N. Rai. 1973. Effect of soil, superphosphate and age at burial of legumes on the changes in inorganic and organic phosphorus during their decomposition. J. Indian Soc. Soil Sci. 21:271–276.

Singh, S., and R.N. Rai. 1975. Effect of salinity, alkalinity, phosphate and age of plants on mineralization of nitrogen from *Sesbania aculeata* and *Melilotus alba* after their incorporation of soil. J. Indian Soc. Soil Sci. 23:122.

Smith, A.M. 1976. Ethylene in soil biology. Ann. Rev. Phytopathol. 14:53–73.

Smith, M.S., W.W. Frye, and J.J. Varco. 1987. Legume winter cover crops. Adv. Soil Sci. 7:95–139.

Sorensen, L.H. 1975. The influence of clay on the rate of decay of amino acid metabolites synthesized in soil during the decomposition of cellulose. Soil Biol. Biochem. 7:171–177.

Srivastva, L.L., N. Ahmed, and B. Mishra. 1984. Studies on nitrogen and carbon status of soils as affected by crop residues, farmyard manure and green manures. In Nitrogen in Soils, Crops and Fertilizers. pp. 277–284. Bull. No. 13, Indian Society of Soil Science, New Delhi.

Srivastava, S.C., and S.M.H. Jafri. 1973. Tracer assessment of mobilization of soil phosphorus due to sunn green manuring. J. Indian Soc. Soil Sci. 21:403–405.

Stewart, B.A. 1987. Preface. Adv. Soil Sci. 7:i–vii. Springer-Verlag, New York.

Stott, D.E., and J.P. Martin. 1989. Organic matter decomposition and retention in arid soils. Arid Soil Res. and Rehabilitation. 3:115–148.

Subbiah, B.V., and N.D. Mannikar. 1964. Selection of green manure crops for the uptake of sub-soil phosphorus studies with P^{32}. Indian J. Agric. Sci. 34:21–27.

Subramani, S., and S. Kannaiyan. 1987. Effect of urea on decomposition of azolla. Int. Rice Res. Newsl. 12(4):57.

Swarup, A. 1987. Effect of presubmergence and green manuring (*Sesbania aculeata*) on nutrition and yield of wetland rice (*Oryza sativa* L.) on a sodic soil. Biol. Fertil. Soils 5:203–208.

Swift, M.J., O.W. Heal, and J.M. Anderson. 1979. Decomposition in Terrestrial Ecosystems. Blackwell, Oxford.

Takeda, K., and C. Furusaka. 1975. Studies on the bacteria isolated anaerobically from paddy field condition. III. Production of fatty acids and ammonia by *Clostridium* species. Soil Sci. Plant Nutr. Tokyo 21:113–118.

Takkar, P.N., and V.K. Nayyar. 1986. Integrated approach to combat micronutrient deficiency. Paper presented in the seminar on growth and moderanization of the fertility industry. Dec. 15–17. FAI, New Delhi.

Thind, H.S., and D.S. Chahal. 1983. Iron equilibria in submerged soils as influenced by green manuring and iron application. J. Agric. Sci. (Camb.) 101:207–221.

Thind, H.S., and D.S. Chahal. 1987. Effect of green manuring (*Sesbania aculeata*) on zinc equilibria in submerged calcareous and noncalcareous soils. Biol. Fertil. Soils 3:179–182.

Thomas, G.V., and M.V. Shantaram. 1984. In situ cultivation and incorporation of green manure legumes in coconut basins. Plant Soil 80:373–380.

Thompson, J.F. 1980. Arginine synthesis, proline synthesis, and related process. In B.J. Miflin (ed.). The Biochemistry of Plants. pp. 375–402. Volume 5, Academic Press, New York.

Thompson, L.M., C.A. Black, and J.A. Zoelluer. 1954. Occurrence and mineralization of organic P in soil with particular reference to association with N, C and pH. Soil Sci. 77:185–196.

Tiwari, K.N., A.N. Pathak, and H. Ram. 1980. Green manuring in combination with fertilizer nitrogen on rice under double cropping system in an alluvial soil. J. Indian Soc. Soil Sci. 28:162–169.

Triplett, G.B., F. Haghiri, and D.M. Van Doren. 1979. Plowing effect of corn yield response to N following alfalfa. Agron. J. 71:801–803.

Tsutsuki, K. 1984. Volatile products and low-molecular weight phenolic products of the anaerobic decomposition of organic matter. In Organic Matter and Rice. pp. 329–343. International Rice Research Institute, Los Banos, Philippines.

Tusneem, M.E., and W.H. Patrick, Jr. 1971. Nitrogen Transformations in Waterlogged Soils, La Agric. Exp. Sta. Bull. 657.

Vallis, I., and R.J. Jones. 1973. Net mineralization of nitrogen in leaves and leaf litter of *Desmodium intortum* and Phaseolus atropurpureus mixed with soil. Soil Biol. Biochem. 5:391–398.

Van Faassen, H.G., and H. Van Dijk. 1987. Manure as a source of nitrogen and phosphorus in soils. In H.G. v. d. Meer et al. (eds). Animal Manure on Grassland and Fodder Crops. pp. 27–45. Martinus Nijhoff Publishers, Dordrecht.

Van Veen, J.A., R. Merckx, and S.C. Van de Geijn. 1989. Plant and soil-related controls of the flow of carbon from roots through soil microbial biomass. Plant Soil 115:179–188.

Varco, J.J. 1986. Tillage Effects on Transformation of Legume and Fertilizer Nitrogen and Crop Recovery of Residue Nitrogen. Ph.D. thesis, University of Kentucky, Lexington.

Varco, J.J., W.W. Frye, M.S. Smith, and J.H. Grove. 1987. Legume nitrogen transformation and recovery by corn as influenced by tillage. In J.F. Power (ed.). The Role of Legumes in Conversation Tillage Systems. pp. 40. Soil Conservation Society of America.

Velu, V., A. Saravanan, and K.M. Ramanathan. 1988. Leaching loss of nitrogen in clay loam soils of Cauvery delta. Oryza 25:374–379.

Venkatakrishnan, S. 1980. Mineralization of green manure (*Sesbania* aculeata, Pers.) nigrogen in sodic and reclaimed soils under flooded conditions. Plant Soil 54:149–152.

Ventura, M., and I. Watanabe. 1990. Azolla and sesbania: Organic fertilizers. Paper presented at the Philippines Environment Conference on 14–18 June, 1990, Manila, Philippines.

Waid, J.S., and A. Chulan. 1981. Release of nitrogen from decomposing legume roots and nodules. In Proc. Symp. on Paddy Soil. pp. 320–322. Science Press, Beijing, Springer Verlag, New York.

Wagger, M.G. 1989. Time of desiccation effects on plant composition and subsequent nitrogen release from several winter annual cover crops. Agron. J. 81:236–241.

Wang, T.S.C., S.Y. Cheng, and H. Tung. 1967a. Dynamics of soil acids. Soil Sci. 104:138–144.

Wang, T.S.C., T.K. Yang, and T.T. Chuan. 1967b. Soil alcohols, their dynamics and their effect upon plant growth. Soil Sci. 104:40–45.

Watanabe, I. 1984. Use of green manures in Northeast Asia. In Organic Matter and Rice. pp. 229–234. International Rice Research Institute, Los Banos, Philippines.

Watanabe, I. 1987. In Azolla Utilization. pp. 197–205. International Rice Research Institute, Los Banos, Philippines.

Watanabe, I., and C.C. Liu, 1990. Improvement of nitrogen-fixing systems and their integration into sustainable rice farming. Trans. 14th Int. Congr. Soil Sci. Commission. III. 134–135.

Watanabe, I., W. Ventura., G.B. Mascarina, and D.L. Eskew. 1989. Fate of Azolla Spp. and urea N applied to wetland rice (*Oryza sativa* L.). Biol. Fertil Soils 8:102–110.

Weeraratna, C.S. 1979. Pattern of N release during decomposition of some green manures in a tropical alluvial soil. Plant Soil 53:287–294.

Wen Qi-xiao, and Yu Tian-ren. 1988. Effect of green manure on physio-chemical properties of irrigated rice soils. In Green Manure in Rice Farming. pp. 275–288. International Rice Research Institute, Los Banos, Philippines.

Westcott, M.P., and D.S. Mikkelson, 1985. Comparative effects of an organic and inorganic nitrogen source in flooded soils. Soil Sci. Soc. Am. J. 49:1470–1475.

Westcott, M.P., and D.S. Mikkelson, 1987. Comparison of organic and inorganic nitrogen sources for rice. Agron. J. 79:937–943.

White, J.L, M. Fried, and A.J. Ohlrogge. 1947. A study of the utilization of phosphorus in green manure crops by the succeeding crops using radiactive phosphorus. Soil Sci. Soc. Am. Proc. 12:174–175.

Williams, S.T., and T.R.G. Gray. 1974. Decomposition of the litter on the soil surface. In C.H. Dickinson and G.J.F. Pugh (eds.). Biology of Plant Litter Decomposition. pp. 611–632. Academic Press, New York.

Williams, W.A., D.S. Mikkelsen, K.E. Mueller, and J.E. Ruckman. 1968. Nitrogen immobilization by rice straw incorporated in lowland rice production. Plant Soil 28:49–60.

Wilson, D.O., and W.L. Hargrove. 1986. Release of nitrogen from crimson clover residue under two tillage systems. Soil Sci. Soc. Am. J. 50:1251–1254.

Wu Mei-ling, and Chen Jia-fang. 1981. A primary study on the specific adsorption of copper ion of paddy soil in southern Jiangsu. Acta Pedol. Sin. 18:234–243.

Yadav, R.L, and Singh, K. 1986. Long term experiments with sugarcane under intensive cropping system and variation in soil fertility. Indian J. Agron. 31:322–325.

Yadvinder-Singh, Bijay-Singh, M.S. Maskina, and O.P. Meelu. 1988. Effect of organic manures, crop residues and green manure (*Sesbania aculeata*) on nitrogen and phosphorus transformations in a sandy loam soil at field capacity and waterlogged conditions. Biol. Fertil. Soils 6:183–187.

Yadvinder-Singh, C.S. Khind, and Bijay-Singh. 1991. Efficient management of leguminous green manures in wetland rice. Adv. Agron. 45:135–189.

Yadvinder-Singh, and E.G. Beauchamp. 1986. Nitrogen mineralization and nitrifier activity in limed and urea treated soils. Commun. Soil Sci. Plant Analysis 7:1369–1381.

Yu, Tian-ren. 1981. Oxidation-reduction properties of paddy soils. In Proceedings of Symposium on Paddy Soils. pp. 95–106. Institute of Soil Science, Academia Sinica, Science Press, Beijing; Springer-Verlag, Berlin, Heidelberg, New York.

Yu, Tian-ren. 1985. Physical Chemistry of Paddy Soils. Science Press, Beijing; Springer-Verlag, Berlin.

Zhu Zhao-liang, Chong-qun Liu, and Bai-fan Jiang. 1984. Mineralization of organic nitrogen, phosphorus, and sulfur in some paddy soils of China. In Organic Matter and Rice. pp. 259–272. International Rice Research Institute, Los Banos, Philippines.

Index

MIX
Papier aus verantwortungsvollen Quellen
Paper from responsible sources
FSC® C105338

If you have any concerns about our products,
you can contact us on
ProductSafety@springernature.com

In case Publisher is established outside the EU,
the EU authorized representative is:
Springer Nature Customer Service Center GmbH
Europaplatz 3, 69115 Heidelberg, Germany

Printed by Libri Plureos GmbH
in Hamburg, Germany